A Primer of
Reliability Theory

A Primer of Reliability Theory

DORIS LLOYD GROSH

Department of Industrial Engineering
Kansas State University

WILEY

John Wiley & Sons

New York Chichester Brisbane Toronto Singapore

To Gene, for his encouragement and occasional cooking,
and to our beloved daughters, Katherine, Barbara, and Margaret.

Library of Congress Cataloging in Publication Data:

Grosh, Doris Lloyd.
    Primer of reliability theory / Doris Lloyd Grosh.
       p.   cm.
    Bibliography : p.
    ISBN 0-471-63820-X
    1. Reliability (Engineering)   I. Title.
TA169.G76 1989
620'.00452–dc19   88-28596
                    CIP

Printed in the United States of America

10 9 8 7 6 5 4 3 2 1

About the Author

Doris Lloyd Grosh studied at Kansas City Junior College, University of Chicago (B.S., mathematics and physics), Universidad Nacional Autonóma de México, and Kansas State College (M.S., mathematics and physics). At Purdue University she married her fellow graduate student, Eugene Grosh, Jr., while pursuing further graduate studies. She spend thirteen years as a full-time housewife and part-time mathematics lecturer at Tulsa University; during this period she was active in community service for the Camp Fire Girls, the League of Women Voters, and the Unitarian Church.

She returned to academic life, with studies at Tulsa University, Oklahoma State University, and Kansas State University (Ph.D., statistics). Since 1968 she has held joint appointments in the Department of Industrial Engineering and the Department of Statistics.

She has been a contributor and referee for *Technometrics, Journal of the American Statistical Society*, and *IEEE Transactions on Reliability*. Organizational memberships have included Society of Women Engineers, American Society for Quality Control, American Statistical Association, American Association of University Professors, MENSA, American Radio Relay League, and the professional honoraries Sigma Xi, Tau Beta Pi, and Pi Mu Epsilon.

She has been a consultant or co-investigator for projects funded by the Nuclear Regulatory Agency, Kansas Department of Health and Environment, and Kansas Department of Transportation.

Preface

It is fair to ask, "Why another reliability book?" The only honest answer I can give is that this is the book I need for the course I teach. If it should serve the needs of other college teachers, that would be a bonus. When I started teaching this subject twenty years ago, the standard textbooks were those by Bazovsky (1961), Shooman (1968), and von Alven (editor, 1964), all of which are excellent. As my knowledge of the subject grew, however, I found that none contained exactly the collection of material I wanted to cover in my class, so I began developing a large collection of handout material, which has evolved into this book.

The choice of material for any university course is always subject to challenge; we teachers can never be sure that we are teaching our students what they will need to know. But there has been sufficient feedback from a fair sprinkling of my former students to know that at least some of these topics have been useful to some of them.

The purpose of the book is to cover mathematical/statistical topics that will help prepare students for reading the reliability literature, particularly the *IEEE Transactions on Reliability*, in which most reliability articles appear. This is not a book about applications and practice; the reader must look elsewhere for those. It is not a source of comprehensive reference material; the bibliographic function has been carried out so splendidly by Mann, Schafer, and Singpurwalla (1974) and Dhillon (1983) that it scarcely needs repeating here.

As for style, I have written in the only way I find comfortable, which is to include a great many intermediate mathematical steps so that students need not wonder how to get from one equation to the next. Some details have been included because, after having myself spent many hours figuring out where some "obvious" or "well-known" result came from, my instincts as a teacher require me to explain it in full detail so my students will not flounder. My reliability students have usually had only one or two semesters of statistics, and some have taken their calculus as long ago as fifteen years (for instance, the mature military officers we sometimes include in our program). I want this book to be usable by such students. Because it contains more details of proof and derivation than is usual, it has also proved useful for independent study. Teachers whose students come to them with weak preparation may want to use this book as a "catch-up" resource.

One advantage I have found in using a text with so many details of development is that I can cover the material more quickly than formerly because the students, seeing the steps all laid out in the book, are following along without slavish notetaking. Nowadays during class I see their attentive eyes rather than the tops of their heads.

In order to maintain some continuity of development, a number of subjects have been placed in chapter appendices. These may quite properly be ignored by many readers. Many of my students come to me with only the vaguest idea about the gamma and beta distributions and knowing nothing at all about order statistics. Chapter 5 is devoted entirely to Epstein's work on life testing, which requires good understanding of these topics, so I precede that chapter with a few class periods devoted to a detour through the material in its appendixes.

Chapter 7 assumes some rudimentary knowledge of Laplace transforms, but the burden of attaining that knowledge must rest with the reader. The topic is lengthy and has been covered so adequately elsewhere that it did not seem reasonable to treat it here. For neophytes, a suggested reference is the first few chapters of Spiegel (1965).

I generated the statistical tables on the mainframe computer at Kansas State University (a National Advanced System 6630) using SAS and IMSL routines. No guarantee of accuracy is made, but the values seem to compare well with those in other published tables.

Thanks are due to a succession of students whose questions and feedback over the years have helped mold this book into its present form. Prabhakar Reddy and Malgorzata Rys helped create some of the exercises. Frank Hwang and Devindra Negi exercised vigilance in catching errors, and Negi has contributed greatly to the solution manual for teachers. Mrs. Joyce Martin has been a superb collaborator in the preparation of the manuscript.

As I review the manuscript preparatory to publication, I am overcome with a sense of its brevity. Every topic could have been treated in double or triple measure. And yet I can barely cover the material in one semester, and it is after all a *primer* of reliability theory, not a treatise. Students who have mastered the material presented here should be ready to further their training with the excellent books by Mann, Schafer, and Singpurwalla, (1974), Kapur and Lamberson (1977), and Bain (1978), and other books published in the past few years.

Readers are encouraged to apprise me of errors, especially those that might create confusion.

Doris Lloyd Grosh
Manhattan, Kansas
November 1988

Contents

Tables

References

Notations and Abbreviations

Index

CHAPTER 1

Introduction

1.1 QUALITY CONTROL AND RELIABILITY

As a beginning student of reliability, you may well be wondering about the contents of the course. You are doubtless aware that it has something to do with applying the laws of probability to hardware units or components. But so also does the study of quality control, which you may already have undertaken. How do the two subjects differ?

Briefly, we could make the following distinction.

1. Quality control procedures were developed to encompass judgments on both measurements and attributes. Measurements can be related to physical characteristics such as washer thickness, distance between holes, electrical resistance, and reflectivity of a polished surface. These are the kind of instantly available data that lend themselves well to analysis by Shewhart x-bar control charts.

Attribute analysis, on the other hand, is concerned with whether an item possesses or does not possess some described characteristic such as a stated size ("go"/"no go" gauging) or a stated degree of perfection (acceptable versus defective product). These data are the kind that are well suited to the p-chart type of control chart.

2. Reliability theory is a subset of quality control; in it the characteristic studied is not some instantly available dimensional measurement but the length of life of the item. Because of the long time periods necessary to gather life data, they are generally far more expensive than simple size measurement data. The analyst may study them more intensively in order to extract as much information as possible from

1

every inspection dollar. Thus there has developed the intensive study of distribution types that we meet in reliability theory, in contrast to the simple control charts of quality control. Yet, let it be remembered that every collection of life data could be converted into either an x-bar chart or a p-chart. Such charts are useful, but they do not answer directly such reliability questions as "What fraction of these elements can I count on to last for 1000 hours, with probability .90?"

In short, the quality control function deals primarily with new products under inspection. Reliability deals with products in service.

1.2 EMPIRICAL RELIABILITY MEASURES

It is important in reliability work to be able to describe quantitatively how well some product holds up over a period of service, or to distinguish how product A, say, compares with product B. In either case, a measure of effectiveness, the reliability, can be obtained by noting the behavior of a large number of "identical" items which are put in service (or on test) at the same time. We can formalize as follows. Let $N_0 = n(0)$ be the number of items put on test at time $t = 0$. Let $n(t)$ be the number of survivors (the number still "alive," or functioning adequately) at time t. A plot of the function $n(t)$ will show the character depicted in Figure 1.1a, the steps corresponding

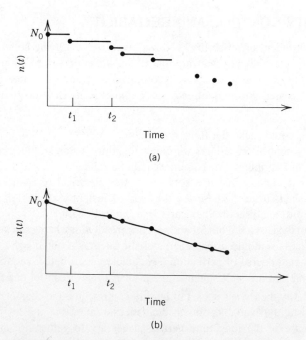

Figure 1.1 The survivor function $n(t)$ for two monitoring situations: (a) continuous monitoring, (b) sporadic monitoring.

to the failure times. Often it is more convenient to work with the ratio or fraction of survivors, defined as

$$R^*(t) = \frac{n(t)}{N_0} \tag{1.1}$$

It is obvious that $R^*(t)$ is a monotone nonincreasing function with $R^*(0) = 1$ and (except in the unlikely event of perfect components) $R^*(\infty) = 0$. Furthermore, if t_i represents the failure times of the components ($i = 1, \ldots, N_0$), then $R^*(t_i)$ will be defined to be continuous from the right, so we should agree that

$$R^*(t_i) = \lim_{\epsilon \to 0} R^*(t_i + \epsilon) \tag{1.2}$$

(But see later for a more aesthetic way of proceeding.) We will call $R^*(t)$ the empirical, or data-based, reliability function. It is a realization, for a particular group of components, of the theoretical reliability function $R(t)$ to be considered in the next chapter.

Every survivor function looks somewhat like Figure 1.1 except for scale, so the inexperienced practitioner may prefer to work with some other function of $n(t)$ that shows more sensitivity to product quality. There are several ways to proceed. The practitioner may wish to consider the increment of the survivor function

$$-\Delta n = n(t) - n(t + \Delta t) \tag{1.3}$$

which is the failure count over the interval Δt. It is of more interest to convert this into a *rate* by dividing it by Δt to get

$$\frac{n(t) - n(t + \Delta t)}{\Delta t} = \frac{-\Delta n}{\Delta t} \tag{1.4}$$

which is the failure count per unit time, and then to "standardize" it by the initial number of units, N_0, yielding the ratio

$$f^*(t) = \frac{n(t_i) - n(t_i + \Delta t)}{\Delta t_{i+1} N_0} = \frac{-\Delta n_{i+1}}{\Delta t_{i+1} N_0} \qquad t_i < t < t_{i+1} \tag{1.5}$$

This is the empirical estimate of probability of failure in the time interval $(t, t + \Delta t)$. It is a realization, for a particular group of components, of the failure density $f(t)$ to be considered in the next chapter.

The failure count defined in Equation 1.3 or the failure count per unit time defined in Equation 1.4 may perhaps be more fruitfully compared, not with the initial count N_0 at the beginning of the test, but with the initial count $n(t)$ at the beginning of the interval of interest. This leads to the function defined as

$$h^*(t) = \frac{n(t_i) - n(t_i + \Delta t)}{\Delta t_{i+1} n(t_i)} = \frac{-\Delta n_{i+1}}{\Delta t_{i+1} n(t_i)} \qquad t_i < t < t_{i+1} \tag{1.6}$$

which will be called the *empirical hazard function*. It is a realization, for a particular group of components, of the theoretical hazard function $h(t)$ to be considered in the next chapter. For many purposes it is considered to be more sensitive than $f^*(t)$ or

$R^*(t)$ for distinguishing between different failure laws, although I have not found this to be true (see Section 2.9 for a discussion of the matter).

To complete the discussion, we must define a fourth function, $F^*(t) = 1 - R^*(t)$. It is called the *empirical unreliability function*, analogous to the theoretical $F(t)$ of the next chapter. In order to use the functions $R^*(t), f^*(t)$, and $h^*(t)$ as they have been defined, it is well to agree on how the data are to be gathered and how the plotting is to be done. There are two cases to consider, based on how the data are gathered. Case 1 can be called continuous monitoring, when the number of units tested is small and the exact time of each failure is recorded. These failure times will be used as the system observation times designated as $\{t_i\}$, and the Δt_i will be the periods between failures, as indicated in Figure 1.2. Specifically, note that we define Δt as

$$\Delta t_{i+1} = t_{i+1} - t_i \qquad i = 0, 1, \ldots$$

The time increment is indexed by the right-hand (terminal) end of the interval. It is obvious that $-\Delta n = n(t_i) - n(t_{i+1}) = 1$ for this situation. Theoretically, $R^*(t)$ is constant over the interval $[t_i, t_{i+1})$ and should be plotted as shown in Figure 1.1a. But the step function shown there is not aesthetically satisfying, so the function $R^*(t)$ is more conveniently defined as in (1.2) only for the end points $t = t_i$, and as

$$R^*(t) = \frac{t_{i+1} - t}{\Delta t_{i+1}} R^*(t_i) + \frac{t - t_i}{\Delta t_{i+1}} R^*(t_{i+1}) \qquad (1.7)$$

over the interval (t_i, t_{i+1}). The alert student will, of course, recognize that Equation 1.7 is merely the straight-line interpolation formula between the points $(t_i, R^*(t_i))$ and $(t_{i+1}, R^*(t_{i_i}))$, as plotted in Figure 1.1b.

We come next to the problem of how to plot $f^*(t)$ and $h^*(t)$ for Case 1. We may use a histogram or bar chart form, not because it seems to make a lot of sense in this particular context, but rather for consistency with the usual way of plotting data from large samples. Or we may choose the frequency polygon type of graph where the ordinate, $f^*(t)$ or $h^*(t)$ is plotted against the center of the Δt interval. Both methods are illustrated in Example 1.1. In any event, formulae 1.5 and 1.6 reduce to

$$f^*(t) = \frac{1}{N_0 \Delta t_{i+1}}, \qquad t_i < t < t_{i+1} \qquad (1.8)$$

and

$$h^*(t) = \frac{1}{n(t_i) \Delta t_{i+1}}, \qquad t_i < t < t_{i+1} \qquad (1.9)$$

with discontinuities at the observation times t_i.

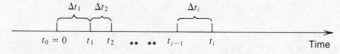

Figure 1.2 Relationship between t_i and Δt_i.

Example 1.1

Ten hypothetical electronic components are placed on life test. Failure times for the components are { 5, 10, 17.5, 30, 40, 55, 67.5, 82.5, 100, 117.5 hours}.

(a) Plot histograms and polygon graphs for $f*(t)$ and $h*(t)$ from the data.

(b) Plot the reliability and unreliability functions $R*(t)$ and $F*(t)$.

For the solution, see Figures 1.3a through 1.3e and Table 1.1.

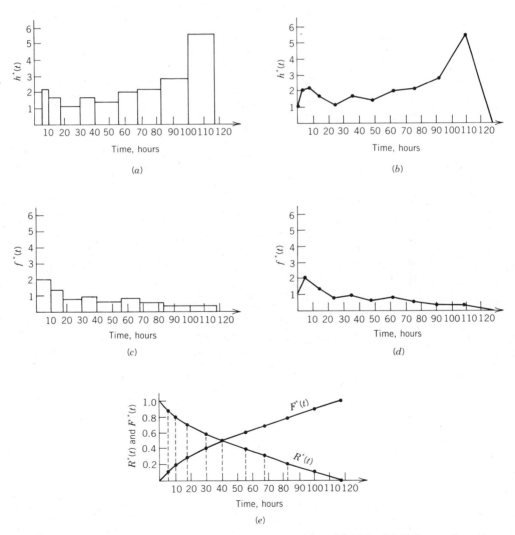

Figure 1.3 Plots for Example 1.1. (a) Histogram plot of $h*(t)$. (b) Polygon plot of $h*(t)$. (c) Histogram plot of $f*(t)$. (d) Polygon plot of $f*(t)$. (e) $F*(t)$ and $R*(t)$.

TABLE 1.1   Calculations for Example 1.1

Failure Number	Operating Time	Failure Density $f^*(t)$	Hazard Rate $h^*(t)$
1	$0 \rightarrow 5$	$1/(10 \times 5) = 0.0200$	$1/(10 \times 5) = 0.0200$
2	$5 \rightarrow 10$	$1/(10 \times 5) = 0.0200$	$1/(9 \times 5) = 0.0222$
3	$10 \rightarrow 17.5$	$1/(10 \times 7.5) = 0.0133$	$1/(8 \times 7.5) = 0.0167$
4	$17.5 \rightarrow 30$	$1/(10 \times 12.5) = 0.0080$	$1/(7 \times 12.5) = 0.0114$
5	$30 \rightarrow 40$	$1/(10 \times 10) = 0.0100$	$1/(6 \times 10) = 0.0167$
6	$40 \rightarrow 55$	$1/(10 \times 15) = 0.0067$	$1/(5 \times 15) = 0.0133$
7	$55 \rightarrow 67.5$	$1/(10 \times 12.5) = 0.0080$	$1/(4 \times 12.5) = 0.0200$
8	$67.5 \rightarrow 82.5$	$1/(10 \times 15) = 0.0067$	$1/(3 \times 15) = 0.0220$
9	$82.5 \rightarrow 100$	$1/(10 \times 17.5) = 0.0057$	$1/(2 \times 17.5) = 0.0286$
10	$100 \rightarrow 117.5$	$1/(10 \times 17.5) = 0.0057$	$1/(1 \times 17.5) = 0.0570$

In the second case, monitoring of the test units is not continuous, but rather the system is observed from time to time, at time points $\{t_i\}$, say (not necessarily equally spaced), at each of which the number of survivors $n(t_i)$ is noted. Formulae 1.1, 1.5, and 1.6 are valid. Case 2 is illustrated in Example 1.2.

Example 1.2

Eight hundred hypothetical components are placed on life test. The system is observed at 3 hours, 6 hours, . . . 30 hours and the number of survivors is noted (see Table 1.2).

TABLE 1.2   Failure Data for 800 Hypothetical Components

Time Interval, Hours	Number of Failures in Interval
$0 \rightarrow 3$	185
$3 \rightarrow 6$	42
$6 \rightarrow 9$	36
$9 \rightarrow 12$	30
$12 \rightarrow 15$	17
$15 \rightarrow 18$	8
$18 \rightarrow 21$	14
$21 \rightarrow 24$	9
$24 \rightarrow 27$	6
$27 \rightarrow 30$	3
Total	350

TABLE 1.3 Calculations for Example 1.2

Time Intervals, hours	Number of Failures in Interval	Failure Density $f^*(t)$	Hazard Rate $h^*(t)$
0 → 3	185	185/(800 × 3) = 0.0771	185/(800 × 3) = 0.0771
3 → 6	42	42/(800 × 3) = 0.0175	42/(615 × 3) = 0.0227
6 → 9	36	36/(800 × 3) = 0.015	36/(573 × 3) = 0.0209
9 → 12	30	30/(800 × 3) = 0.0125	30/(537 × 3) = 0.0175
12 → 15	17	17/(800 × 3) = 0.0071	17/(507 × 3) = 0.0112
15 → 18	8	8/(800 × 3) = 0.0033	8/(490 × 3) = 0.0054
18 → 21	14	14/(800 × 3) = 0.0058	14/(482 × 3) = 0.0097
21 → 24	9	9/(800 × 3) = 0.00375	9/(468 × 3) = 0.0064
24 → 27	6	6/(800 × 3) = 0.0025	6/(459 × 3) = 0.0044
27 → 30	3	3/(800 × 3) = 0.0013	3/(453 × 3) = 0.0022

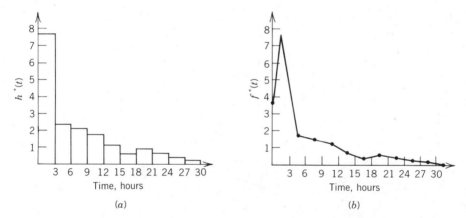

Figure 1.4 Plots for Example 1.2. (a) Histogram plot of $h^*(t)$ (b) Polygon plot of $h^*(t)$

(a) Plot $f^*(t)$ and $h^*(t)$ for the data given (two plotting methods for each).

(b) Plot the reliability and unreliability functions $R^*(t)$ and $F^*(t)$.

Solution: The calculations are in Table 1.3, and the plots are shown in Figure 1.4.

EXERCISES 1.2

1. Ten units are placed on life test, and the individual failure times are {7, 17, 25, 32, 40, 48, 50, 55, 61, and 65 hours}.

(a) Plot $f^*(t)$ and $h^*(t)$ for the data given (two plotting methods for each).

(b) Plot the reliability and unreliability functions $R^*(t)$ and $F^*(t)$.

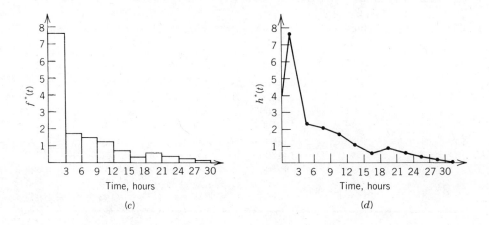

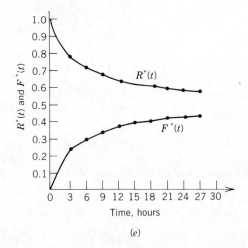

Figure 1.4 (Continued)   (c) Histogram plot of $f^*(t)$ (d) Polygon plot of $f^*(t)$ (e) $F^*(t)$ and $R^*(t)$.

2. An 11,000-hour life test on a sample of 80 transistors produced the following data.

(a) Plot $f^*(t)$ and $h^*(t)$ for the data given (two plotting methods for each).

(b) Plot the reliability and unreliability functions $R^*(t)$ and $F^*(t)$.

Time Interval, hours	Number of Failures in Interval
0 → 1000	20
1000 → 2000	15
2000 → 3000	10
3000 → 4000	5
4000 → 5000	6
5000 → 6000	5
6000 → 7000	5
7000 → 8000	6
8000 → 9000	4
9000 → 10,000	2
10,000 → 11,000	2
Total	80

3. Repeat Exercise 1 for the following failure times:

{4, 8, 14, 16, 22, 28, 39, 43, 57, 59, 65, 67, 77, 80, 85, 86, 88, 93, 99}

4. Repeat Exercise 2 for the following data set.

Time Interval, hours	Number of Failures in Interval
0 → 10	10
10^+ → 20	7
20^+ → 30	14
30^+ → 40	14
40^+ → 50	14
50^+ → 60	15
60^+ → 70	3
70^+ → 80	6
80^+ → 90	8
90^+ → 100	9

5. Repeat Exercise 1 for the following failure times:

{3, 31, 46, 55, 70, 80, 84, 111, 122, 154, 168, 193, 194, 199, 298}

6. Repeat Exercise 2 for the following data set.

Time Interval, Upper Limit	Number of Failures in Interval
20	16
40	11
60	10
80	16
100	12
125	16
150	7
175	8
200	3
300	1

7. Repeat Exercise 2 for the following data set.

Time Interval, hours	Number of Failures in Interval
$70 \rightarrow 80$	2
$80^+ \rightarrow 85$	2
$85^+ \rightarrow 90$	6
$90^+ \rightarrow 95$	18
$95^+ \rightarrow 100$	18
$100^+ \rightarrow 105$	13
$105^+ \rightarrow 110$	19
$110^+ \rightarrow 115$	14
$115^+ \rightarrow 120$	5
$120^+ \rightarrow 125$	2
$125^+ \rightarrow 130$	1

CHAPTER 2

Reliability Distributions

2.1 BASIC FORMULATIONS

In the preceding chapter we considered how an engineer might have gathered performance data and summarized it in graphical form. Here we expand those basic ideas into a more theoretical formulation.

Let T be a nonnegative continuous random variable that represents the useful life (or length of life, or time to failure) of a component (or unit or piece of equipment). The failure law for the component can be described in any of several ways. Perhaps the most fundamental formulation is in terms of $F(t)$, the cumulative distribution function (or CDF) defined as the probability that the unit "lives" for at most time t, and which we write as

$$F(t) = P(T \leq t) \qquad (2.1)$$

This is also referred to as the *unreliability* function. An equivalent and sometimes more useful formulation is the *reliability* function $R(t)$, the probability that the component lives for more than time t, which is designated as

$$\begin{aligned} R(t) &= P(T > t) \\ &= 1 - P(T \leq t) \\ &= 1 - F(t) \qquad (2.2) \end{aligned}$$

It is traditional also to describe the failure law in terms of the density function

$$f(t) = F'(t) \qquad (2.3)$$

which must have the properties that

$$f(t) \geq 0$$

$$\int_0^\infty f(t)dt = 1 \tag{2.4}$$

It follows that the probability of the component failing between times t_1 and t_2 is given by

$$P(t_1 < T \leq t_2) = \int_{t_1}^{t_2} f(t)dt$$

$$= F(t_2) - F(t_1), \tag{2.5}$$

and that

$$F(t) = \int_0^t f(x)dx \tag{2.6}$$

It will be assumed throughout these discussions that the random variable T is well behaved in the sense that $F(t)$ is continuous, so that there are no "lumps" of probability to be accounted for separately. As an immediate result of this assumption, it follows that

$$P(T = t_0) = P(t_0 < T \leq t_0) = \int_{t_0}^{t_0} f(t)dt = 0 \tag{2.7}$$

and therefore

$$P(t_1 \leq T < t_2) = P(t_1 \leq T \leq t_2)$$

$$= P(t_1 < T \leq t_2) \tag{2.8}$$

$$= P(t_1 < T < t_2)$$

It is obvious from these definitions that $F(t)$ is a monotone nondecreasing function of t with

$$F(0) = 0$$

$$F(\infty) = 1$$

whereas $R(t)$ is a monotone nonincreasing function of t with

$$R(0) = 1$$

$$R(\infty) = 0$$

It is important to note that the function $f(t)$, although it describes how the failure probability is spread over time, is not itself a probability, but a *probability density*.

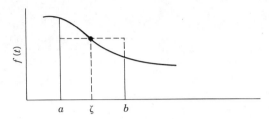

Figure 2.1 The mean value theorem illustrated.

It can be used to compute the approximate probability over a short interval when the exact calculation is inconvenient or analytically intractable. To justify this, let us invoke the mean value theorem (MVT) which states that for a continuous function $f(t)$ we can write $P(a < T \le b)$ as

$$\int_a^b f(t)dt = (b - a)f(\xi) \tag{2.9}$$

where ξ is in the interval (a, b), and can thus be written as $\xi = a + m(b - a)$, where $m < 1$. The relationship 2.9 is exact for some (probably unknown) interior point ξ (see Figure 2.1). If the interval length $b - a = \Delta t$ is small, a reasonably good approximation may be achieved by using the midpoint $t^* = (a + b)/2$, instead of the unknown ξ. Then the exact equality in (2.9) is replaced by the approximation

$$\int_a^b f(t)dt \approx (b - a)f(t^*) = f(t^*)\Delta t \tag{2.10}$$

For illustration, let us consider two examples.

Example 2.1

Consider the normal $N(100, 625)$ distribution with mean $\mu = 100$ and standard deviation $\sigma = 25$. The random variable thus defined is, strictly speaking, not nonnegative, but the amount of probability to the left of $T = 0$ is negligible because there is almost zero area under the density curve to the left of $T = \mu - 4\sigma$. We can approximate some desired probability, say $P(104 \le T \le 106)$, by

$$f(105)(106 - 104) = \frac{1}{\sqrt{2\pi}(25)} \exp\left[-\frac{1}{2}\left(\frac{105 - 100}{25}\right)^2\right] \tag{2}$$

$$= 0.313$$

This is in good agreement with the answer obtained in the usual way from a normal table by calculating

$$P(104 \le T \le 106) = P\left(\frac{104 - 100}{25} \le Z \le \frac{106 - 100}{25}\right)$$

$$= P(.16 \le Z \le .24) = .5948 - .5636 = .0312$$

where Z is the $N(0, 1)$ or "standard normal" variable.

Example 2.2

Consider a component whose length of life in hours is well described by the negative exponential density function

$$f(t) = .002e^{-.002t} \qquad \text{for } t \ge 0$$

The probability of failure between hours 510 and 515 can be obtained approximately as

$$f(512.5)(515 - 510) = .002e^{-.002(512.5)}(5)$$

$$= .00359$$

The exact value, here easily obtained by integration, is

$$F(515) - F(510) = e^{-.002(510)} - e^{-.002(515)} = .00359$$

The quantity $f(t)dt$ used in these two examples is frequently referred to as a *probability element*. This is dimensionally correct, since dt has the units (time) and $f(t)$ is interpreted as failure probability per unit time.

There is still a fourth convenient formulation for describing the failure law in terms of conditional probabilities. For justification, consider the following example.

Example 2.3

A machine has a useful life well described by the normal $N(500, 100^2)$ distribution.

1. What is the probability that a new machine of this type will last at least 600 hours?
2. What is the probability that a machine of this type that has already functioned for 500 hours will function for at least 100 hours more?

To the beginning student, these two questions may appear to be the same, but they are not. To answer question 1, the required probability is

$$P(T > 600) = P\left(\frac{T - 500}{100} > \frac{600 - 500}{100}\right) = P(Z > 1)$$

$$= 1 - .8413 = .1587$$

To answer question 2, we must recall the rule for conditional probabilities, where

$$P(A|B) = \frac{P(A \text{ and } B)}{P(B)}$$

For the present case this becomes

$$P(T > 600|T > 500) = \frac{P(T > 600, T > 500)}{P(T > 500)}$$

$$= \frac{P(T > 600)}{P(T > 500)} = \frac{.1587}{.5} = .3174$$

It can be seen from this example that it is convenient to introduce a conditional density function, defined as the failure probability per unit time at time t, *given* that failure has not yet occurred at time t. This function is called the *hazard rate, hazard function, instantaneous failure rate*, or often simply *failure rate* (as opposed to failure density) and is defined as

$$h(t) = \frac{f(t)}{R(t)} = \frac{f(t)}{1 - F(t)} \tag{2.11}$$

We have, then, four different but equivalent and interrelated functions for describing a statistical failure law for a component. The relationship between them can be described graphically in Figure 2.2. As shown in the graph, $f(t)$ is the height of the density curve, $F(t)$ is the left-hand tail area, and $R(t)$ is the right-hand tail area. The quantity $h(t)$, which is not easy to display graphically, is the height of the ordinate divided by the right-hand tail area. The concept of the hazard rate (as opposed to failure density) is a bit slippery to grasp. It is often defined in terms like these: "The failure rate per unit time at time t, conditional on survival until time t." But of course if a component has not survived until time t, it clearly has no probability associated with failing or surviving for any additional period.

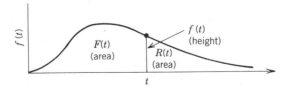

Figure 2.2 Relationships between the four performance functions.

Beginning reliability students often have a hard time understanding the difference between $f(t)$ and $h(t)$. Frequent reference to Example 2.3 will help. Or here is another example. Consider the difference between the following questions regarding a newborn child.

1. What is the probability that she will die in her sixty-first year? $f(t)$
2. What is the probability that if she reaches age 60 she will die in the following year? $h(t)$

Question 1 uses the density function $f(t)$ and question 2 uses the hazard function $h(t)$.

If any of the four quantities $f(t)$, $F(t)$, $R(t)$, and $h(t)$ are given, the other three may be obtained from it. This is easy to do in the three following cases.

Case 1   Assume $f(t)$ is given. Then the other three functions can be computed by

$$F(t) = \int_0^t f(x)dx \tag{2.12a}$$

$$R(t) = \int_t^\infty f(x)dx = 1 - F(t) \tag{2.12b}$$

$$h(t) = \frac{f(t)}{R(t)} \tag{2.12c}$$

Case 2   Assume $F(t)$ is given. Then the other three functions are computable as

$$f(t) = F'(t) \tag{2.13a}$$
$$R(t) = 1 - F(t) \tag{2.13b}$$
$$h(t) = \frac{F'(t)}{1 - F(t)} \tag{2.13c}$$

Case 3   Assume $R(t)$ is given. Then the other three functions are computable as

$$F(t) = 1 - R(t) \tag{2.14a}$$
$$f(t) = -R'(t) \tag{2.14b}$$
$$h(t) = \frac{f(t)}{R(t)} = \frac{-R'(t)}{R(t)} \tag{2.14c}$$

Case 4   The case in which only $h(t)$ is given requires considerably more work. We begin with the relationship 2.14c. Treating it as a differential equation to be solved, we multiply both sides by dt to get

$$\frac{R'(t)dt}{R(t)} = -h(t)dt \tag{2.15}$$

This is a variables separable differential equation, and the solution can be obtained using either the definite integral or the indefinite integral approach. We will do it both ways for illustrative purposes. In the definite integral approach the integration is carried out over the range from 0 to t. This necessitates the use of a dummy variable of integration, since it is not permissible to use the same letter for both the "running variable" and the end point of the integral. (See Appendix 2C for a discussion of dummy variables and their role.) Thus we write

$$\int_0^t \frac{R'(x)dx}{R(x)} = -\int_0^t h(x)dx$$

The left-hand side yields

$$\ln R(x)\Big|_0^t = \ln R(t) - \ln R(0)$$

But $R(0) = 1$, so $\ln R(0) = 0$, yielding

$$\ln R(t) = -\int_0^t h(x)\,dx = -H(t) \qquad H(t) = \int_0^t h(x)\,dx$$

say, or

$$R(t) = \exp\left[-\int_0^t h(x)dx\right] = e^{-H(t)} \qquad \text{\large✳} \qquad (2.16a)$$

From this we obtain

$$F(t) = 1 - \exp\left[-\int_0^t h(x)dx\right] = 1 - e^{-H(t)} \qquad \text{\large✳} \qquad (2.16b)$$

and, since $f(t) = h(t)R(t)$, it follows that

$$f(t) = h(t)\exp\left[-\int_0^t h(x)dx\right] = h(t)e^{-H(t)} \qquad \text{\large✳} \qquad (2.16c)$$

The quantity $H(t)$ is called the *cumulative hazard function*.

If we had carried out the operation as an indefinite integral, we would have needed to indicate by $H^*(t)$ any antiderivative of $h(t)$, so that the differential equation 2.14c

$$\frac{R'(t)dt}{R(t)} = -h(t)\,dt$$

would yield

$$\ln R(t) = -H^*(t) + c \qquad (2.17)$$

The constant of integration can be evaluated with the boundary condition that $R(0) = 1$, so that

$$0 = -H^*(0) + c$$

yielding $c = H^*(0)$. Then

$$\ln R(t) = -[H^*(t) - H^*(0)] = -\int_0^t h(x)\,dx$$

as before.

It is interesting to note that (2.16a) implies that the area under the hazard rate curve must be infinite. This is because $R(\infty) = 0$ will not be true unless $\int_0^\infty h(x)dx = H(\infty) = \infty$.

To illustrate the use of Equation 2.16c, we present some examples.

Example 2.4

Assume that $h(t) = \lambda$ (a constant) and find $f(t)$, $F(t)$, and $R(t)$.

Solution: We compute

$$H(t) = \int_0^t h(x)dx = \int_0^t \lambda\,dx = \lambda t$$

Then it follows immediately that (for $t \geq 0$) the four descriptive functions are

$$h(t) = \lambda \tag{2.18a}$$

$$R(t) = e^{-\lambda t} \tag{2.18b}$$

$$F(t) = 1 - e^{-\lambda t} \tag{2.18c}$$

$$f(t) = \lambda e^{-\lambda t} \tag{2.18d}$$

This CFR (constant failure rate) law is one of the most important in reliability work, for reasons that will become apparent as we proceed. The corresponding failure law is called the *negative exponential* or sometimes simply the *exponential* (since there exists no positive exponential law). In this work we use the notations

$$X \sim \text{NGEX}(\lambda)$$

$$X \sim \text{CFR}(\lambda)$$

to stand for the random variable or failure law described by Equations 2.18a–d.

Example 2.5

Assume that $h(t)$ is constant for the interval $0 \leq t \leq t_0$ and then changes to a different constant for the rest of the time interval. This could correspond to a change in operating conditions of the unit at time t_0. The failure rate is described by the piecewise function

$$h(t) = c_0 \qquad \text{for } 0 \le t \le t_0 \tag{2.19}$$
$$= c_1 \qquad \text{for } t > t_0$$

Because $h(t)$ is defined in piecewise manner, it should be no surprise that the other three failure law functions must also be defined piecewise. The operations are carried out in two steps, as follows.

(a) *For* $0 \le t \le t_0$. Since $h(t) = c_0$ over this interval, we can use the results of the previous example, replacing the λ which appears there by c_0.

(b) *For* $t > t_0$. This interval requires more care. The quantity in the exponent of $R(t)$ is

$$\int_0^t h(x)\,dx = \int_0^{t_0} c_0\,dx + \int_{t_0}^t c_1\,dx$$
$$= c_0 t_0 + c_1(t - t_0)$$

Thus

$$R(t) = e^{-c_0 t_0} e^{-c_1(t - t_0)} \tag{2.20}$$

and

$$F(t) = 1 - e^{-c_0 t_0} e^{-c_1(t - t_0)} \tag{2.21}$$

Using (2.18d) to calculate $f(t)$, we must keep in mind that the value of $h(t)$ must be the one that is appropriate to the interval, namely c_1.

Thus, combining the results from both intervals, we have

$$R(t) = e^{-c_0 t} \qquad \text{for } 0 \le t \le t_0$$
$$= e^{-c_0 t_0} e^{-c_1(t - t_0)} \qquad \text{for } t > t_0 \tag{2.22}$$

and

$$f(t) = c_0 e^{-c_0 t} \qquad \text{for } 0 \le t \le t_0$$
$$= c_1 e^{-c_0 t_0} e^{-c_1(t - t_0)} \qquad \text{for } t > t_0 \tag{2.23}$$

Since $F(t)$ is easily obtained as the complement of $R(t)$, we dispense with its formulation.

Figure 2.3 shows the graphs for this example where $c_0 < c_1$ and where $c_0 > c_1$. Notice that $R(t)$ is continuous and $f(t)$ is discontinuous. We could have expected the latter result, of course, since $f(t)$ is proportional to $h(t)$ and any discontinuity in $h(t)$ will be reflected in $f(t)$ and vice versa. Since integration is a smoothing operation, it should be no surprise that $R(t)$ is a continuous function. In fact, if it were not, there would be a "lump" of probability at any discontinuity, a case we ruled out of consideration.

Of particular interest in the preceding example is the special case where $c_0 = 0$. This leads to the following result, known as the *shifted exponential* distribution.

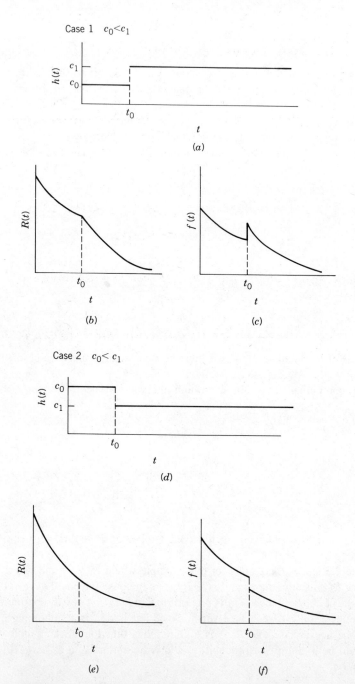

Figure 2.3 Graphs for Example 2.5. (a) $h(t)$ for $c_0 < c_1$. (b) $R(t)$ for $c_0 < c_1$. (c) $f(t)$ for $c_0 < c_1$. (d) $h(t)$ for $c_0 > c_1$. (e) $R(t)$ for $c_0 > c_1$. (f) $f(t)$ for $c_0 > c_1$.

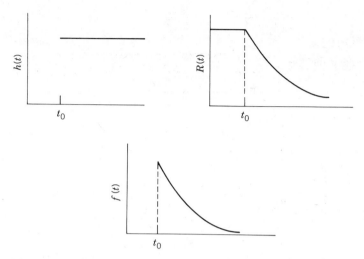

Figure 2.4 Graphs for Example 2.6, the shifted exponential.

Example 2.6

We assume that $h(t)$ has the form

$$h(t) = 0 \qquad \text{for } 0 \le t \le t_0 \tag{2.24}$$
$$= c \qquad \text{for } t > t_0$$

which results in

$$R(t) = 1 \qquad \text{for } 0 \le t \le t_0 \tag{2.25}$$
$$= e^{-c(t-t_0)} \qquad \text{for } t > t_0$$

and

$$f(t) = 0 \qquad \text{for } 0 \le t \le t_0 \tag{2.26}$$
$$= ce^{-c(t-t_0)} \qquad \text{for } t > t_0$$

These are shown graphically in Figure 2.4. It is difficult to imagine a component that might display this type of behavior, but the shifted negative exponential distribution is a useful model from the customer's point of view when the component is protected by an ironclad guarantee providing for immediate replacement of any item that fails before time t_0.

In another distribution which is sometimes useful, the failure rate is constant for a while and then increases linearly (IFR) for the balance of its life.

Example 2.7

Let $h(t)$ have the form

$$h(t) = c \qquad\qquad \text{for } 0 \le t \le t_0 \qquad\qquad (2.27)$$
$$= c + m(t - t_0) \qquad \text{for } t > t_0, \ m > 0$$

The calculation of $R(t)$ and $f(t)$ for $0 \le t \le t_0$ proceeds exactly as in Examples 2.5 and 2.6. For the interval $t > t_0$ we have

$$H(t) = \int_0^t h(x)dx = \int_0^{t_0} c\,dx + \int_{t_0}^t [c + m(x - t_0)]dx$$
$$= ct_0 + c(t - t_0) + m(t - t_0)^2/2$$
$$= ct + m(t - t_0)^2/2$$

Hence

$$R(t) = e^{-ct} \qquad\qquad \text{for } 0 \le t \le t_0$$
$$= e^{-ct}e^{-m(t-t_0)^2/2} \qquad \text{for } t > t_0 \qquad\qquad (2.28)$$

and

$$f(t) = ce^{-ct} \qquad\qquad \text{for } 0 \le t \le t_0$$
$$= [c + m(t - t_0)]e^{-ct - m(t-t_0)^2/2} \qquad \text{for } t > t_0 \qquad\qquad (2.29)$$

These functions are displayed in Figure 2.5.

A distribution of considerable interest in reliability work is the following.

Example 2.8   The Weibull Distribution

Let $h(t)$ have the general power function or polynomial form

$$h(t) = c_1 t^{c_2}$$

where c_1 is positive and c_2 may be positive, negative, or zero. The resulting curve shapes for $h(t)$ are increasing failure rate, decreasing failure rate, and constant failure rate (IFR, DFR, CFR).

The quantity needed for the exponent of $R(t)$ is

$$H(t) = \int_0^t h(x)dx = c_1 \int_0^t x^{c_2}dx = \frac{c_1 t^{c_2+1}}{c_2 + 1}$$

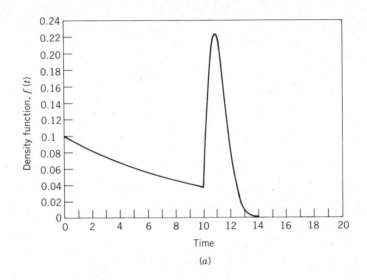

(a)

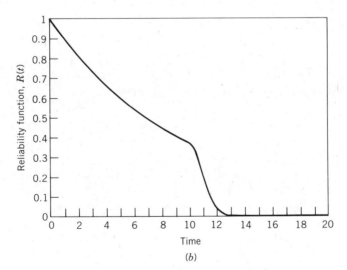

(b)

Figure 2.5   Graphs for the piecewise case of Example 2.7.

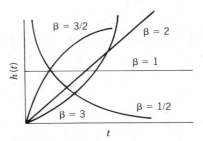

Figure 2.6 Hazard functions for the Weibull distribution.

Because this form is algebraically awkward to use, it is traditional to make the substitutions

$$c_1 = \frac{\beta}{\theta^\beta}$$

$$c_2 + 1 = \beta \text{ (the "Weibull slope")}$$

This leads to the more attractive formulations

$$h(t) = \frac{\beta}{\theta}\left(\frac{t}{\theta}\right)^{\beta-1} \tag{2.30a}$$

and

$$H(t) = \left(\frac{t}{\theta}\right)^\beta$$

which lead via the application of (2.16a) and (2.16c) to

$$R(t) = e^{-(t/\theta)^\beta} \tag{2.30b}$$

and

$$f(t) = \frac{\beta}{\theta}\left(\frac{t}{\theta}\right)^{\beta-1} e^{-(t/\theta)^\beta} \tag{2.30c}$$

When $\beta = 1$, the negative exponential distribution results. The hazard curves resulting from various values of β are shown in Figure 2.6. The Weibull distribution is taken up again in more detail in Section 2.7.

EXERCISES 2.1

1. The purpose of this exercise is to cultivate a feel for the use of the mean value theorem (MVT) and for the idea of a probability element. Evaluate the desired probability for the density function $f(t) = 6e^{-6t}$, with $t > 0$, by three different

methods. (i) Use exact calculations (integration or table). (ii) Use MVT at the left end of interval, and show percentage error. (iii) Use MVT at the center of the interval and show percentage error.

(a) $P(.1 \le T \le .2)$

(b) $P(.3 \le T \le .35)$

2. Repeat the previous exercise for the following situation. Assume the normally distributed variable $X \sim N(50, 25)$ and find

(a) $P(51.2 \le X \le 51.5)$

(b) $P(49 \le X \le 50)$.

3. Repeat Exercise 1 for the density function $f(t) = .001e^{-.001t}$ and find

(a) $P(1000 \le T \le 1010)$

(b) $P(1000 \le T \le 1500)$

Based on Exercises 1 through 3, what do you conclude about the validity of the MVT approximation for large intervals as opposed to small intervals?

4. Repeat Exercise 1 for the normally distributed variable $X \sim N(10, 9)$ for

(a) $P(12 \le X \le 12.1)$

(b) $P(9.1 \le X \le 10.5)$

5. Repeat Exercise 1 for the Weibull variable for which $\theta = 10$ and $\beta = 3$, to find

(a) $P(10 \le X \le 10.1)$

(b) $P(10 \le X \le 10.5)$

6. Find the functions $f(t)$ and $R(t)$ for the hazard function

$$h(t) = 0 \qquad \text{for } 0 \le t \le 5$$
$$= 13 \qquad \text{for } 5 < t$$

Calculate and plot $f(t)$ and $R(t)$.

7. Repeat Exercise 6 for the hazard function

$$h(t) = 0.5 \qquad \text{for } 0 \le t \le 2$$
$$= 0.7 \qquad \text{for } 2 < t$$

8. Repeat Exercise 6 for the hazard function

$$h(t) = 0 \qquad\qquad \text{for } 0 \le t \le 10$$
$$= 0.1(t - 10) \qquad \text{for } 10 < t$$

9. Repeat Exercise 6 for the hazard function

$$h(t) = 0.01 \qquad \text{for } 0 \le t \le 50$$
$$= 0.02 \qquad \text{for } 50 < t$$

10. Repeat Exercise 6 for the hazard function

$$h(t) = .5 \qquad\qquad\qquad \text{for } 0 \le t \le 2$$
$$= .5 + 1.2[(t - 2)] \qquad \text{for } 2 < t$$

11. Repeat Exercise 6 for the hazard function

$$h(t) = 0.5 \qquad\qquad\qquad \text{for } 0 \le t \le 1$$
$$= 0.5 + 0.1(t - 1) \qquad \text{for } 1 < t$$

12. Repeat Exercise 6 for the hazard function

$$h(t) = .2 - .1t \qquad\qquad \text{for } 0 \le t \le 1$$
$$= .1 \qquad\qquad\qquad \text{for } 1 \le t \le 3$$
$$= .1 + .2(t - 3) \qquad \text{for } 3 \le t$$

13. Repeat Exercise 6 for the hazard function

$$h(t) = a - m_1 t \qquad\qquad \text{for } 0 \le t \le t_0$$
$$= b \qquad\qquad\qquad \text{for } t_0 < t \le t_1$$
$$= b + m_2(t - t_1) \qquad \text{for } t_1 < t$$

Select a value for m_1 that makes the hazard function continuous at t_0.

2.2   THE HAZARD FUNCTION

The student who has gone through a course or two in statistics may have noticed that when introducing a statistical distribution, authors generally concentrate attention on the density function (or for discrete distributions, on the probability mass function). For reliability work it is sometimes more pertinent to focus on the hazard function $h(t)$.

We have already seen that the negative exponential distribution is associated with constant failure rate (CFR), whereas the Weibull family can be IFR, CFR, or DFR, as shown in Figure 2.6.

Although the normal distribution is theoretically inappropriate for reliability work (since the random variable is not nonnegative), it is in fact frequently useful for describing the lifetimes of components, so it is of interest to examine the behavior of the hazard function. Unfortunately, this cannot be done analytically, since the reliability function $R(t)$ cannot be expressed in closed form. Canceling the common factor $\sqrt{2\pi}\sigma$ in numerator and denominator, we have

$$h(t) = \frac{\exp[-(t - \mu)^2/2\sigma^2]}{\int_t^\infty \exp[-(x - \mu)^2/2\sigma^2]\,dx} \tag{2.31}$$

With the help of a good normal table (or pocket calculator), the student can construct the function numerically. This was done in the construction of Figure 2.7, which

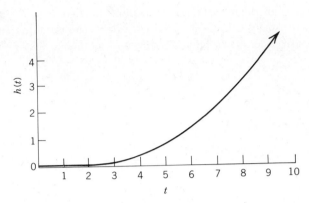

Figure 2.7 The hazard function for the $N(5,1)$ distribution.

depicts $h(t)$ for the $N(5,1)$ distribution. Since the values are essentially scale-independent, it is of more general applicability than the small parameter values would imply.

It is traditional to present a general model of a "typical" real-life hazard function, known in the literature as the *bathtub* curve, and shown in Figure 2.8. As a matter of fact, the bathtub curve is probably an idealization, since real components almost never display the type of behavior depicted. Human beings probably fit the model best. Region A corresponds to the infant mortality period, when the death rate is high for newborns, with weak "units" failing quickly, and when each week of life successfully lived seems to be an indication of the strength necessary to live additional weeks. The time t_1 marks the beginning of adulthood, when deaths may be attributed primarily to chance causes. Region B, then, is called the *chance failure* region, and its association with constant failure rate will be examined in more detail when we consider the Poisson process in Section 2.3. The time t_2 marks the beginning of the old-age, or wearout, period of Region C.

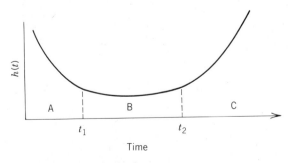

Figure 2.8 The bathtub curve.

   In addition to the four functions for describing the failure law—$f(t)$, $F(t)$, $R(t)$, and $h(t)$—it is usually convenient to have available the mean life and a variability measure. As usual, we define the mean life or mean time to failure (MTTF) as

$$\mu = E(T) = \int_0^\infty t f(t)\, dt \tag{2.32}$$

and, for measure of variability, the variance as

$$\sigma^2 = E(T - \mu)^2 = \int_0^\infty (t - \mu)^2 f(t)\, dt \tag{2.33a}$$

$$= E(T^2) - \mu^2 \tag{2.33b}$$

The variance has dimension (time)^2, so a more convenient measure is σ, the standard deviation, which is the root-mean-square deviation of failure times, measured from the MTTF μ. For some distributions the calculation of the MTTF from Equation 2.32 is tedious, so it is useful to have an alternative technique, which is presented in the following theorem.

THEOREM 2.1

For well-behaved distributions, the MTTF μ is the area under the reliability curve; that is,

$$\mu = \int_0^\infty R(t)\, dt \tag{2.34}$$

Proof A trick is used, integrating the right-hand side of (2.34) by parts. Recalling that

$$\int_a^b u\, dv = uv\Big|_a^b - \int_a^b v\, du \tag{2.35}$$

we write

$$u = R(t) \qquad\qquad v = t$$
$$du = R'(t)\, dt \qquad dv = dt$$

so that

$$\int_0^\infty R(t)\, dt = t R(t)\Big|_0^\infty - \int_0^\infty R'(t)t\, dt \tag{2.36}$$

The first term on the right-hand side is

$$\lim_{t \to \infty} t R(t) - (0)R(0)$$

where the limit is indeterminate. We assume, and this is what is meant by "well-behaved" in the statement of the theorem, that $R(t)$ approaches zero faster than $1/t$ as $t \to \infty$ (not a very onerous restriction). Thus the first term vanishes. For the second term in (2.36), we recall that $-R'(t) = f(t)$ (see Equation 2.14b) so that

$$\int_0^\infty R(t)\, dt = \int_0^\infty t f(t)\, dt = \mu$$

as desired. A more elegant proof of this result is presented in Appendix 2E.

EXERCISES 2.2

1. Are the following functions valid hazard models?
   (a) e^{-4t} (b) e^{4t} (c) $4t^2$ (d) $5t^3$ (e) $3t^{-2}$ (f) e^{-4t}/t^2

2. Use Theorem 2.1 to compute the mean of the NGEX(λ) distribution. Compute it again with the standard method. Which method do you prefer?

3. Master the details of proof for Theorem 2.1.

4. Use Table T2 and your pocket calculator (or electronic spreadsheet) to create and then plot the hazard function for the $N(6, 1)$ distribution.
   Hint. It should look very much like Figure 2.7.

5. Use l'Hôpital's rule (review your undergraduate calculus book if necessary) to demonstrate that the hazard function for the normal case does indeed approach linearity for large arguments.

6. Suppose that the break points between the three regions of the bathtub curve are at times t_1 and t_2, that the slopes of the three regions are $-m_1$, 0, and m_2 respectively, and that the constant portion of the curve has height C. Derive the piecewise formulae for $h(t)$ over the three ranges.

7. For Exercise 6 calculate the piecewise formulae for the cumulative hazard function $H(t)$ and use an electronic spreadsheet program to plot it for some convenient constant values of C, m_1, m_2, t_1, and t_2. Plot $R(t)$ and $f(t)$ for the same data using Equations 2.16a and 2.16c.

8. (For the algebraically fearless.) Derive the analytic (algebraic) expressions for $R(t)$ and $f(t)$ for the previous problem.

2.3 THE POISSON PROCESS

We have frequent occasion in reliability studies to work with renewal processes, as typified by the following situation. Consider a collection of identical units available for performing some function. At time $t = 0$ a unit from the collection is put into service and allowed to function until it fails at time t_1, when it is instantaneously replaced by another unit from the collection, which in turn functions until its failure at time t_2, whereupon it is immediately replaced, and so on. This can be depicted as a sequence of failures ("arrivals") on the time axis, as in Figure 2.9. The arrivals occur at random times depending on some probability law about which various assumptions can be made. Let $N(t)$ be the integer-valued random variable that designates the number of arrivals (failures) by time t. It is evident that $N(t + s) - N(t)$ is the number of arrivals that occur in an interval of length s beginning at time t. It is desired to formulate an expression for

$$P_n(t) = P[N(t) = n] \tag{2.37}$$

the probability of n arrivals by time t. The most common and most fruitful set of assumptions to make about the arrival process are the so-called Poisson postulates which follow.

POSTULATE 1

$$P_0(0) = 1 \quad \text{or} \quad N(0) = 0 \tag{2.38}$$

In other words, the count is initiated to zero at time zero.

POSTULATE 2

For a very small interval of time $\Delta t\,(=h)$, it is true that

$$P[N(t + h) - N(t) = 1] = \lambda h + o_1(h) \tag{2.39}$$

where $o_1(h)$ is a differential of higher order, that is, a quantity that approaches zero much faster than h, when h approaches zero. The postulate states that the probability of a single arrival in a small interval of time h is (almost) directly proportional to the length of interval, and λ is the constant of proportionality, called the intensity of the process, or the arrival rate.

POSTULATE 3

For a small interval of time h, it is true that

$$P[N(t + h) - N(t) \geq 2] = o_2(h) \tag{2.40}$$

where $o_2(h)$ is again a differential of higher order. The postulate states that in a very small interval there is essentially zero probability of two or more arrivals; that is, simultaneous arrivals are impossible.

Figure 2.9

POSTULATE 4

For any times t, t', and s it is true that

$$P[N(t + s) - N(t) = n] = P[N(t' + s) - N(t') = n] \qquad (2.41)$$

The postulate states that the probability of any given number of arrivals in any period of fixed length s is the same, no matter when the period begins. In other words, the process is stationary over time. (The postulate does not claim that the actual number of arrivals will be equal for different periods.)

POSTULATE 5

For any times t, s, t', and s' for which $t' \geq t + s$, it is true that

$$\text{cov}[N(t + s) - N(t), N(t' + s') - N(t')] = 0 \qquad (2.42)$$

The postulate states that the arrival counts for any two nonoverlapping intervals are independent random variables; what happens in one interval has no influence on what happens in any subsequent interval.

These five postulates—which the beginner may regard as rather general and innocuous—constitute in fact a powerful and restrictive set of assumptions which leads uniquely to a single, well-defined probability distribution, as we prove in the following paragraphs.

Postulates 2 and 3 (Equations 2.39 and 2.40) taken together imply that for a small interval h

$$P[N(t + h) - N(t) = 0] = 1 - \lambda h \qquad (2.43)$$

(The quantities $o_1(h)$ and $o_2(h)$ are ignored, since they are essentially zero.)

Now consider a small interval of time h, starting at time t, and consider the state of the system (the total number of arrivals) at the beginning and end of the interval. There are two cases to consider.

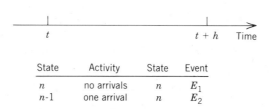

Figure 2.10 The event states for a Poisson process.

Case 1: General Case   As depicted in Figure 2.10, there are two disjoint events, E_1 and E_2, that can result in having n arrivals by the end of the interval; all other possibilities are ruled out by Postulate 3. Thus

$$P_n(t + h) = P[E_1 \text{ or } E_2]$$

$$= P[E_1] + P[E_2]$$

$$= P[\text{the count was } n \text{ at time } t, \text{ and no arrivals}$$
$$\text{occurred in interval } h] + P[\text{the count was}$$
$$n - 1 \text{ at time } t, \text{ and one arrival occurred in}$$
$$\text{interval } h] \tag{2.44}$$

By Postulates 4 and 5, the events within brackets in the two terms are independent, so the probability of the compound event in each bracket is simply the product of the associated probabilities. Thus

$$P_n(t + h) = P[\text{the count was } n \text{ at time } t] P[\text{no arrival occurred in interval } h]$$
$$+ P[\text{the count was } n - 1 \text{ at time } t] P[\text{one arrival occurred in interval } h]$$

Thus

$$P_n(t + h) = P_n(t)(1 - \lambda h) + P_{n-1}(t)\lambda h \tag{2.45}$$

Multiplying out the parentheses on the right-hand side, transposing the resulting first term, and then dividing the equation by h leads to

$$\frac{P_n(t + h) - P_n(t)}{h} = -\lambda P_n(t) + \lambda P_{n-1}(t)$$

If the interval h shrinks to zero, we see that the left-hand side defines the derivative of the probability with respect to time. The result is

$$\frac{dP_n(t)}{dt} = P'_n(t) = -\lambda P_n(t) + \lambda P_{n-1}(t), \qquad n = 1, 2, 3, \ldots \tag{2.46}$$

Case 2: Boundary Case   As depicted in Figure 2.11, there is only one way for the count to be zero at time $t + h$, and that is for the count to have been zero at time t, with no arrivals during the interval h. This gives

$$P_0(t + h) = P(E_3)$$

$$= P[\text{the count was } 0 \text{ at time } t, \text{ and no arrivals occurred in } h]$$

$$= P[\text{the count was } 0 \text{ at time } t] P[\text{no arrivals occurred in interval } h]$$

$$P_0(t + h) = P_0(t)(1 - \lambda h)$$

Transposing the first term, as before, dividing by h, and taking the limit as $h \to 0$, we get

$$\frac{dP_0(t)}{dt} = P'_0(t) = -\lambda P_0(t) \tag{2.47}$$

Figure 2.11 The event states for a Poisson process, boundary case.

We see that Equations 2.46 and 2.47 represent a set of linked differential equations in $P_0(t)$, $P_1(t)$, $P_2(t)$, ... Such a set of difference–differential equations is tedious to solve by elementary methods. The solution to (2.47) is easy to obtain, since it is of the variables separable form. The resulting solution,

$$P_0(t) = e^{-\lambda t} \tag{2.48}$$

can be inserted into (2.46) with $n = 1$, resulting in the linear differential equation

$$P_1'(t) + \lambda P_1(t) = \lambda e^{-\lambda t}$$

which can in turn be solved without much difficulty for $P_1(t)$. The expression obtained for $P_1(t)$ can then be substituted in Equation 2.46 with $n = 2$, and so on. After a few steps, the pattern emerges for the general form of the solution.

The student with adequate background in Laplace transforms will find that the solution is far more easily obtained using that technique. This is done in Appendix 2B. Here we simply state without proof that the joint solution for Equations 2.46 and 2.47 has the form

$$P_n(t) = \frac{e^{-\lambda t}(\lambda t)^n}{n!}, \qquad n = 0, 1, 2, \ldots \tag{2.49}$$

of which (2.48) is just a special case. It is easy to verify by differentiation of (2.49) that it does indeed satisfy (2.46), just as differentiation of (2.48) yields (2.47).

It is well known that the Poisson distribution has mean and variance both equal to its parameter. In the present case this implies that if a Poisson process with intensity λ is observed for a length of time t, the mean and variance of the number of arrivals $N(t)$ in that interval are

$$E[N(t)] = \lambda t \tag{2.50}$$

and

$$\text{Var}[N(t)] = \lambda t \tag{2.51}$$

Example 2.9

A Poisson process has intensity (average arrival rate) $\lambda = 8$ arrivals per hour.

(a) What is the probability that in the next hour there will be exactly five arrivals?

(b) What is the probability that in the next half hour there will be at most three arrivals?

(c) What is the expected number of arrivals in the next two hours? In the next 20 minutes?

Solutions:

(a) We have $\lambda t = (8)(1) = 8$. Then

$$P(5 \text{ arrivals}) = P_5(1) = \frac{e^{-8}8^5}{5!} = .092$$

(b) For a half-hour period, we have $\lambda t = (1/2)8 = 4$.

$$P(\text{at most 3 arrivals}) = P_0(1/2) + P_1(1/2) + P_2(1/2) + P_3(1/2)$$

$$= e^{-4}(1 + 4 + \frac{4^2}{2!} + \frac{4^3}{3!}) = .433$$

(c) The answers are $\lambda t = 16$ in the first case and $\lambda t = 8/3$ in the second case.

The following are important illustrations of counting processes that are Poisson in character, or nearly so.

1. The number of cars per unit time passing a particular observation point.
2. The number of customers per unit time entering a grocery store, a checkout line, a barbershop, and so on.
3. The number of breakdowns per unit time of a piece of equipment, if it is due to the failure of a component that can be immediately repaired or replaced (this is the paradigm renewal process),
4. The number of defects per unit length of wire, fence, or cloth.
5. The number of galaxies per unit volume of space.

There exists an important relation between the Poisson arrival process just described and the negative exponential distribution considered in Section 2.1. Consider the Poisson process depicted by the time axis shown in Figure 2.12, with arrival times $t_1, t_2, t_3, \ldots$. Often the times between arrivals are of interest, since these may represent trouble-free periods between breakdowns, usable lengths of wire, or breaks

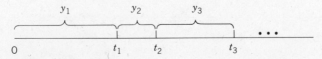

Figure 2.12   A typical Poisson process and interarrival times $\{y_k\}$.

in traffic flow when pedestrians can cross a street or cars can merge with traffic, and so on. Thus it may be that the *interinterval* times are of as much interest as the times $\{t_i\}$ themselves. It is necessary to be very careful about notation here, since there are so many different times involved in the discussion. The arrival times designated as $\{t_i\}$ are the times ("epochs" is the word used by some for these instants) that would be observed on a stopwatch started at some instant labeled $t = 0$ and then never reset. The periods *between* the Poisson arrivals are the elapsed times that would be observed on a stopwatch reset to zero at each Poisson arrival. How shall we designate them? Some authors call them $\{\tau_i\}$, some call them $\{t_i'\}$. Both notations, in my experience, have caused student confusion, so I have designated them as $\{y_i\}$. They are all realizations of the same continuous random variable which is the length of time between successive arrivals; let us call it T.

Does it bother you that $\{y_i\}$ are observed values of T? If so, then call it Y, but remember that this Y is the same length-of-life variable we called T in Section 2.1. Be warned: at some future time (specifically, in Chapter 5) we will want to regard the times between arrivals, or lifetimes, as the fundamental random variables to be designated as $\{t_1\}$, and the Poisson arrival instants as the secondary or derived variables to be assigned some other designation. I reserve the right to change notation when it seems appropriate ("it means what I declare it to mean"). There should be no trouble if close attention is paid to the definitions at each stage. Since T is a (nonnegative) random variable, it has its own density function $f(y)$ and its own CDF (cumulative distribution function or unreliability function) $F(y)$.

Again, we are in a situation for which the notation is awkward. If you are a beginner, you will feel confused that the random variable T has density $f(y)$ and CDF $F(y)$. But recall that y is just a numerical value, not a random variable. You would not be disturbed if we wrote $F(7) = P(T \le 7)$. The notation $F(y) = P(T \le y)$ represents a similar situation. To digress for a moment, consider an example that may help clarify the issue. Consider two distinct random variables X and Y, each with its own probability law. Let

X = number of boxes of SUDSO sold per day,

Y = number of boxes of CLEANO sold per day.

We define the CDFs for the two variables as $F(x)$ and $G(y)$, where $F(x) = P(X \le x)$ and $G(y) = P(Y \le y)$. Suppose we wish to consider the probability that on a certain day the sales of the two products are equal. If the common sales level is specified, say 2000 boxes, we feel no awkwardness at writing $F(2000) = G(2000)$. But what if the common sales level is not specified? Then the best we can do is to write $F(x) = G(x)$ or $F(y) = G(y)$ and, in either case, one or the other of the CDFs seems to be related to the wrong argument. This is our situation here; with practice we can adjust to such awkward statements. At any rate, the form of the CDF may be determined as follows.

Let two observers A and B be stationed (at time $t = 0$) to watch a Poisson process whose intensity is λ, and let them continue to watch until the first arrival occurs at time $t_1(=y_1)$. Before that happens, at some time $t < t_1$ each observer is asked for a report. Assume that A has been equipped with a stopwatch for recording the arrival time. If queried about the state of the system at time t, he may say, "I don't know what the first interarrival time is going to be, but I do know that it will be greater than the present time t." In other words, $Y > t$. The second observer B has been equipped with a counter: her job is simply to count the number of arrivals. At time t (before the first arrival), if queried about the state of the system, she can respond, "I haven't counted anybody yet"—that is, $N(t) = 0$. Now these statements

$$T > t$$

and

$$N(t) = 0$$

are simply two descriptions of the same event—no arrivals by time t. Hence they have the same probability and we can write

$$P(T > t) = P(N(t) = 0) \tag{2.52}$$

In terms of the unknown probability law $F(\cdot)$ for interarrival times that was mentioned earlier, the left-hand side of (2.52) is

$$P(T > t) = 1 - P(T \le t) = 1 - F(t) \tag{2.53}$$

whereas the right-hand side for a Poisson process of intensity λ is

$$P[N(t) = 0] = P_0(t) = e^{-\lambda t} \tag{2.54}$$

Thus the first interarrival time has the property that, equating (2.53) and (2.54),

$$1 - F(t) = e^{-\lambda t} \tag{2.55}$$

and thus

$$f(t) = \lambda e^{-\lambda t} \tag{2.56}$$

Because of Postulate 4 (the stationarity postulate), the second interarrival time must obey the same law, as must the third, and so on. We can summarize these results in the following theorem.

THEOREM 2.2

The interarrival times of a Poisson process with intensity λ are distributed negative exponentially with the same parameter λ.

Comment. As was seen in Exercise 2.2.3, the mean of the NGEX(λ) distribution is $1/\lambda$. By Theorem 2.2 a Poisson process with $\lambda = 10$ arrivals per hour will give rise to a distribution of interarrival times whose mean is 0.1 hour. This accords well with our intuition. The part of the theorem that is counterintuitive is that the shape of the distribution should be negative exponential. One can only respond by reminding the student that not everything true is obvious and that not everything that is obvious is true.

Another important characteristic of the negative exponential distribution can now be discussed, one that depends on the form of its density function but is related to its origin in the Poisson process. Let us imagine a hitchhiker who positions himself by the side of a lonely road where he has heard that the traffic flow is Poisson with mean rate (in his lane) of $\lambda = 6$ cars per hour. He understands correctly that on the average one car will go past every 10 minutes. After standing in the cold for 30 minutes, he may think, "Well, I haven't seen any cars during my long wait, so by the law of averages there is extremely high probability that a car will come in the next minute or two." Is he correct? Alas, no. Let us see why.

We are concerned with, say, the probability that he must wait at least 10 minutes *more*, *given that* he has already waited for 30 minutes. If we let

$T =$ the waiting time until the next car,

$S =$ the time he has already waited,

$t =$ the extra time he must wait, starting now

we want to know

$$P(T > t + S \mid T > S) \tag{2.57}$$

where $S = 30$ and $t = 10$. Recalling the rule for conditional probabilities, that

$$P(A|B) = \frac{P(A \text{ and } B)}{P(B)}$$

we write (2.57) in the form

$$\frac{P(T > t + S \text{ and } T > S)}{P(T > S)} \tag{2.58}$$

Now the intersection (*and*) of the event $\{A: T > t + S\}$ with the event $\{B: T > S\}$ is the event $\{A: T > T + S\}$ itself. Here A implies B; if a Venn diagram were drawn, the A circle would be entirely inside the B circle. Hence (2.58) can be written as

$$\frac{P(T > t + S)}{P(T > S)}$$

and applying (2.55) to evaluate the probabilities, we get

$$P(T > t + S \mid T > S) = \frac{e^{-\lambda(t+s)}}{e^{-\lambda S}}$$
$$= e^{-\lambda t} \tag{2.59}$$

This equation can be put in the following words. The probability of waiting t *more* minutes, given that there has already occurred a wait of length S of any size (including zero), depends not at all on the preliminary wait, but only on the added period of interest. This is the *no-memory*, or *forgetfulness*, property of the negative exponential distribution. It says that components obeying this failure law "have no history"; they are always as good as new until they fail. Fuses work like this, don't they? When they fail, it is not because they have worn out, but because of a surge of current—like a Poisson arrival. It is for this reason that the middle portion (B) of the bathtub curve in Figure 2.8 is said to correspond to chance (random) failures. Components in this portion of the curve will fail—not because they are worn out, but because of outside influences (jolts) that occur in a Poisson manner. Such components, if they could talk, might be given to preaching slogans like "Every day is a brand new beginning" or "What's past is past—the future begins now." And for them it is true.

These components are in contrast with those in the wear-out portion of the curve, which behave like the components of Example 2.3. The normal distribution considered there does display a memory property, since

$$P(T > 100) \neq P(T > 600 \mid T > 500)$$

EXERCISES 2.3

1. Master the details of deriving Equations 2.47 and 2.48 from the Poisson postulates 2.39 through 2.42.

2. Master the details of the proof that a Poisson process has exponential interarrival times.

3. Master the details of deriving the no-memory property of the exponential distribution.

4. A Poisson process has intensity $\lambda = 60$ arrivals per hour.
   (a) What is the expected number of arrivals in the next half hour? What is the probability that the actual number of arrivals in the next half hour will in fact equal the answer you stated?
   (b) Answer the questions in part a for a fifteen-minute interval.
   (c) What is the probability that in the next ten minutes there will be exactly five arrivals? At least five arrivals?

5. Assuming the Poisson distribution of defects, if a certain type of fencing has on the average two defects per 100 linear feet, what is the probability that the next 200 feet from the roll will be defect-free?

6. In the previous problem, if the previous 50 feet of fence sold had no defects, what is the probability that the next 100 feet will have no defect? At least one defect?

7. A Poisson process has intensity (arrival rate) $\lambda = 3$ arrivals per minute.
   (a) What is the probability of no arrivals in a one-minute interval.
   (b) What is the probability of three arrivals in the next ten minutes?

(c) How many arrivals can be expected in the next ten minutes?

(d) What is the probability of three arrivals in the next ten minutes, if there have already been ten in the previous ten minutes?

(e) What is the average length of time with no arrivals?

(f) What is the probability of no arrivals during this average interval?

8. You live on a country road whose average traffic rate in each direction is 30 cars per hour. A hitchhiker is standing by the side of the road waiting for a lift. He knows that 99.9 percent of people will not stop for a hitchhiker under any circumstances.

(a) What is the effective arrival rate for him?

(b) How long could he expect to wait for a possible ride?

(c) What is the probability of his having to wait at least 30 minutes for a ride?

2.4 MORE ABOUT THE POISSON AND NEGATIVE EXPONENTIAL DISTRIBUTIONS

An important question asked by the reliability practitioner is how to ascertain whether or not some renewal (arrival) process of interest is really Poisson or not? There are a number of goodness-of-fit tests that may be performed.

Test I In the classical χ^2 goodness-of-fit test studied in basic statistics courses, the process is watched for many periods of constant length, say T_0, and the arrival count for each period constitutes one observation. If the process is watched for n periods, the corresponding observed counts $\{f_1, \ldots, f_n\}$ may be tested for Poissonness using the χ^2 goodness-of-fit test. There are two drawbacks to this test. One is that the observation periods must be of equal length. Another is that a large number of observations are required in order to have a reasonable number of degrees of freedom.

An advantage of the test is that counts are sometimes relatively simple to obtain. In bygone years, students hired as summer workers by the highway department used to sit at the side of the road pressing a "clicker" (counting device) as cars went by; nowadays an automatic counter imbedded in or laid across the road does the same job. If the process under study is a working system based on some equipment, the number of failed components turned in for replacement per day can serve as a failure count.

Example 2.10

The number of defects per roll of wire in a day's production of 200 rolls is as follows.

Defect count	0	1	2	3	4	5	6	7	8	
Number of rolls	7	32	50	42	35	20	5	6	3	200

Is it reasonable to use the Poisson distribution as the model for failure count per roll? (Use significance level .05.)

Solution: Since no value for the Poisson parameter λ is suggested, we use $\hat{\lambda}$ as estimated by the sample mean; that is, $\bar{x} = (\Sigma f_i x_i)/200 = 594/200 = 2.97$. Table 2.1 shows the needed calculations, the p_i values being obtained from the formula

$$p_i = f(x_i) = e^{-2.97}(2.97)^x/x!$$

The so-called computed χ^2, defined as

$$Q = \sum_{i=1}^{k} \frac{(f_i - e_i)^2}{e_i}$$

$$= \sum_{i=1}^{k} \frac{f_i^2}{e_i} - n$$

has the value 5.14, which is to be compared with $\chi^2_{.05}(7) = 14.067$. Q lies in the acceptance region for the hypothesis test, so we cannot reject the Poisson model at the .05 level. Notice that the last expected cell frequency of 2.4 was not considered small enough to require lumping with the adjacent cell. The degrees of freedom for the test for $k = 9$ classes are $k - 1 - 1 = 7$.

TABLE 2.1   Work Table for Example 2.10

x_i	p_i	$e_i = 200\,p_i$	f_i
0	.051	10.2	7
1	.152	30.4	32
2	.226	45.2	50
3	.224	44.8	42
4	.166	33.2	35
5	.099	19.8	20
6	.049	9.8	5
7	.021	4.2	6
≥ 8	.012	2.4	3
	1.000	200.0	200

Test II   Another classical χ^2 goodness-of-fit test is based on the interarrival times. We assume that the process is observed for one or more periods of time and that the arrival times $\{t_i\}$ are noted and recorded. The interarrival times $\{y_i\}$ are then constructed from the relationship $y_i = t_i - t_{i-1}$ where $t_0 = 0$. If the data come from more than one observation period, they can be "spliced" together by ignoring the

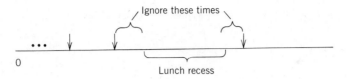

Figure 2.13 An interrupted Poisson process.

incomplete interarrival time at the end of period one and the beginning of period two, as depicted in Figure 2.13. It can be proved[1] that the incomplete periods have a different distribution than the complete ones. The observed $\{y_i\}$ are then tested for exponentiality using the χ^2 goodness-of-fit test. Although the total observation time in this test may be substantially less than that for Test I, a great number of observations are still needed to provide a reasonable number of degrees of freedom. Furthermore, it may be considerably more difficult to record arrival times than merely counts.

Example 2.11

A renewal process was observed until 100 failures had been noted. The individual lifetimes of the items were recorded and then categorized as shown in the leftmost two columns of Table 2.2. Is it reasonable to believe that the observed process might have been Poisson with failure rate $\lambda = 0.01$ failures per hour?

TABLE 2.2 Work Table for Example 2.11

Life Length	Count	$F(t)$	p_i	e_i
0 to 20 hours	20	.181	.181	18.1
20 to 40 hours	16	.330	.149	14.9
40 to 60 hours	12	.451	.121	12.1
60 to 80 hours	10	.551	.100	10.0
80 to 100 hours	5	.632	.081	8.1
100 to 150 hours	17	.777	.145	14.5
150 to 200 hours	7	.865	.088	8.8
200 to 300 hours	10	.950	.085	8.5
300 to 400 hours	3	.982	.032	3.2
More than 400 hours	0	1.000	.018	1.8
	100		1.000	100.0

---

[1] See Feller, 1966, page 11.

Solution: The rightmost columns contain the calculations for the χ^2 goodness-of-fit test on the interarrival times using the model

$$f(t) = \lambda e^{-\lambda t}$$
$$F(t) = 1 - e^{-\lambda t}$$

The cell probabilities p_i are computed from $p_i = F(t_i) - F(t_{i-1})$.
The test statistic is

$$Q = 20^2/18.1 + \ldots + 10^2/8.5 + 3^2/5.0 - 100$$
$$= 3.33$$

with $9-1=8$ degrees of freedom. Clearly there is no need to reject the null hypothesis. (Notice the unequal classes in this example.)

The next test is a very nice one for small amounts of data. It depends on a pair of theorems that are not well known and deserve wider dissemination. Their proofs are not provided until later (see Appendix 5D) because they depend on the properties of order statistics, which will be considered in Chapter 5.

THEOREM 2.3

If a Poisson process is monitored for a predetermined period of time T_0, and if there are n arrivals during that period, the n arrival times $\{t_i\}$ ($i = 1, \ldots, n$) look like (or have the same distribution as) the ordered observations from a random sample of size n from the continuous uniform $CUD(0,T_0)$ distribution.

THEOREM 2.4

If a Poisson process is monitored for a predetermined number n_0 of arrivals, the previous arrival times $\{t_i\}$ ($i = 1, \ldots, n_0-1$) look like (or have the same distribution as) the ordered observations from a random sample of size $n_0 - 1$ from the continuous uniform distribution $CUD(0,t_{n_0})$.

The beauty of these theorems is that they allow the experimenter to "eyeball" a plot of the data and decide intuitively whether the assumption of Poissonness is appropriate. Of course, some experience with samples from CUDs is necessary for dependable judgments. (For a more formal treatment, a Kolmogorov–Smirnov test can be performed.) In Figure 2.14 we show a collection of simulated Poisson processes, obtained as follows.

A set of random numbers can be used to simulate a Poisson process with the parameter λ by recognizing that the interarrival times $\{y_i\}$ have CDF

$$F(y) = 1 - e^{-\lambda y} \tag{2.60}$$

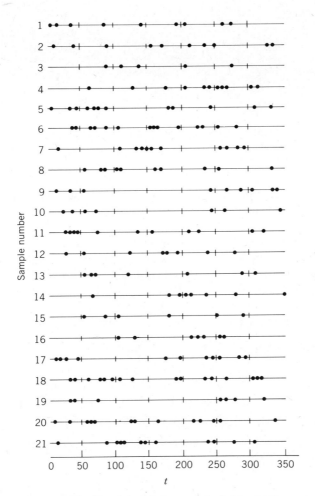

Figure 2.14 Simulated Poisson arrival processes.

Since $0 \le F(y) \le 1$, we simulate various y_i by using a random number u_i from CUD(0, 1) as a sample value for $F(y)$. (Often we use u_i for $1 - F(y)$, since $1 - u_i$ is just as random as u_i itself.) This process of simulation can be explained somewhat intuitively by reference to Figure 2.15. Table 2.3 contains the details of the construction of the simulated Poisson process times $\{t_i\}$. The mean of the simulated interarrival times is $\bar{y} = 666.0$ and the standard deviation is $S = 500.9$.

This sample shows how "undependable" the negative exponential distribution is, in terms of parameter estimates. The form of the density function, $f(t) = \lambda e^{-\lambda t}$, means that although there is a preponderance of small values there is a "heavy tail" that corresponds to a nontrivial number of very large values. If a particular set of random numbers lacks large values, the sample values generated will markedly underestimate the true mean $1/\lambda$. To see better how this works, notice that for even a "perfect"

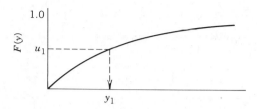

Figure 2.15 The graphical interpretation of Equation 2.60.

distribution of random numbers like $\{ .1, .2, \ldots , .9 \}$, the value of $\bar{y}$ is only .8802 of the theoretical population mean.

To test the simulated arrival times of Table 2.3 for Poissonness, we can use the Kolmogorov–Smirnov test. For small samples like the present one, it is an appropriate alternative to the χ^2 goodness-of-fit tests illustrated earlier in this section. The test is based on measuring the largest vertical distance between the empirical CDF step function $S_n(t)$ for a data set and the theoretical CDF $F(t)$ which is being tested. That largest distance may occur just to the "front" or just to the "rear" of a step in the plot of $S_n(t)$, so we must compute the double set of differences,

$$\left| S_n(t_{i-1}) - F(t_i) \right| , \left| S_n(t_i) - F(t_i) \right| , \quad i = 1, \ldots, n$$

and find the largest value D among all these. Based on Theorem 2.4, we regard the values in the last column of Table 2.3 as ordered random values from a continuous uniform distribution over the interval (0, 3330) for a sample size of four. More formally, to test Poissonness we test the equivalent hypothesis

$$
\begin{aligned}
H_0 : F(t) &= t/3330 &\quad &\text{for } 0 \leq t \leq 3330 \\
&= 0 &\quad &\text{for } t < 0 \\
&= 1 &\quad &\text{for } t > 3330
\end{aligned}
$$

$H_1 : F(t)$ is not as given

TABLE 2.3 Simulation of a Poisson Process with $\lambda = .001$ (Using $F(x) = u_i$)

Random Number u_i	Interarrival Time y_i	Poisson Arrival Time $t_i = \sum_{k=1}^{i} y_k$
.7644	1446	1446
.2339	266	1712
.2186	247	1959
.4018	514	2473
.5756	857	3330

TABLE 2.4 Kolmogorov–Smirnov Test for the Simulated Sample of Table 2.3

Poisson Arrival Time	Index	$F(t_i)$	$S_n(t_i)$	$S_n(t_{i-1})$	$\|F(t_i) - S_n(t_i)\|$	$\|F(t_i) - S_n(t_{i-1})\|$
1446	1	.43	.25	0	.18	.43
1712	2	.51	.50	.25	.01	.26
1959	3	.59	.75	.50	.16	.09
2473	4	.74	1.00	.75	.01	.26

$D = \max\{.18, .01, .16, .01, .43, .26, .09, .26\} = .43$

As always, the empirical CDF is

$$S_n(t_i) = \frac{i}{n}$$

The calculations are shown in Table 2.4. The observed largest distance $D = .43$ is compared with $D_{.05} = .624$ for $n = 4$ from Table T10. The observed $D = .43$ is less than the critical value $D_{.05}$, so the proposed null hypothesis cannot be rejected at the $\alpha = .05$ level.

EXERCISES 2.4

1. (a) Use the random numbers of Table T9 (or the random number generator of your electronic spreadsheet or pocket calculator) to create a set of twenty simulated CUD(0,1) observations. Solve formula 2.60 for y and use the result to create twenty simulated lifetimes with $\lambda = .01$ (so mean life is 100). Call these generated lifetimes $\{y_i\}$ and use them as interarrival times to create a Poisson process of arrival instants. Plot the Poisson arrival instants on a time axis. Do the points look fairly evenly spread over the range of observation? This exercise is quick and easy to do using an electronic spreadsheet.

   (b) Use the Kolmogorov–Smirnov test to decide whether your created arrival times can be considered Poisson.

2. Construct the plot for the example in Tables 2.3 and 2.4, showing the straight-line CDF $F(t)$ and the step function $S_n(t)$. Verify, by "eyeballing" the set of "front" and "rear" differences in the last two columns of Table 2.4.

   Comment. The student who is adept with electronic spreadsheets can create this plot using a microcomputer. Can you? List the t_i values in column A, including the endpoints of the interval. Insert before each observation some extra values that are infinitesimally smaller, so that each t_i is nested tightly below by at least one other number. In column B assign the $Sn(t)$ value i/n and in column C the computed $F(t)$ value. Graph these three columns as TYPE XY. If the hypothesized CDF is not linear, many more values are needed to display its shape; the DATA/FILL command may be used, then supplemented with inserted rows to provide the needed nesting of observations.

3. Repeat Exercise 1 using a different set of random numbers.

4. Repeat Exercise 1a ten times and plot the resulting simulated Poisson processes on the same sheet of graph paper, to get a feel for what a Poisson process looks like. Your result should be similar to Figure 2.14.

Comment.   If you do not like hand calculations and hand plotting, you can learn to do this exercise using an electronic spreadsheet. There is a trick to getting the individual samples to appear one over the other. Clearly each sample must have a different *y*-value associated with it. Can you figure out how to do it? Putter around with it for a while and if necessary, come back here for more information.

Answer.   If you couldn't figure it out, don't feel badly; one of my very bright students had to give me a hint. Suppose you have created two Poisson processes in column A and column B (say) of your spreadsheet. Insert extra columns to the right (say) of columns A and B so that now your samples appear in columns A and C. In column B insert the constant value 0 all the way down to the end of your sample, and in column D insert the value 1 all the way down. These are the *y*-values for the two samples.

Move the two-column block C–D to the area below the A–B sample block. Column A is now twice as long and contains all the arrival times for both samples; column B contains their corresponding *y*-values. A DATA/SORT procedure on column A will intermesh the arrival times in natural sorted order. Now when you use the GRAPH command (symbols only!), sample 2 will appear one unit above sample 1. Clearly this procedure can be adapted to any number of samples. For better visibility you might want to lift sample 1 above the abscissa a little bit by using .1 (say) as its *y*-value instead of zero.

2.5   THE NEGATIVE EXPONENTIAL DISTRIBUTION AS A FAILURE LAW MODEL

Throughout its development the theory of reliability has been based heavily on the negative exponential failure law, primarily because of its mathematical tractability. It is the appropriate model for used-good-as-new components, like fuses and many other electronic parts, because of the forgetfulness property described in Section 2.3. The author was for a long time chagrined to spend excessive time in class discussing a distribution whose mode is at zero—a most counterintuitive situation. Reconciliation to the use of the negative exponential can be achieved based on the following considerations.

First, it may be reasonable to assume that for a particular mission of interest, the failure rate of an equipment is—if not constant—then fairly well bounded by a lower limit λ_1 and an upper limit λ_2, so that $\lambda_1 < h(t) < \lambda_2$. Then it will be true that the corresponding reliability function R(t) is bounded by

$$e^{-\lambda_2 t} \le R(t) \le e^{-\lambda_1 t} \tag{2.61}$$

Sometimes a lower bound on reliability may be all that is needed.

To assuage the feeling of nervousness that may be aroused by dealing with a distribution whose mode is at zero, the author has found it fruitful to study an analogous discrete distribution, the geometric. This is the *waiting-time* distribution associated with Bernoulli trials. A series of experiments constitute a set of Bernoulli trials if the following conditions hold.

1. The outcomes of each trial are a dichotomy of mutually exclusive events— success or failure, heads or tails, defective or good, and so on.
2. The trials are independent.
3. There is constant probability for each outcome from trial to trial.

If, in addition to the Bernoulli conditions, there is a predetermined number of trials, the usual binomial distribution is generated. If experimentation is continued until some fixed number of successes is observed, however, the result is a waiting-time distribution, here the negative binomial or Pascal distribution. To be specific, let

k = the desired number of successes (a predetermined constant)

X = The number of trials needed to produce the k successes (a random variable)

p = the probability of success at each trial

$q = 1 - p$

Then the probability mass function (pmf) for the random variable X is

$$f(x:k,p) = \binom{k-1}{x-1} p^k q^{x-k}, \quad x = k, k+1, \ldots \quad (2.62)$$

The name *negative binomial* is associated with this distribution because (2.62) is the $(x - k + 1)^{\text{th}}$ term in the expansion of $(1/p - q/p)^{-k}$, just as the binomial mass function

$$b(x:n,p) = \binom{n}{x} p^x q^{n-x}, \quad x = 0, 1, \ldots, n \quad (2.63)$$

is the xth term in the expansion of the binomial expression $(q + p)^n$. The mean of the negative binomial distribution is k/p and its variance is kq/p^2.

Suppose now that $k = 1$, so that we are waiting for the first success. The probability function reduces to

$$g(x:p) = q^{x-1}p, \quad x = 1, 2, \ldots, \quad (2.64)$$

a set of monotone decreasing values, so that just as in the negative exponential distribution, the smallest value possible is the most likely value! This is true no matter how small p is. Even though a success has a probability of only .001, we are more likely to achieve the first success on the first trial than on the second. Although the expected number of trials to reach success is $1/p = 1000$, we are more likely to achieve the success on the first trial than on the 1000th. Specifically, $g(1, .001) = .001$ whereas $g(100, .001) = .00037$.

The similarity between the geometric distribution and the negative exponential distribution does not end with the smallest-as-mode property. Although it was not emphasized earlier, the negative exponential is also a waiting time distribution—the waiting time until a Poisson arrival. The geometric distribution is the unique memoryless discrete distribution, just as the negative exponential is the unique memoryless continuous distribution. The proof is similar to that of Section 2.3. Consider that for the geometric distribution

$$P(X > x) = g(x + 1) + g(x + 2) + g(x + 3) + \dots$$

$$= q^x p + q^{x+1} p + q^{x+2} p + \dots$$

$$= q^x p(1 + q + q^2 + \dots)$$

$$= q^x p/(1 - q)$$

$$= q^x \tag{2.65}$$

Then the probability of having to wait more than s more trials for a success, given that one has already waited more than x trials, is

$$P(X > s + x \mid X > x) = \frac{P(X > s + x \text{ and } X > x)}{P(X > x)}$$

$$= \frac{P(X > s + x)}{P(X > x)}$$

$$= \frac{q^{s+x}}{q^x}$$

$$= q^s$$

$$= P(X > s) \tag{2.66}$$

This result is exactly analogous to that of Equation 2.59, but this time it seems more natural because we have more experience in this area. For example, consider the case where an unbiased coin is being tossed until the first head appears. Most of us know that even though we may have tossed the coin four or five times with a solid run of tails, the coin doesn't remember the past and it isn't going to try harder to turn up a head for us to satisfy some vague notions we may have about "laws of probability."

A final similarity between the geometric and exponential distributions which may be of interest is the form of the moments (mean and variance) as well as of the functions themselves, when we recognize that any positive number can be written as some power of the transcendental number e. In particular, for $0 \le q \le 1$, we can write $q = e^{-\lambda}$ for some $0 \le \lambda < \infty$. The striking similarities between the two distributions are laid out in Table 2.5. The analogies are not perfect but are very suggestive, and our better intuition about the geometric distribution may be of service in helping us to get a feel for the negative exponential distribution.

We have seen two different methods for recognizing the exponential failure law. The first is to use the data analysis technique of Chapter 1 to see whether the resulting

TABLE 2.5 Comparisons Between the Negative Exponential and the Geometric Distributions

Aspect of Distribution	Negative Exponential	Geometric
Random variable	Amount of elapsed time until an "arrival"	Number of trials until a "success"
Type	Continuous, $x > 0$	Discrete $x = 1, 2, \ldots$
Awaited event	A Poisson arrival	A Bernoulli success
Density (mass) function	$\lambda e^{-\lambda x} = \lambda(e^{-\lambda})^x$	$q^{x-1}p = \dfrac{p}{q}\, q^x$
Right-hand tail $P(X > x)$	$e^{-\lambda x} = (e^{-\lambda})^x$	q^x
Mean	$1/\lambda$	$1/p$
Variance	$1/\lambda^2$	q/p^2
Distribution shape	Time	Trial

bar graph plot of $f^*(t)$ has the appropriate exponential appearance. This technique is appropriate when a number of units have been put in service or on test simultaneously and failure times noted. Unfortunately, as will be demonstrated in Section 2.9, the method is not very discriminating. A second technique is appropriate when observing a renewal process, and depends on the uniform appearance of the actual failure times, described in Theorems 2.3 and 2.4. If the interarrival times from the renewal process are then calculated and ranked in ascending order, the data are in a form where the first test could be applied.

Epstein (1960a,b) suggests a third (graphical) procedure for checking exponentiality. It depends on the formulae

$$f(t) = \lambda e^{-\lambda t} \tag{2.67}$$

and

$$F(t) = 1 - e^{-\lambda t} \tag{2.68}$$

so that

$$\lambda t = -\ln[1 - F(t)] \tag{2.69}$$

or

$$\lambda t = \ln[1 - F(t)]^{-1}$$

If we plot the ordinate y_i defined by

$$y_i \equiv \ln[1 - F(t)]^{-1} \tag{2.70}$$

against t_i, the result is a straight line with slope λ. Of course, it is assumed that $F(t)$ is not known (otherwise we wouldn't be making the test), so we use the estimate

$$\hat{F}(t_i) = i/(n + 1) \tag{2.71}$$

We use this estimate because $E[F(T_{(i)})] = i/(n + 1)$. (Section 2.10 gives additional explanation.) Then y_i in Equation 2.70 assumes the form

$$y_i = \ln\left(\frac{n + 1}{n + 1 - i}\right) \tag{2.72}$$

Table 2.6 shows simulated sample data and Figures 2.16a and b give the associated graphs. The parameter λ may be estimated graphically as the slope of the line, or arithmetically as $1/\bar{t}$ (provided that $A = 0$).

Remark. Geometrically, a line with positive slope may have any of the three forms shown in Figure 2.17. The zero-intercept case is the one we have been discussing. It is possible that a set of data could plot like Figure 2.17c with t-intercept of A, say. Then the model is not Equation 2.69, but

$$f(t) = \lambda e^{-\lambda(t-A)}, \qquad t \geq A \tag{2.73}$$

TABLE 2.6 Two Examples Patterned after Epstein (1960b)

Forty-nine items were placed on life test and the test was terminated after all items had failed. The observed failure times were as given.

			Example 2.12a			
0.2	8.9	30.1	53.2	62.6	119.7	212.0
0.5	9.0	31.8	53.9	74.0	130.4	234.5
1.2	12.6	34.4	55.3	78.8	135.4	250.2
1.8	12.8	39.2	55.4	103.2	157.9	265.8
1.9	19.7	39.4	57.4	108.6	158.9	373.1
3.0	20.3	40.7	58.7	109.7	175.4	403.2
7.4	23.2	40.9	59.7	115.0	191.4	463.7
			Example 2.12b			
212	331	489	726	1072	1715	2455
221	363	527	800	1299	1716	2546
223	425	610	862	1327	2039	2824
248	432	653	868	1521	2132	2912
264	442	692	869	1627	2193	3841
269	451	714	945	1659	2214	3943
292	461	724	947	1673	2293	4188

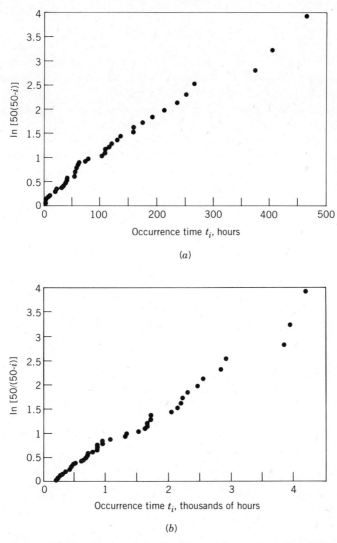

Figure 2.16 Plots of data from Example 2.12 in Table 2.5.
(a) Data from Example 2.12a. (b) Data from Example 2.12b.

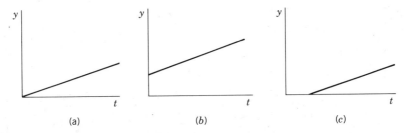

Figure 2.17 Lines with positive slope. (a) Zero intercept. (b) Positive intercept. (c) Negative intercept.

This is the *shifted exponential* we met in Section 2.1. Presumably it corresponds to some kind of presorting process by means of which all early failures less than A have been eliminated. (How this could be done is a mystery.)

The configuration of Figure 2.17b is not possible, because there can be no jump in the plot from 0 to d, say, on the y axis; the value $t = 0$ for the zero[th] observation corresponds to $y = 0$, so the plot must go through the origin. But it is possible to realize something similar to Figure 2.17b, namely the type of behavior shown in Figure 2.18. This situation can arise when there is contamination of the population by a few duds. The population is then a *mixture* of the main population with the failure law given by Equation 2.69, and a small portion of poor components with the failure law

$$f_1(t) = be^{-bt} \tag{2.74}$$

where $b > \lambda$. The resulting mixture density has the form

$$f^*(t) = cf_1(t) + (1 - c)f(t) \tag{2.75}$$

where $0 \le c \le 1$ is the fraction of contamination. The mixture CDF is

$$F^*(t) = cF_1(t) + (1 - c)F(t) \tag{2.76}$$

Changing signs and adding unity to both sides of the equation gives

$$1 - F^*(t) = c[1 - F_1(t)] + (1 - c)[1 - F(t)] \tag{2.77}$$
$$= ce^{-bt} + (1 - c)e^{-\lambda t}$$

Thus

$$y = \ln[1 - F^*(t)]^{-1} \tag{2.78}$$

has the form

$$y = \ln[ce^{-bt} + (1 - c)e^{-\lambda t}]^{-1} \tag{2.79}$$

which, upon multiplication of the bracket by $e^{\lambda t}e^{-\lambda t}$, results in

$$y = \ln e^{\lambda t}[1 - c + ce^{-(b-\lambda)t}]^{-1} \tag{2.80}$$
$$= \ln e^{\lambda t} - \ln[1 - c(1 - e^{-(b-\lambda)t})] \tag{2.81}$$
$$= \lambda t - \ln D \tag{2.82}$$

where the quantity D (for denominator) is a positive fraction, so that (2.82) can be written as

$$y = \lambda t + d$$

with $d = -\ln D$. As $t \to \infty$, D approaches $1 - c$, so for a small amount of contamination (2.82) is very nearly

$$y \simeq \lambda t$$

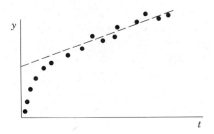

Figure 2.18 Contamination by duds.

A third example, similar to one provided by Epstein (1960a,b), is depicted in Figure 2.19; Table 2.7 shows an example of this type of behavior. The monograph by Jensen and Peterson (1982) is recommended for a study of burn-in and contamination by duds.

The graphical method of detecting the shifted exponential or contamination by duds is appealing, but I do not recommend it with much enthusiasm. Inspection of Table 2.6, Example 2.12b, and Table 2.7 makes it clear to the experienced user that the data are contrived, and that no plotting is necessary to decide that something is "fishy" about the numbers. In fact, it has been difficult for me to create by simulation any data sets of these two types that could be picked out by graphical methods without being more readily recognized by simply eyeing the numbers.

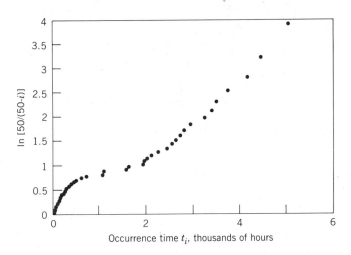

Figure 2.19 Plot of the data in Table 2.6 depicting contamination.

TABLE 2.7 Contamination by Duds

Forty-nine items were placed on life test and the test was run until all items had failed. The observed failure times were as given.

40	92	192	373	1100	2258	3232
52	92	192	398	1586	2437	3392
55	125	252	462	1637	2548	3480
61	145	272	502	1933	2628	2628
72	161	275	633	1956	2725	2725
76	167	285	732	2014	2802	2802
86	170	312	1080	2107	2928	2928

EXERCISES 2.5

1. Use Epstein's graphical procedure to determine the exponentiality of the failure law of the following ranked interarrival times:

   {5, 9, 12, 24, 27, 34, 39, 39, 45, 65, 66, 83, 99, 102, 103, 106, 108, 173}

2. Repeat Exercise 1 for the data set
   {0, 1, 1, 1, 1, 1, 2, 2, 3, 4, 6, 8, 10, 15, 16}

3. Repeat Exercise 1 for the data set
   {5, 14, 15, 20, 26, 32, 40, 42, 44, 47, 61, 67, 78, 106, 122, 140, 167, 191, 204, 210}

4. Repeat Exercise 1 for the data set
   {3, 5, 44, 47, 50, 52, 79, 81, 97, 142}

5. Repeat Exercise 1 for the data set
   {22, 33, 54, 82, 86, 97, 120, 143, 209, 252}

6. Repeat Exercise 1 for the data set
   {2, 4, 6, 10, 16, 31, 41, 43, 49, 75, 81, 109, 123, 129, 142}

7. Repeat Exercise 1 for the data set
   {76, 220, 243, 385, 521, 771, 871, 935, 1002, 1059, 1773, 1941}

8. Repeat Exercise 1 for the data set you created in Exercise 2.4.1.

2.6 THE GAMMA DISTRIBUTION

In future chapters there will be need for working with the gamma distribution which is defined by the density function

$$f(x) = Kx^{\alpha-1}e^{-x/\beta}, \qquad \alpha, \beta > 0, x \geq 0 \tag{2.83}$$

where K is a normalizing constant chosen to give unit area under the curve. A random variable with density given by (2.83) will hereafter be written as

$$X \sim G(\alpha, \beta) \qquad (2.84)$$

The student who is not well acquainted with the gamma function $\Gamma(\cdot)$ is encouraged to consult Appendix 2A before proceeding further.

It is of interest to answer the following questions for the distribution in (2.83).

1. What is the value of the normalizing constant K?
2. What is the mean μ of the distribution?
3. What is the variance σ^2 of the distribution?

Question 1 requires the evaluation of $\int_0^\infty f(x)\,dx$, question 2 requires $\int_0^\infty xf(x)\,dx$ and question 3 requires $\int_0^\infty x^2 f(x)\,dx$. It is therefore convenient to derive a general expression for the rth noncentral moment

$$E(X^r) = \int_0^\infty x^r f(x)\,dx$$

and then set $r = 0$, 1, and 2 in turn. Specifically, we want to evaluate

$$E(X^r) = K \int_0^\infty x^{r+\alpha-1} e^{-x/\beta}\,dx \qquad (2.85)$$

The substitution $y = x/\beta$ converts this into

$$E(X^r) = K\beta^{r+\alpha} \int_0^\infty y^{r+\alpha-1} e^{-y}\,dy$$

$$= K\beta^{\alpha+r} \Gamma(\alpha+r) \qquad (2.86)$$

Now setting $r = 0$ we get

$$E(1) = K\beta^\alpha \Gamma(\alpha) = 1$$

whereby

$$K = [\Gamma(\alpha)\beta^\alpha]^{-1} \qquad (2.87)$$

Setting $r = 1$ and recalling the recursion formula $\Gamma(\alpha+1) = \alpha\Gamma(\alpha)$, we get

$$E(X) = \mu = K\beta^{\alpha+1}\Gamma(\alpha+1)$$

$$= \frac{\beta^{\alpha+1}\alpha\Gamma(\alpha)}{\beta^\alpha\Gamma(\alpha)} \qquad (2.88)$$

$$= \alpha\beta$$

Setting $r = 2$ and recalling that $\Gamma(\alpha+2) = (\alpha+1)\alpha\Gamma(\alpha)$ gives

$$E(X^2) = \alpha(\alpha+1)\beta^2 \qquad (2.89)$$

so that

$$\sigma^2 = E(X^2) - E^2(X)$$
$$= \alpha(\alpha + 1)\beta^2 - (\alpha\beta)^2$$
$$= \alpha\beta^2 \qquad (2.90)$$

The shape of the curve depends on the value of α, hence its usual designation as the *shape parameter* of the distribution. This can be seen as follows. For $\alpha = 1$ expression 2.83 reduces to the negative exponential distribution. For all other values of α, the first two derivatives are

$$f'(x) = K e^{-x/\beta} x^{\alpha-2} [-x/\beta + (\alpha - 1)] \qquad (2.91)$$

$$f''(x) = K e^{-x/\beta} x^{\alpha-3} [(-x/\beta + \alpha - 1)^2 - (\alpha - 1)] \qquad (2.92)$$

Equating (2.91) to zero yields horizontal tangents at ∞, at zero for $\alpha > 2$, and at $x = (\alpha - 1)\beta$ for $\alpha > 1$. Thus if a mode exists, it is at

$$x_0 = (\alpha - 1)\beta \qquad (2.93)$$

Equating (2.92) to zero yields inflection points at

$$x = (\alpha - 1)\beta \pm \beta \sqrt{\alpha - 1} \qquad (2.94)$$

It is seen from (2.83) that $f(x) \to 0$ as $x \to \infty$ for all values of α. The character of the curve is determined by what happens for small values of x, particularly near $x = 0$. There are five cases to consider. These are illustrated in Figure 2.20.

Case 1   For $\alpha < 1$ we have $f(0) = \infty$ and $f'(0) = \infty$. The curve has no mode and it approaches the vertical axis asymptotically. There is no inflection point.

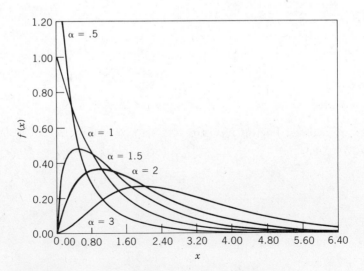

Figure 2.20   Selected gamma density functions.

Case 2 For $\alpha = 1$ we have $f(0) = 1/\beta$ and $f'(0) = -1/\beta^2$. This is the negative exponential case. There is no inflection point.

Case 3 For $1 < \alpha < 2$ we have $f(0) = 0$ and $f'(0) = \infty$. The curve begins at the origin with vertical tangent, increases to its modal value, then decreases. Since $\sqrt{\alpha - 1} > \alpha - 1$ for $a < 2$, there is only one inflection point, the one to the right of the mode.

Case 4 For $\alpha = 2$ we have $f(0) = 0$ and $f'(0) = 1/\beta^2$. The curve comes out from the origin at a positive angle, increases to its modal value, then decreases. There is a degenerate inflection point at the origin and a genuine one to the right of the mode.

Case 5 For $\alpha > 2$ we have $f(0) = 0$ and $f'(0) = 0$. The curve comes out from the origin horizontally and has two genuine inflection points.

Based on the analysis just given, it is possible to sketch any gamma density function with a minimum of "point plotting." Notice that the basic curve shape is determined entirely by the *shape parameter* and not at all by β, which is merely a *scale parameter*.

A distribution of particular interest is that for which $\alpha = \nu/2$ ($\nu = 1, 2, 3, \ldots$) and $\beta = 2$. Such a distribution (or random variable) is said to be a chi-square with ν degrees of freedom. We can write it briefly as

$$G(\nu/2, 2) = \chi^2(\nu) \tag{2.95}$$

Example 2.13

Plot, with a minimum of calculation, the density function for $X \sim G(4, 1/2)$.

Solution: The mean is at $\alpha\beta = 2$, the mode is at $(\alpha - 1)\beta = 3/2$, and the standard deviation is $\beta\sqrt{\alpha} = 1$. There are inflection points at $\beta(\alpha - 1 \pm \sqrt{\alpha - 1}) = (3 \pm \sqrt{3})/2 = 1.5 \pm .866 = 0.634$ and 2.366. Since $\alpha = 4$ the curve is horizontal at the origin. The form of $f(x)$ is $(8/3)x^3 e^{-2x}$ so that $f(.634) = .19$, $f(3/2) = .45$, $f(2) = .39$, $f(2.366) = .31$, $f(2.5) = .28$, $f(3) = .18$, $f(4) = .06$, $f(5) = .02$. Figure 2.21 shows the completed sketch for the example.

An important property of the gamma distribution, which will not be proved here since it depends on moment generating functions, is the additivity property.

THEOREM 2.5

If X_1 is distributed $G(\alpha_1, \beta)$ and if X_2 is distributed $G(\alpha_2, \beta)$ and if X_1 and X_2 are independent, then the sum $X_1 + X_2$ is distributed $G(\alpha_1 + \alpha_2, \beta)$. This can be stated more briefly as

$$G(\alpha_1, \beta) + G(\alpha_2, \beta) = G(\alpha_1 + \alpha_2, \beta)$$

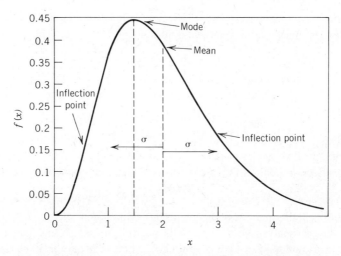

Figure 2.21   The density function for the $G(4, 1/2)$ distribution.

Notice that the scale parameters must be equal for the additivity property to hold. An important corollary is that the sum of independent chi-square variables is also a chi-square variable with the summed degrees of freedom.

One important result of Theorem 2.5 is that it leads to the recognition of the gamma distribution as a waiting-time distribution. Consider a component that is subject to shocks arriving as a Poisson process with intensity λ. Suppose that the component is subject to degradation at each shock with complete failure at the kth shock. The time to failure of the equipment is of interest. Because the shocks arrive in a Poisson manner, the intershock times $T_i (i = 1, \ldots, k)$ are distributed as

$$T_i \sim \text{NGEX}(\lambda) \tag{2.96}$$

or equivalently

$$T_i \sim G(1, 1/\lambda) \tag{2.97}$$

Then the total time to failure for the equipment is $T = T_1 + \ldots + T_k$, which is distributed as

$$T \sim G(k, 1/\lambda) \tag{2.98}$$

This form of the gamma distribution with integer-valued shape parameter is often called the *Erlang* distribution.

It is easy to verify that the expressions $F(t)$, $R(t)$, and $h(t)$ for a $G(\alpha, \beta)$ variable are not expressible in simple form. It is interesting that such a failure law does not have the CFR property that holds for failure after a single shock. For evaluating the reliability of a unit with gamma failure law, we need the incomplete gamma function

$$F(x; \alpha, \beta) = [\Gamma(\alpha)\beta^\alpha]^{-1} \int_0^x y^{\alpha-1} e^{-y/\beta} dy \tag{2.99}$$

There are several unattractive options. We can (a) consult the abbreviated Table T4 in the appendix or a volume of incomplete gamma functions (a well-equipped scientific library will have one), or (b) use a computer package such as IMSL, or (c) use the special relationship of the following theorem for the Erlang distribution.

THEOREM 2.6   THE ERLANG–POISSON RELATIONSHIP

Designating $F(x; k, \beta)$ as the left-hand tail area of the Erlang $G(k, \beta)$ distribution,

$$F(x; k, \beta) = [\Gamma(k)\,\beta^k]^{-1} \int_0^x y^{k-1} e^{-y/\beta}\, dy \tag{2.100}$$

and poi$(j; \lambda')$ as the Poisson probability term,

$$\text{poi}(j; \lambda') = \frac{e^{-\lambda'}(\lambda')^j}{j!} \tag{2.101}$$

it is true that, for $\lambda' = x/\beta$,

$$F(x; k, \beta) = \sum_{j=k}^{\infty} \text{poi}(j; \lambda') \tag{2.102}$$

In words, the left-hand tail probability of an Erlang distribution is equal to the right-hand tail of a suitably chosen Poisson distribution.

Proof   The technique is to use repeated integration by parts in (2.100) with

$$u = e^{-y/\beta} \qquad\qquad v = y^k/k$$
$$du = -(1/\beta)e^{-y/\beta}\, dy \qquad dv = y^{k-1}\, dy$$

Then

$$F(x; k, \beta) = \frac{1}{\Gamma(k)\,\beta^k}\left[\frac{x^k e^{-y/\beta}}{k}\Big|_0^x + \frac{1}{\beta k}\int_0^x y^k e^{-y/\beta}\, dy\right]$$

$$= \frac{1}{\Gamma(k)\,\beta^k}\frac{x^k e^{-x/\beta}}{k} + \frac{1}{k\Gamma(k)\,\beta^{k+1}}\int_0^x y^k e^{-y/\beta}\, dy \tag{2.103}$$

Since $k\Gamma(k) = \Gamma(k + 1)$, Equation 2.103 assumes the form of a recursion formula

$$F(x; k, \beta) = \text{poi}(k; x/\beta) + F(x; k + 1, \beta) \tag{2.104}$$

Repeating the process indefinitely leads to the theorem, provided that the remainder term after r steps, say, converges to zero with increasing r. That it does indeed do so is evident intuitively from the fact that as the shape parameter α increases, the mean

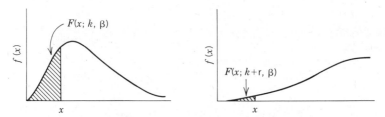

Figure 2.22   The convergence of the remainder term in Equation 2.104.

of the distribution increases proportionately, pulling the distribution to the right. The fraction of area under the curve to the left of a fixed x value does therefore decrease to zero, as illustrated in Figure 2.22.

Example 2.14

Suppose that a piece of equipment fails at the fifth shock, where the shocks are arriving in Poisson manner with intensity $\lambda = 0.01$ shocks per hour. Then the failure law is $G(5,100)$, and the mean life for such a piece of equipment is 500 hours. The reliability for a 200-hour mission is

$$R(200) = 1 - F(200; 5, 100)$$

$$= 1 - \sum_{j=5}^{\infty} \text{poi}(j; 2)$$

$$= \sum_{j=0}^{4} \text{poi}(j; 2) = .947$$

The reliability for a 700-hour mission is

$$R(700) = 1 - F(700; 5, 100)$$

$$= \sum_{j=0}^{4} \text{poi}(j; 7) = .173$$

In a similar way, the student may verify that $R(500) = .440$, reminding us again that any equipment has high reliability only for missions that are much less than its expected life. For this reason the concepts of maintainability and availability (depending on repair) are important in reliability theory and practice and will be considered in Chapter 7.

Example 2.15

Suppose in our example that the unit in question is one we manufacture in our factory, and that to increase customer acceptance, we desire to provide a money-back guarantee. For economic reasons we want no more than 5 percent of our customers to be dissatisfied. How long can the guarantee period x_g be? We set $F(x_g: 5, 100) = .05$, or equivalently, set $\sum_{j=0}^{4} \text{poi}(j : \lambda') = .95$ and determine by interpolation that $\lambda' = 1.97$ and hence $x_g = \beta\lambda' = 197$ hours. In a future discussion, in Section 5.3, we will encounter a somewhat more convenient technique for solving this guarantee problem.

The hazard function for the general $G(\alpha, \beta)$ distribution is not convenient to use, but for the $G(k, \beta)$ Erlang case Theorem 2.6 provides a form that is easy to program for a pocket calculator or electronic spreadsheet. We demonstrate for $k = 5$. The hazard function is

$$h(t) = \frac{f(t; 5, \beta)}{1 - F(t; 5, \beta)} = \frac{f(t: 5, \beta)}{\sum_{j=0}^{4} \text{poi}(j; t/\beta)}$$

$$= \frac{t^4 e^{-t/\beta}}{\Gamma(5)\beta^5}\left[\sum_{j=0}^{4} \frac{e^{-t/\beta}(t/\beta)^j}{j!}\right]^{-1}$$

Letting $a = t/\beta$ for convenience of notation,

$$h(a\beta) = \frac{1}{\beta}\frac{a^4}{4!}\left(1 + a + \frac{a^2}{2!} + \frac{a^3}{3!} + \frac{a^4}{4!}\right)^{-1}$$

$$= \frac{1}{\beta}\left(\frac{a^4}{24 + 24a + 12a^2 + 4a^3 + a^4}\right)$$

Figure 2.23 displays this example as well as others, for the special case when $\beta = 1$. Notice that as $t \to \infty$ it is true that $h(t) \to 1/\beta = \lambda$, the arrival rate for the intershock process.

Thus the $G(\alpha, \beta)$ hazard function is seen to be IFR over the useful lifetime of such a piece of equipment, approaching very slowly to CFR. This is in contrast to the normal distribution, whose density has the "uphill-max-downhill" character but whose hazard function approaches a linear rate of increase, and the Weibull, which has a density curve much like the gamma in appearance but whose hazard function increases polynomially. As a convenience to the student, we offer the hazard functions for $G(\alpha, \beta)$ with $\beta = 1$ whose graphs are given in Figure 2.23.

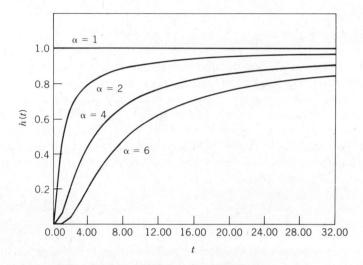

Figure 2.23   Gamma distribution hazard curves.

For $\alpha = 2$, $h(t) = \dfrac{t}{t + 1}$

For $\alpha = 3$, $h(t) = \dfrac{t^2}{2 + 2t + t^2}$

For $\alpha = 4$, $h(t) = \dfrac{t^3}{6 + 6t + 3t^2 + t^3}$

For $\alpha = 6$, $h(t) = \dfrac{t^5}{120 + 120t + 60t^2 + 20t^3 + 5t^4 + t^5}$

EXERCISES 2.6

1. Practice several times the derivations of K, μ, and σ^2 for the $G(\alpha, \beta)$ distribution.

2. Using the results of Equations 2.88, 2.90, and 2.95, find the mean and variance of the $\chi^2(\nu)$ distribution.

3. Analyze the curves for the following random variables and plot them, using not more than four computed values of $f(x)$.

(a) $X \sim G(3/2, 3)$

(b) $X \sim G(1/2, 3)$

(c) $X \sim \chi^2(8)$

4. Consider a unit with a length-of-life distribution that is $G(8, 20)$, in hours.

  (a) Compute the mean, variance, and standard deviation of the distribution.

  (b) What is the most likely time of failure?

  (c) What is the reliability for a mission of length 100 hours?

  (d) What is the reliability for a mission of length 150 hours?

  (e) What is the reliability for a mission of length 240 hours?

  (f) What fraction of the distribution lies within two standard deviations of the mean?

5. Repeat the previous problem for the $G(5, 100)$ distribution using the mission times 200 and 400 respectively in parts d and e.

6. Repeat Exercise 4 for the $G(3, 20)$ distribution using 60 hours in part c, 30 hours in part d, and 10 hours in part e.

7. Find the 5 percent guarantee value (the value that 95 percent of the population outlives) for the $G(8, 20)$ distribution of Exercise 4.

8. Repeat Exercise 7 for the $G(3, 20)$ distribution.

2.7 THE WEIBULL DISTRIBUTION

The Weibull distribution was introduced in Example 2.8 in terms of its hazard function

$$h(t) = \left(\frac{\beta}{\theta}\right)\left(\frac{t}{\theta}\right)^{\beta-1} \tag{2.105}$$

which leads to the density function

$$f(t) = \left(\frac{\beta}{\theta}\right)\left(\frac{t}{\theta}\right)^{\beta-1} \exp\left[-\left(\frac{t}{\theta}\right)^{\beta}\right] \tag{2.106}$$

and the reliability function

$$R(t) = \exp\left[-\left(\frac{t}{\theta}\right)^{\beta}\right] \tag{2.107}$$

We will indicate the distribution by writing $T \sim \text{WEI}(\theta, \beta)$. As noted previously, the distribution is IFR, CFR, or DFR depending on whether the shape parameter β is greater than, equal to, or less than unity. The negative exponential density is thus a special case of the Weibull (for $\beta = 1$), just as it is of the gamma distribution. The value $\beta = 2$, which corresponds to the Rayleigh distribution, is also a demarcation between two different types of IFR behavior, those that are concave upward ($\beta > 2$) and those that are concave downward ($1 < \beta < 2$).

In order to determine the density function curve shapes for the various values of β, it is convenient to look at the derivative

$$f'(t) = \frac{\beta}{\theta^2} e^{-V^\beta} V^{\beta-2} (-\beta V^\beta + \beta - 1) \tag{2.108}$$

where V stands for t/θ. The exponential factor guarantees that $\lim_{t\to\infty} f(t) = 0$ and that the curves all approach the t-axis asymptotically. Equating to zero the last factor in (2.108), we see that there is a mode (maximum value) at

$$t_0 = \theta \left(\frac{\beta - 1}{\beta} \right)^{1/\beta} \tag{2.109}$$

provided that $\beta > 1$. The second derivative is

$$f''(t) = \frac{\beta}{\theta^3} e^{-V^\beta} V^{\beta-3} \left[\beta^2 V^{2\beta} - 3\beta(\beta - 1)V^\beta + (\beta - 1)(\beta - 2) \right] \tag{2.110}$$

When equated to zero this derivative yields inflection points at the roots of

$$\beta^2 V^{2\beta} - 3(\beta - 1)\beta V^\beta + (\beta - 1)(\beta - 2) = 0 \tag{2.111}$$

that is to say, at the roots of

$$V^\beta = \frac{3(\beta - 1) \pm \sqrt{(\beta - 1)(5\beta - 1)}}{2\beta} \tag{2.112}$$

It can be seen that inflection points will exist only if $\beta > 1$, and that there will be an inflection point below the mode only if $\beta > 2$ since $\beta = 2$ is a value that makes (2.112) vanish. By considering $\lim_{t\to 0} f(t)$ and $\lim_{t\to 0} f'(t)$, we see that there are

TABLE 2.8   Summary of Weibull Curve Shape Criteria

Case	β-value	$f(0)$	$f'(0)$	Character of Curve at $t = 0$
1	$0 < \beta < 1$	∞	$-\infty$	Comes down asymptotically from ∞. No mode, no inflection points.
2	$\beta = 1$	$\frac{\beta}{\theta}$	<0	Negative exponential case. No inflection points.
3	$1 < \beta < 2$	0	∞	Comes out vertically from the origin. Has mode. Has one inflection point to the right of the mode.
4	$\beta = 2$	0	>0	Comes out from the origin at a positive angle. Has mode. Has degenerate inflection point at t $= 0$ plus genuine one to the right of the mode
5	$2 < \beta < \infty$	0	0	Comes out from the origin horizontally. Has mode. Has two inflection points, one on each side of the mode

five distinct curve shapes, depending on the value β. These are exactly analogous to those described more fully in Section 2.6 for the gamma distribution, so they are only summarized briefly here in Table 2.8. It is obvious why β is called the shape parameter for the Weibull distribution, just as α was for the gamma distribution. Figure 2.24 shows typical curves for $\theta = 1$.

It should be mentioned that other notations for the Weibull distribution are common in the literature; the one used here seems to be the simplest. The parameter θ is clearly a scale parameter.

A common three-parameter generalization of the Weibull is the shifted distribution with

$$R(t) = \exp\left[-\left(\frac{t - \eta}{\theta - \eta}\right)^{\beta}\right] \qquad (2.113)$$

where the location parameter η is the nonzero lower limit of the support region. We will not deal with it here, but it is discussed fairly widely in the literature.

The determination of the mean and variance of the distribution is accomplished through calculating the value of the rth noncentral moment

$$\mu_r' = E(T^r) = \int_0^\infty t^r f(t)\, dt$$

$$= \frac{\beta}{\theta^\beta} \int_0^\infty t^{r + \beta - 1} \exp\left[-\left(\frac{t}{\theta}\right)^\beta\right] dt \qquad (2.114)$$

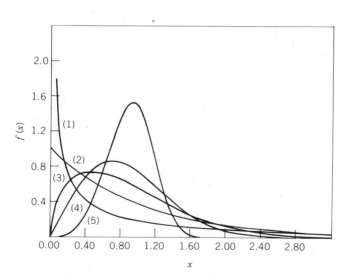

Figure 2.24 Selected Weibull density functions. Parameter values, with $\theta = 1$ in all cases, are (1) $\beta = 0.5$, (2) $\beta = 1$, (3) $\beta = 1.5$, (4) $\beta = 2$, (5) $\beta = 4$.

The substitution $y = (t/\theta)^\beta$ converts (2.114) into a gamma-function type of integral, to wit

$$\mu'_r = \theta^r \int_0^\infty y^{r/\beta} e^{-y} \, dy$$

$$= \theta^r \Gamma(1 + r/\beta) \tag{2.115}$$

Setting $r = 1$ yields

$$\mu = \theta\Gamma(1 + 1/\beta) \tag{2.116}$$

Setting $r = 2$ and using $\sigma^2 = \mu'_2 - \mu^2$ yields

$$\sigma^2 = \theta^2[\Gamma(1 + 2/\beta) - \Gamma^2(1 + 1/\beta)] \tag{2.117}$$

Example 2.16

Let us examine the behavior of the collection of Weibull members WEI$(1,\beta)$ for $\beta = 1, 2, 3, 4, 10, 100$. Using Equations 2.116, 2.117, and 2.109, we calculate the values shown in Table 2.9.

Clearly, as $\beta \to \infty$, $\mu = \Gamma(1 + 1/\beta)$ approaches $\Gamma(1) = 1$ and the variances appear to approach zero. The limiting form for the mode (for $\theta = 1$) can be written as

$$\lim_{\beta \to \infty} (1 - 1/\beta)^{1/\beta} = 1$$

TABLE 2.9 Summary of Results of Example 2.16

β	μ	σ^2	Mode
1	1	1	0
2	.89	.21	.71
3	.8933	.11	.87
4	.9063	.065	.93
10	.9513	.013	.99
100	.9943	.0002	.9998995

The estimation of the parameters θ and β is not an easy task and must be carried out either graphically or by means of some sophisticated numerical technique.[2] Despite this difficulty, the Weibull is a popular model for life distributions, probably because it can be used to model either burn-in or wear-out—though not simultaneously—and also because its reliability function is so mathematically tractable. In order to decide

---

[2] See Bain (1978) and Mann, Schafer, and Singpurwalla (1974).

whether a set of data is well modeled by the Weibull distribution, a CDF plotting procedure similar to that illustrated in Section 2.5 for the exponential distribution may be used. The CDF

$$F(t) = 1 - \exp\left[-\left(\frac{t}{\theta}\right)^{\beta}\right]$$

is linearized by taking natural logarithms twice to get

$$\ln \ln \frac{1}{1 - F(t)} = \beta \ln t - \beta \ln \theta \tag{2.118}$$

From this it is clear why β is often referred to as the *Weibull slope*.

Example 2.17

To illustrate the use of the graphical method, consider the ordered set of lifetimes

$$\{57, 65, 69, 71, 74, 75, 83, 86, 90, 91, 97, 101, 109, 113, 138\}$$

generated from the WEI(100, 3) population. The graphical analysis was done using an electronic spreadsheet that has the capability of generating graphs and computing regression coefficients. The worksheet is shown in Table 2.10. Column A is generated

TABLE 2.10   Calculations for Weibull Example 2.17

A Index	B Times	C ln(time)	D Ordinates	E Line
1	57	4.05	−2.74	−2.27
2	65	4.18	−2.01	−1.70
3	69	4.23	−1.57	−1.48
4	71	4.27	−1.25	−1.33
5	74	4.31	−0.98	−1.12
6	75	4.31	−0.76	−1.12
7	83	4.42	−0.55	−0.64
8	86	4.46	−0.37	−0.49
9	90	4.50	−0.19	−0.31
10	91	4.51	−0.02	−0.26
11	97	4.57	0.15	0.02
12	101	4.62	0.33	0.21
13	109	4.69	0.52	0.54
14	113	4.73	0.73	0.70
15	138	4.93	1.02	1.57

Mean = 88.01   Standard deviation = 20.49

Regression output:
Constant   −20.045
R squared   0.94
X coefficient(s)   4.388 (i.e., slope for least squares line)

with the DATA-FILL command, column B contains the sorted lifetimes, column C contains the natural logarithms of the lifetimes, column D contains the ordinates for plotting (based on $\hat{F}(t_i) = i/(n + 1)$) which are

$$y_i = \ln \ln \frac{16}{16 - i} \tag{2.119}$$

and column E contains the computations for the least squares regression line

$$\hat{y}_i = a + b \ln t_i$$

The least squares coefficients are computed by the spreadsheet command REGRESSION, whose partial output is shown at the bottom of the table. For this sample the estimated slope is $\hat{b} = 4.388$ (the true value is 3) and the intercept is $-\hat{\beta} \ln \hat{\theta}$, which yields the estimate $\hat{\theta} = 96.16$ (the true value is 100). The plot is shown in Figure 2.25. Two other simulated samples from the same population appear in the exercises at the end of this section.

Before the invention of the electronic spreadsheet, practitioners relied on special Weibull graph papers which had a logarithmic scale on one axis and a doubly logarithmic scale on the other axis. The "best" line was drawn in by eye and the Weibull slope was read off an auxiliary scale on the margin of the paper. The justification for the process is as follows.

Originally, the value of b could have been estimated from the actual slope of the line by reading the coordinates of two points and then computing the slope as "rise over run." The crudeness of that method is replaced by introducing a special "estimation point" and an auxiliary β-scale line. The estimation point is that point

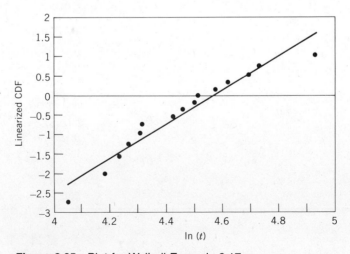

Figure 2.25   Plot for Weibull Example 2.17.

where $F(t) = 1 - e^{-1} = .632$ so that $\ln \ln 1/R(t) = 0$. This point, projected vertically downward onto the time axis, yields the estimator $\hat{\theta}$, since $\ln \ln 1/R(t) = 0$ has the solution $\beta(\ln t - \ln \theta) = 0$, whereby $t = \theta$. The estimation point is used as one of the two "read-off" points of the line, the other being chosen arbitrarily. It is traditional to relocate the estimation point (to get it out of the way) by placing it at the left of the paper and then to supply a calibrated β-scale axis at the top of the paper. An auxiliary line is drawn through the estimation point, parallel (or perpendicular, depending on the brand of paper) to the "eyeball" line. The point where the parallel line intersects the β-scale axis provides the β-estimate. The use of the parallel (or perpendicular) line obviates the need for reading the coordinates of the second point.

Clearly, different analysts examining a given data set can arrive at quite different lines. This renders the whole process questionable and is one reason why Gorski (1968) rails at those who use it—as well as those who adopt the Weibull model without good cause. The availability of electronic spreadsheets has rendered the hand plotting technique obsolete, so we do not discuss it further; those interested in it may consult Nelson (1967) or Kapur and Lamberson (1977).

Another method for estimating θ from the least-squares-obtained slope $\hat{\beta}$ is to use the relationship

$$\mu = \theta \Gamma(1 + 1/\beta)$$

in the method-of-moments form

$$\hat{\theta} = \frac{\bar{x}}{\Gamma(1 + 1/\hat{\beta})} \tag{2.120}$$

In our example problem, using $\hat{\beta} = 4.388$ and Table T5 of the gamma function yields

$$\hat{\theta} = \frac{88.00}{\Gamma(1.228)} \approx \frac{88.00}{0.9108} = 96.62$$

EXERCISES 2.7

1. For the Weibull family, plot the density curves for $\theta = 2$ and $\beta = 1/3, 1/2, 2, 3$. Compute the mean, mode, and variance for each of the cases.

2. Here is a data set for you to practice on:

   { 50, 58, 59, 61, 63, 92, 93, 98, 105, 108, 120, 121, 140, 162, 172}

   (a) Analyze the data as was done in Table 2.9, and decide whether the Weibull distribution is an appropriate model.

   (b) If it is, estimate the parameters β and θ.

   (c) Test to see whether the Weibull fit is better than exponential fit.

3. Repeat Exercise 2 for the data set

   { 25, 27, 67, 68, 71, 78, 81, 85, 97, 98, 106, 109, 123, 141, 154}

4. Use Table T9 (or the RAND function of your calculator or electronic spreadsheet) to simulate a Weibull sample of size 20 with parameters $\theta = 50$, $\beta = 3$. (Do this by solving Equation 2.118 for t and replacing $F(t)$ by the CUD(0,1) random values.) Apply the procedure of Example 2.17 to see how well the simulated sample estimates the parameters.

5. Repeat Exercise 4 several times using a different set of random numbers each time. Use the plots to become familiar with the amount of variability to be expected when using the graphical method of Example 2.17 and Figure 2.25.

2.8   THE LOGNORMAL DISTRIBUTION

The lognormal distribution is introduced here because it seems to have an increasingly important role in reliability work. According to the AT&T company (1983), the lifetimes of semiconductor components after burn-in are well described by the lognormal distribution, as are the repair times for certain types of equipment. In the past the distribution has been used mostly by economists. The monograph by Aitchison and Brown (1957) is recommended, and here we (reluctantly) use their notation.

A random variable X is said to have the lognormal (or logarithmico–normal) distribution designated as $\Lambda(\mu, \sigma^2)$, if $Y = \ln X$ has the normal distribution $N(\mu, \sigma^2)$. Thus the density function for X is

$$f(x) = \frac{1}{x\sigma\sqrt{2\pi}} \exp\left[\frac{-(\ln x - \mu)^2}{2\sigma^2}\right], \qquad x > 0 \qquad (2.121)$$

There are some minor problems with the usual notation and terminology. First, the variable X would be better described as *antilognormal*, since X is the antilog of a normally distributed variable. Second, the parameters μ and σ^2 are the moments of the "auxiliary distribution" Y, not of X, the variable under consideration. We will follow Aitchison and Brown by using the notation $a = E(X)$ and $b^2 = \text{Var}(X)$.

The moments of the distribution are calculated in straightforward manner as follows. Equating the derivative to zero gives

$$f'(x) = -Ke^{-Q}\frac{(\ln x - \mu + \sigma^2)}{x^2\sigma^2} \qquad (2.122)$$

where

$$K = (\sigma\sqrt{2\pi})^{-1}$$

and

$$Q = \frac{(\ln x - \mu)^2}{2\sigma^2} \qquad (2.123)$$

gives the modal value

$$x_0 = e^{\mu - \sigma^2}$$

The general noncentral moment $E(X^r)$ is obtained by a change of variable integration using $y = \ln x$, to wit

$$E(X^r) = (\sigma\sqrt{2\pi})^{-1} \int_0^\infty x^r e^{-Q} \frac{dx}{x}$$

$$= (\sigma\sqrt{2\pi})^{-1} \int_{-\infty}^\infty e^{yr} e^{-Q} \, dy \qquad (2.124)$$

The exponent $yr - Q$ can be written in the form

$$\frac{-[(y - \mu)^2 - 2\sigma^2 yr]}{2\sigma^2}$$

which, when the square in y is completed, becomes

$$\frac{-[y - (\mu + r\sigma^2)]^2}{2\sigma^2} + \mu r + \frac{r^2 \sigma^2}{2}$$

Upon integrating (2.124) in this form, we get

$$E(X^r) = e^{\mu r + r^2 \sigma^2/2} \qquad (2.125)$$

Those familiar with moment generating functions will realize that (2.125) could have been obtained by recognizing that

$$E(X^r) = E(e^{Yr})$$

is the moment generating function of the normally distributed Y, where Y is the auxiliary variable in the transform space. Setting $r = 1$ in (2.125) yields the mean

$$E(X) = a = e^{\mu + \sigma^2/2} \qquad (2.126)$$

Setting $r = 2$ and using $\text{Var}(X) = E(X^2) - E^2(X)$, we get

$$\text{Var}(X) = b^2 = e^{2\mu + 2\sigma^2} - (e^{\mu + \sigma^2/2})^2 \qquad (2.127)$$

$$= e^{2\mu + 2\sigma^2} - e^{2\mu + \sigma^2}$$

$$= e^{2\mu}(e^{2\sigma^2} - e^{\sigma^2})$$

$$= a^2(e^{\sigma^2} - 1)$$

$$= a^2 \eta^2 \qquad (2.128)$$

where $\eta^2 = e^{\sigma^2} - 1$. The quantity η is seen to be the coefficient of variation of the distribution. The third central moment can be verified to be (using the notation of Aitchison and Brown)

$$\lambda_3 = a^3(\eta^6 + 3\eta^4) \qquad (2.129)$$

and the fourth central moment is

$$\lambda_4 = a^4(\eta^{12} + 6\eta^{10} + 15\eta^8 + 16\eta^6 + 3\eta^4) \tag{2.130}$$

The resulting dimensionless skewness and kurtosis indices are

$$\alpha_3 = \eta^3 + 3\eta \tag{2.131}$$

and

$$\alpha_4 = 3 + \eta^8 + 6\eta^6 + 15\eta^4 + 16\eta^2 \tag{2.132}$$

Thus the skewness and kurtosis both increase strongly with $\eta = b/a$.

In the exercises at the end of this section, you are asked to simulate a random sample from the lognormal distribution. To do this, a set of standard normal deviates is needed. One way to create them is with the Box-Mueller transformation which uses independent CUD(0,1) variates U_1 and U_2 to generate standard normal pairs Z_1 and Z_2 from the formulae

$$Z_1 = \sqrt{-2\ln U_1}\cos 2\pi U_2 \tag{2.133a}$$

$$Z_2 = \sqrt{-2\ln U_1}\sin 2\pi U_2 \tag{2.133b}$$

Even though Z_1 and Z_2 are not mathematically independent, their distribution for large samples behave as if they were statistically independent. If you feel doubtful about this, ignore (2.133b) and use (2.133a) only; in that case, you will create only half as many standard deviates as the number of uniform deviates. Values for U_1 and U_2 are available in Table T11, or you can use your electronic spreadsheet. To create the random samples with your spreadsheet, use the RAND, COPY, and RANGE VALUE (or equivalent) commands to generate two columns of CUD(0,1) numbers. Put these in columns A and B, say. Then write the Box-Mueller formulae (2.133a,b) into columns C and D respectively, using U_1 from column A and U_2 from column B. If you concatenate the resulting values from columns C and B into one large sample, you will find that you have created an adequate simulated sample from the $N(0, 1)$ population. For general $N(\mu, \sigma^2)$ variates the usual transformation $x = \mu + \sigma z$ is applied.

To compute a particular quantile for the lognormal distribution, we let x_γ be the quantile defined by

$$\gamma = P(X \le x_\gamma)$$

$$\gamma = P(\ln X \le \ln x_\gamma)$$

$$= P(Y \le \ln x_\gamma)$$

$$= P\left(\frac{Y - \mu}{\sigma} \le \frac{\ln x_\gamma - \mu}{\sigma}\right)$$

$$= P\left(Z \le \frac{\ln x_\gamma - \mu}{\sigma}\right) \tag{2.134}$$

where $Z \sim N(0, 1)$. Then we see that

$$\frac{\ln x_\gamma - \mu}{\sigma} = z_{1-\gamma} \qquad (2.135a)$$

Notice that there is some awkwardness in notation here because the subscripts don't seem to match. The number x_γ is a left hand critical point of the X distribution, whereas the corresponding left hand critical point of the standard normal distribution is indicated as $z_{1-\gamma}$ due to the convention of indexing most tabled variables from the upper end of the curve. We can also write (2.135a) in the equivalent form

$$\frac{\ln x_\gamma - \mu}{\sigma} = -z_\gamma \qquad (2.135b)$$

Then (2.135a) and (2.135b) can be written as

$$x_\gamma = e^{\mu + \sigma z_{1-\gamma}} \qquad (2.136a)$$

or

$$x_\gamma = e^{\mu - \sigma z_\gamma} \qquad (2.136b)$$

If simulated lognormal samples are to be generated as indicted above, the preferred form for the transformation is simply

$$x = e^{\mu + \sigma z} \qquad (2.136c)$$

In particular

$$x_{.5} = e^\mu \qquad (2.137)$$

$$x_{.16} \approx e^{\mu - \sigma} \qquad (2.138)$$

$$x_{.84} \approx e^{\mu + \sigma} \qquad (2.139)$$

Dividing (2.137) by (2.138) yields

$$\sigma = \ln(X_{.5}/X_{.16}) \qquad (2.140)$$

which can be used to obtain an estimate of σ by using the data quantiles, just as (2.137) can be used to estimate μ from the sample median. The maximum likelihood estimators of μ and σ^2 are

$$\hat\mu = \left(\sum_{i=1}^{n} \ln x_i \right)/n \qquad (2.141)$$

and

$$\hat\sigma^2 = \left[\sum_{i=1}^{n} (\ln x_i - \hat\mu)^2 \right]/n \qquad (2.142)$$

Figures 2.26 through 2.29 show graphs of $f(x)$ for selected values for σ, using the median of the distribution e^μ as reference scale. Notice that $f(x)$ has the indeterminate

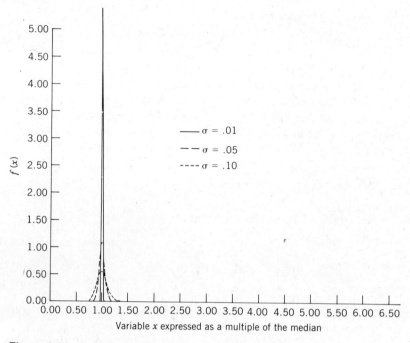

Figure 2.26 Lognormal densities for very small values of σ and $\mu = 2$.

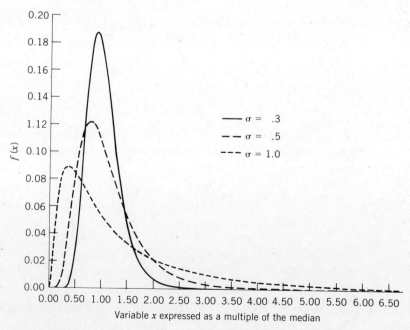

Figure 2.27 Lognormal densities for small values of σ and $\mu = 2$.

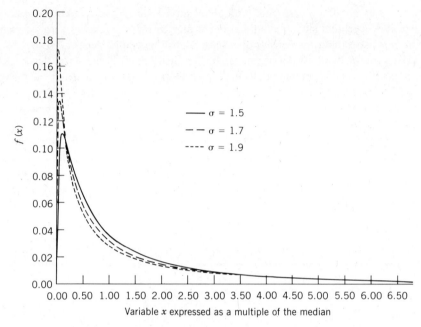

Figure 2.28 Lognormal densities for medium values of σ and $\mu = 2$.

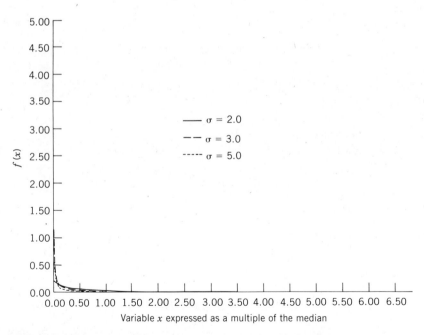

Figure 2.29 Lognormal densities for large values of σ and $\mu = 2$.

form $0 \div 0$ at $x = 0$ and l'Hôpital's rule is of no use (try it and see why). It may be argued, however, that since the normal curve of $y = \ln x$ approaches zero as y tends toward $-\infty$, so should the density of $x = e^y$ vanish as x approaches zero.

The value μ is a location parameter for the distribution in the sense that it alone determines the median or midpoint of the distribution. As is clear from the curves, the small mode generated by large σ is also associated with large variance, great skewness, and slow convergence to the horizontal axis. Notice the different vertical scales for the figures.

The interval of greatest probability, which lies near the mode, is determined by both μ and σ, so that both can truly be called location parameters. The dependence of α_3 and α_4 on η as shown in (2.131) and (2.132) make it reasonable to call η as well as σ a shape parameter. Table 2.11 displays the values of all moments for $\mu = 2$ and selected values of σ.

TABLE 2.11   Moments for Selected Lognormal Distributions with $\mu = 2$

σ	0.1	0.5	1.0	1.5	2.0
Mean	7.43	8.37	12.18	22.76	54.6
Median	7.39	7.39	7.39	7.39	7.39
Mode	7.32	5.75	2.72	0.78	0.14
Standard deviation	0.74	4.46	15.97	66.30	339.7
Coefficient of variation	0.10	0.53	1.31	2.91	7.32
Skewness	0.30	1.75	6.18	33.5	414.4
Kurtosis	0.16	5.90	110.90	10075.3	–

EXERCISES 2.8

1. Compute the mean, variance, mode, and coefficient of variation for the lognormal distribution which has $\mu = 3$ and $\sigma = 0.5$. Plot the density function for the distribution.

2. Repeat Exercise 1 for $\mu = 3$ and $\sigma = 1.0$.

3. Simulate a random sample of size 50 from the distribution of Exercise 1, using the Box-Mueller transformation given in (2.133a,b).

2.9   PLOTTING FOR MODEL SELECTION

It was formerly believed that an important use of the hazard function was to distinguish between various failure models such as gamma, Weibull, exponential, and so on.

When we contemplate the strikingly different curve shapes of the various hazard functions, such a belief appears reasonable. Naturally, we would expect random variability to distort an empirical hazard function, as defined in Section 1.1, from the theoretical curves shown earlier in this chapter. With the goal of demonstrating to students the kind of variability to expect, this writer tried creating by simulation a collection of examples for classroom use. The results were a severe disappointment.

The procedure was as follows. A FORTRAN program was written using an IMSL (International Mathematical and Statistical Library) random variable generator. Samples of the chosen size were created and sorted, and the empirical hazard function was computed from the usual Equation 1.6 for nongrouped failure times, to wit

$$h^*(t_i) = [n(t_i)\,\Delta t_i]^{-1}$$

The FORTRAN data set consisting of the triplet $[i, t_i, h^*(t_i)]$ was then passed by an appropriate job control language (JCL) into a Statistical Analysis System (SAS) program, which has a simple and cheap plotting routine. Since large numbers of samples were to be created, I wanted to avoid the more elegant but expensive Calcomp plots. It was not deemed worthwhile to convert the plots into histograms, as in Section 1.2, so the results have the appearance of scatter diagrams.

Four distributions were studied: exponential, normal, gamma, and Weibull. The intention was to compile a notebook showing many examples of each, so that students could learn the patterns of variation and would henceforth be able to select the proper model by eye. The plan failed because few of the plots looked anything like their theoretical curves. Figures 2.30 through 2.33 present a typical set of the plots.

If hazard function plots will not distinguish between model types, what will? The answer is old-fashioned probability plotting, the kind that uses appropriate scaling to convert the CDF into a straight line as we did in Section 2.5, and for which probability paper is widely available from the TEAM Company (Technical and Engineering Aids in Management) and the Codex Company. For our simulated samples the amount of work needed for hand plotting would have been prohibitive, so all samples were plotted using the IMSL subroutine USPRP which gives straight-line CDF plots for six models: normal, lognormal, half-normal, exponential, Weibull, and extreme value. The results were satisfying. A few selected plots for an exponential sample are shown in Figures 2.34 through 2.39. Although perfect selectivity was not possible with the twenty samples of each type generated, the discrimination was certainly much better than with the hazard function plots. The plots make it fairly clear that the normal, the lognormal, and the extreme-value models should all be rejected. The exponential, the Weibull, and the half-normal could all be accepted. Of course the Weibull should be accepted since the exponential is a special Weibull case. The acceptance of the half-normal should be no great surprise since its density function shares with the exponential the property of starting high and decreasing monotonically.

We cannot display here the number of plots necessary for developing a feeling for the simulation results. To encourage the interested reader to experiment, we provide in Appendix 2F the computer codes that were used.

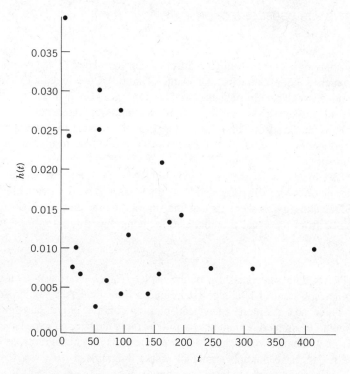

Figure 2.30 Empirical hazard function for simulated sample from exponential distribution with $\lambda = 100$.

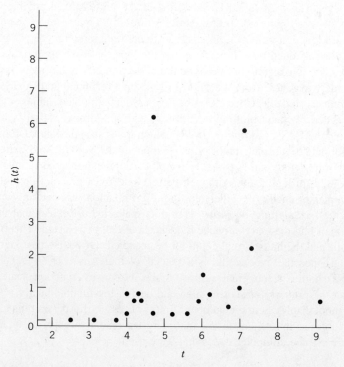

Figure 2.31 Empirical hazard function for simulated sample from gamma distribution ($\alpha = 6$, $\beta = 1$).

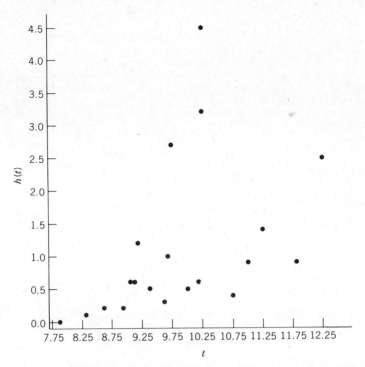

Figure 2.32 Empirical hazard function for simulated sample from normal $N(10,1)$ distribution.

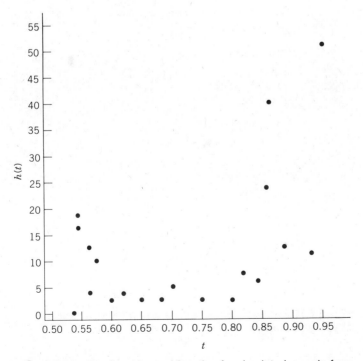

Figure 2.33 Empirical hazard function for simulated sample from Weibull distribution with $\theta = .76$, $\beta = 6$.

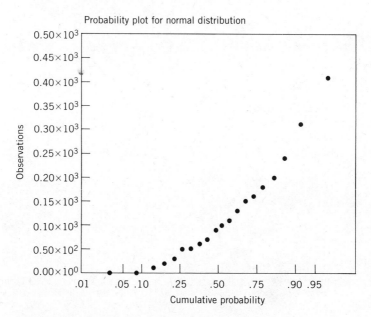

Figure 2.34   An exponential sample tested for normality using IMSL routine USPRP.

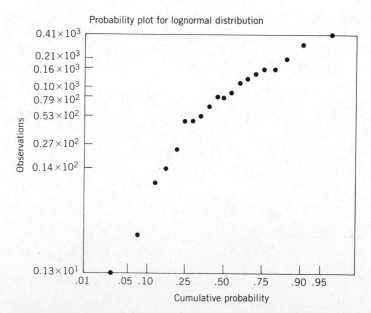

Figure 2.35   An exponential sample tested for lognormality using IMSL routine USPRP.

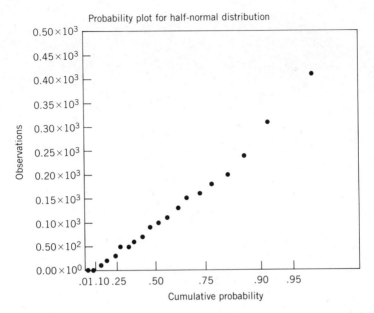

Figure 2.36 An exponential sample tested for half-normality using IMSL routine USPRP.

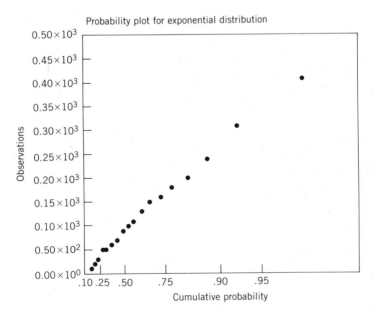

Figure 2.37 An exponential sample tested for exponentiality using IMSL routine USPRP.

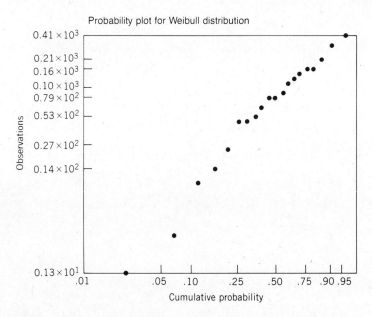

Figure 2.38 An exponential sample tested for Weibullness using IMSL routine USPRP.

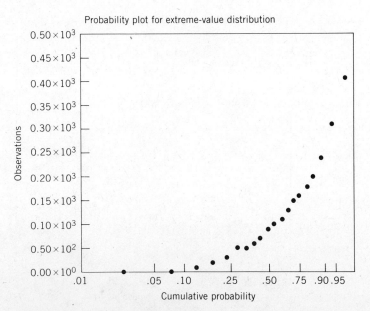

Figure 2.39 An exponential sample tested against the extreme-value model using IMSL routine USPRP.

2.10 PLOTTING POSITIONS

In order to follow parts of the discussion of this section, some understanding is needed of the beta distribution (Appendix 2A), order statistics (Appendix 5A), and transformation of variables (Appendix 5C). Readers who want to postpone or omit this section may do so without penalty, since it is not needed for subsequent chapters.

In Sections 2.5 and 2.7, where Epstein's graphical procedure for testing exponentially was illustrated, the quantities

$$\hat{F}(t_i) = \frac{i}{n+1} \ , \qquad i = 1, \ldots, n \qquad (2.143)$$

were used to compute ordinates corresponding to the abscissa values $\{t_i\}$. In the FORTRAN IMSL routine USPRP referred to in Section 2.9 and Appendix 2F, the values used for $\hat{F}(t)$ were

$$\hat{F}(t_i) = \frac{3i-1}{3n+1} \ , \qquad i = 1, \ldots n \qquad (2.144)$$

for some plots and

$$\hat{F}(t_i) = \frac{i - \frac{1}{2}}{n} \ , \qquad i = 1, \ldots, n \qquad (2.145)$$

for others. Some authors discuss the so-called median plotting positions, which require a table indexed by both i and n. Students may well wonder why there is such an abundance of choices for $W_i = \hat{F}(X_{(i)})$ as plotting positions for the observed failure times $\{t_i\}$. In response, we consider the problem from several points of view.

The most elementary approach is to create an ogive or a linearized ogive using as plotting positions the sample CDF values given by

$$W_i = S(t_{(i)}) = \frac{i}{n} \ , \qquad i = 1, \ldots, n \qquad (2.146)$$

Readers who have used the Kolmogorov–Smirnov test will recognize the procedure. It is little used in this application because of its bias. Instead, it is often replaced by the "midpoint" plotting positions given by (2.145), which can be justified as follows. The smallest of n observations may be thought of as representative of the smallest $(1/n)^{\text{th}}$ fraction of the population. The midpoint of the corresponding CDF interval is $0.5/n$. The second smallest observation may be thought of as representative of the $(2/n)^{\text{th}}$ fraction of the population. The midpoint of the corresponding CDF interval is

$$\frac{1}{2}\left(\frac{1}{n} + \frac{2}{n}\right) = \frac{1.5}{n} = \frac{2-0.5}{n}$$

This line of reasoning leads to the general formula (2.145).

Both (2.146) and (2.145) ignore the statistical distribution of the values $F(X_{(i)})(i = 1, \ldots, n)$, which we develop here.

For a random sample $(X_1, X_2, \ldots, X_n)$ from a population with density $f(x)$, the complete order statistic is defined as the ranked n-tuple

$$X_{(1)}, X_{(2)}, \ldots, X_{(n)} \tag{2.147}$$

where

$$X_{(1)} \leq X_{(2)} \leq \ldots \leq X_{(n)} \tag{2.148}$$

The joint density of the complete order statistic is given by

$$n! f(x_{(1)}) f(x_{(2)}) \ldots f(x_{(n)}) \tag{2.149}$$

The marginal density of the ith partial order statistic is

$$g_i(x_{(i)}) = \frac{n!}{(i-1)!(n-i)!} [F(x_{(i)})]^{i-1} f(x_{(i)}) [1 - F(x_{(i)})]^{n-i} \tag{2.150}$$

If we designate the CDF as $W_i = F(X_{(i)})$, we see that W_i, or the *probability integral transform*, is itself a random variable, since it is a function of the random variable $X_{(i)}$. As shown in Appendix 5C, the rule for transformation of variables leads to the density of W_i given by

$$h_i(W_i) = g_i(X_{(i)}) \left| \frac{dX_{(i)}}{dW_i} \right| \tag{2.151}$$

where the Jacobian is

$$\left| \frac{dX_{(i)}}{dW_i} \right| = \left[\frac{dW_i}{dX_{(i)}} \right]^{-1} = \left[\frac{dF(X_{(i)})}{dX_{(i)}} \right]^{-1} = \left[f(X_{(i)}) \right]^{-1} \tag{2.152}$$

and this cancels the third factor in (2.150). Thus the density function for W_i is

$$h_i(w) = \frac{n!}{(i-1)!(n-i)!} w^{i-1} (1-w)^{n-i}, \quad 0 \leq w \leq 1 \tag{2.153}$$

This function is recognizable (see Equation 2A.19) as the beta density with parameters i and $n + 1 - i$. Thus the plotting position W_i for $t_{(i)}$ should be chosen as the value that seems most appropriate for estimating a $be(i, n + 1 - i)$ variable. There are various ways of doing this. The easiest way is to use the mean of the distribution, leading to the value previously set forth in Equation 2.71, to wit

$$E(\hat{F}_i) = \frac{i}{i + (n - i + 1)} = \frac{i}{n + 1} \tag{2.154}$$

The problem with the mean plotting positions given by (2.154) is that for very small and very large i, the beta distributions for the $\{W_i\}$ are quite skew (see Figure 2.40), and thus the mean is not regarded as a representative value for the distribution. For such asymmetrical distributions, the median is considered more typical of the distribution than the mean, and is thus the generally preferred plotting point. In the years BC ("before computers") the determination of the medians was mathematically

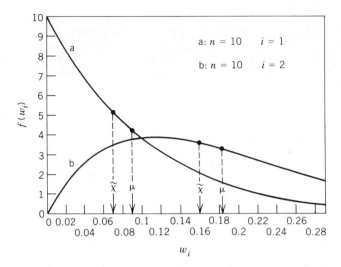

Figure 2.40 The skewness of selected plotting point variables.

intractable except for $i = 1$ and $i = n - 1$. It is left to the student to verify that for $n = 10$ and $i = 1$ the mean value of the W_1 distribution is $1/11 = .09091$, compared with the median value $.06697$—a relative difference of some 26 percent in magnitude (or 36 percent, depending on which value is regarded as correct).

A pioneer worker in the area of plotting points, Leonard Johnson (1951), used the easily obtainable exact median ranks for $i = 1$ and $i = n$ (see Exercise 2.10.1) and then interpolated linearly to obtain approximate intermediate values for $i = 2, \ldots, n-1$. Table 2.12 shows a comparison of his interpolated values with the somewhat more correct values obtained by computer using the FORTRAN IMSL routine MDBETI, along with the exact value obtained from repeated iterations with Newton's method.

Since the IMSL-computed values are so close to the exact median rank values, it is recommended that the former be used in any situation in which a high degree of accuracy is desired. Table 2.13 shows an appropriate FORTRAN code which can be used to generate any needed set of median ranks. The value $P = .5$ can also be altered to any desired value, such as .1 and .9, to create tolerance intervals for the ranks. This is discussed by Johnson (1964).

The formula

$$W_i = \frac{i - .3}{n + .4} \tag{2.155}$$

is a popular approximation formula for the median rank values and should suffice for practical work. It is recommended for any computer application for which a simple formula is desired. The formula

$$\frac{i - 3/8}{n + 1/4} \tag{2.156}$$

TABLE 2.12   Exact and Approximate Median Ranks for $n = 10$

Index i	Johnson's Interpolated Value	IMSL Value	Exact Value	$\dfrac{i - .3}{n + .4}$
1	.06697	.06697	.06697	.06731
2	.16320	.16227	.16226	.16346
3	.25943	.25857	.25857	.25962
4	.35566	.35510	.35510	.35577
5	.45189	.45170	.45169	.45192
6	.54811	.54830	.54831	.54808
7	.64434	.64491	.64490	.64423
8	.74057	.74143	.74143	.74038
9	.83680	.83773	.83774	.83654
10	.93303	.93303	.93303	.93269

is another popular one, as is (2.144). The reader who finds tables of median rank plotting points in other books can use Table 2.12 as a comparison standard for determining which set of plotting points is being offered therein.

If the plot is being made merely to decide on the suitability of the model, by seeing whether a particular scaling on the CDF axis yields a straight line, it is probably immaterial which set of plotting positions is used. If parameter estimation is the purpose, more care is needed and the Rys report (1986) using a minimum squared error criterion may be of interest, as may the references cited on page 301 of Kapur and Lamberson (1977).

The degree and direction of the bias that arise from using means versus medians as plotting positions depends on what model is being tested. We have seen that for small i the mean exceeds the median, and that for large i the reverse is true. Keep in mind, however, that it is not the plotting positions themselves but some function of them which is the ordinate; equivalently, special paper that effects a transformation of variables may be used. For the exponential distribution the ordinate is (see Equation 2.70)

$$Y_i = \ln(1 - W_i)^{-1} \tag{2.157}$$

whereas for the Weibull distribution it is

$$Y_i = \ln \ln(1 - W_i)^{-1} \tag{2.158}$$

which is plotted against $\ln t_{(i)}$. Figure 2.41 shows the difference that results from the two choices—the means versus the approximate medians—for a simulated exponential sample from the NGEX(100) distribution. The data set for Figure 2.41 is

$$\{21, 27, 41, 51, 62, 63, 77, 77, 114, 125, 129, 186, 229, 243, 270\}$$

The estimate of λ from the data is $\overline{X} = 114.4$. The estimate of λ from the least squares line using Equation 2.143 (line a) is $\hat{\lambda} = 106.78$ and from line b based on Equation 2.155 is $\hat{\lambda} = 97.56$.

TABLE 2.13 FORTRAN Program for Computing Median Ranks

```
C  THIS PROGRAM REQUIRES THE IMSL LIBRARY
      REAL P,X
      NUM = 20
C
C  THIS IS THE MAXIMUM SAMPLE SIZE IN THE CALCULATION SET
C
      DO 105 N = 1, NUM
      WRITE (6,91) N
91    FORMAT (20X, 'THE SAMPLE SIZE IS  ',I2)
      WRITE (6,92)
92    FORMAT (20X, 'INDEX', 5X, 'MEDIAN RANK')
      DO 100 J = 1, N
      A = REAL (J)
      B = REAL (N − J + 1)
      P = .5
C
C  THIS IS THE CUMULATIVE PROBABILITY VALUE FOR THE MEDIAN.
C  FOR TOLERANCE LIMITS IT MAY BE CHANGED TO ANY DESIRED LEVEL
C  ALPHA AND ITS COMPLEMENT (1—ALPHA).
C
      CALL MDBETI(P,A,B,X,IER)
      WRITE (6,93) J,X
93    FORMAT (22X,I2,10X,F7.6)
100   CONTINUE
105   CONTINUE
      STOP
      END
/*
//LKED.SYSLIB   DD
//             DD
//             DD
//   DD  DSN = SYS1.IMSL.SPFLIB,DISP = SHR
//GO.SYSIN   DD  *
/*
```

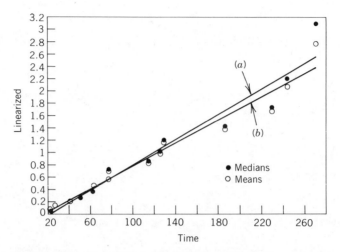

Figure 2.41   Comparison of plotting positions for an exponential sample: a, least squares line for medians plotting; b, least squares line for means plotting.

Figure 2.42 illustrates the same type of comparison for the Weibull example of Table 2.10. Line a, the least squares line using the means, yielded $\hat{\beta} = 4.39$ and $\hat{\theta} = 96.36$, whereas line b based on the approximate medians of Equation 2.155 yields $\hat{\beta} = 4.76$ and $\hat{\theta} = 96.03$.

Notice that for both the exponential and the Weibull the approximate medians-based line is rotated slightly counter-clockwise from the means-based line. The difference is noticeable but not astounding; the resulting parameter estimates are somewhat

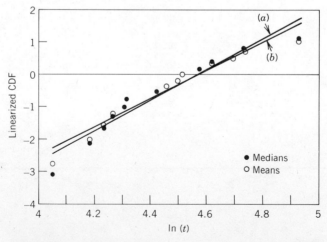

Figure 2.42   Comparison of plotting positions for the Weibull sample of Table 2.10: a, least squares line for median plotting; b, least squares line for means plotting.

different but probably less than differences caused by sampling variation. It is probably safe for the practitioner to use without guilt any set of plotting positions that seems most appealing.

EXERCISES 2.10

1. Set $n = 10$ and $i = 1$ in Equation 2.153, and integrate the result to obtain the CDF $H(W_1)$. Then set $H(W_1) = .5$ and verify that the median of the W_1 distribution is $\tilde{W}_1 = 1 - (.5)^{.1} = .06697$, as compared with the mean value $1/11 = .09091$. Repeat the exercise for $i = 10$ and show that the median value of the W_{10} distribution is the complement $\tilde{W}_{10} = (.5)^{.1} = .93303$, as compared with the mean value $10/11 = .90909$.

2. Set $i = 2$ (or 9) and verify that the determination of the median value is impossible using elementary mathematical techniques. If you know how, use some search technique such as Newton's method, regula falsi, Gauss–Seidel, or a similar algorithm to verify that the median is .16227 (or .83773) compared with mean .18182 (or .81818).

   Hint. An easy way to carry out an iterative procedure is to use an electronic spreadsheet. The iteration formula can be copied automatically as many times as necessary for convergence.

3. Verify Johnson's interpolated values in Table 2.11.

4. Verify at least one of the exact solutions in Table 2.11 using Newton's method.

   Hint. With Johnson's result as starting value, two or three iterations should be enough for eight-place accuracy.

5. Use the set of simulated failure times {3, 4, 17, 20, 48, 88, 104, 159, 164, 170} in the exponential model to plot the data as was done in Figure 2.16 for the data of Table 2.6. If possible, use an electronic spreadsheet with plotting and regression capability. First, use the mean plotting positions from Equation 2.143 and then, on the same graph, use the approximate median plotting positions from (2.155). Draw the least squares line through each set of points. Is there a noticeable difference between the lines? Estimate the slope from both graphs and use it to estimate the failure rate λ.

   Comment. The data were simulated from a CFR population with mean 100.

6. (a) Use the Weibull simulated data from Exercise 2.7.2 to plot superimposed graphs as you did in Exercise 5. (If possible, do this by means of an electronic spreadsheet.) Compute the β and θ estimates for both cases. Which are closer to the true values of $\beta = 3$ and $\theta = 100$?

   (b) Do the same for the data from Exercise 2.7.3.

7. Use a random number generator or Table T9 to simulate several Weibull samples. Plot the resulting data and compute the parameter estimates from your plotted data, just as you did in Exercise 6.

APPENDIX 2A

Gamma and Beta Functions

2A.1 THE GAMMA FUNCTION

Let us define $\Gamma(\alpha)$ as

$$\Gamma(\alpha) = \int_0^\infty x^{\alpha-1} e^{-x}\, dx \qquad (2A.1)$$

For integer values of α we can evaluate this function using successive integrations by parts with the following substitutions:

$$u = x^{\alpha-1} \qquad\qquad v = -e^{-x}$$
$$du = (\alpha - 1)x^{\alpha-2}\, dx \qquad dv = e^{-x}\, dx$$

Then

$$\Gamma(\alpha) = -e^{-x} x^{\alpha-1}\Big|_0^\infty + (\alpha - 1)\int_0^\infty x^{\alpha-2} e^{-x}\, dx \qquad (2A.2)$$

$$= (\alpha - 1)\int_0^\infty x^{\alpha-2} e^{-x}\, dx$$

$$= (\alpha - 1)\Gamma(\alpha - 1) \qquad (2A.3)$$

Keep up this process, reducing the exponent of the polynomial factor at each step. This gives

$$\Gamma(\alpha) = (\alpha - 1)(\alpha - 2)\Gamma(\alpha - 2)$$
$$= (\alpha - 1)(\alpha - 2)(\alpha - 3)\Gamma(\alpha - 3)$$

If α is an integer, we finally reach a terminating step

$$\Gamma(1) = \int_0^\infty x^{1-1} e^{-x}\, dx = \int_0^\infty e^{-x}\, dx = 1$$

Hence

$$\Gamma(\alpha) = (\alpha - 1)! \qquad (2A.4)$$

90

If α is not an integer, we keep integrating by parts until we reach an argument between 1 and 2, because these are given in tables such as Table T5. Thus $\Gamma(4.2) = (3.2)(2.2)(1.2)\Gamma(1.2) = (3.2)(2.2)(1.2)(.9182) = 7.7561$.

Two other ways of writing the gamma function are useful in particular applications. The first is obtained by substituting $x = z^2$ in Equation 2A.1. The result, after rearranging, is

$$\Gamma(\alpha) = 2 \int_0^\infty z^{2\alpha-1} e^{-z^2} \, dz \tag{2A.5}$$

The second alternative form is obtained by substituting $x = z^2/2$ in (2A.1), which yields

$$\Gamma(\alpha) = 2^{1-\alpha} \int_0^\infty z^{2\alpha-1} e^{-z^2/2} \, dz \tag{2A.6}$$

This form is especially useful when working with the normal distribution. Either of these forms can be used to prove that $\Gamma(\frac{1}{2}) = \sqrt{\pi}$. It involves a trick, as follows:

$$\Gamma\left(\frac{1}{2}\right)\Gamma\left(\frac{1}{2}\right) = \sqrt{2} \int_0^\infty z^0 e^{-z^2/2} \, dz \, \sqrt{2} \int_0^\infty w^0 e^{-w^2/2} \, dw \tag{2A.7}$$

$$= 2 \int_0^\infty \int_0^\infty e^{-(z^2+w^2)/2} \, dz \, dw \tag{2A.8}$$

$$= 2 \int_0^{\pi/2} \int_0^\infty e^{-r^2/2} r \, dr \, d\theta \tag{2A.9}$$

$$= 2 \int_0^{\pi/2} d\theta \int_0^\infty e^{-r^2/2} r \, dr$$

$$= 2(\pi/2)\left(-e^{-\infty} + e^{-0}\right) = \pi \tag{2A.10}$$

Hence

$$\Gamma\left(\frac{1}{2}\right) = \sqrt{\pi}, \qquad \text{Q.E.D.} \tag{2A.11}$$

To apply this equation to finding the area under the standard normal curve, consider the integral

$$K \int_{-\infty}^\infty e^{-z^2/2} dz = 1 \tag{2A.12}$$

where K is to be evaluated. Since the integrand is an even function,

$$K \int_{-\infty}^{\infty} e^{-z^2/2}\, dz = 2K \int_{0}^{\infty} e^{-z^2/2} dz \qquad (2A.13)$$

$$= \sqrt{2}K \left(\sqrt{2} \int_{0}^{\infty} e^{-z^2/2} dz \right)$$

$$= \sqrt{2}K\, \Gamma\left(\frac{1}{2}\right) = K\sqrt{2\pi} \equiv 1 \qquad (2A.14)$$

Thus the "normalizing constant" for the standard normal curve is

$$K = 1/\sqrt{2\pi}, \qquad \text{Q.E.D.} \qquad (2A.15)$$

More about the Gamma Function

THEOREM: $\Gamma(0) = \infty$

Proof

$$\Gamma(0) = \int_{0}^{\infty} \frac{e^{-x}}{x}\, dx = \int_{0}^{.5} \frac{e^{-x}}{x}\, dx + \int_{.5}^{\infty} \frac{e^{-x}}{x}\, dx = Q_1 + Q_2$$

Now if $0 < x < .5$, then $-.5 < -x < 0$, so that we have

$$e^{-.5} < e^{-x} < e^{0} = 1$$

and thus

$$\frac{e^{-x}}{x} > \frac{e^{-.5}}{x}$$

Then

$$Q_1 = \int_{0}^{.5} \frac{e^{-x}\, dx}{x} > e^{-.5} \int_{0}^{.5} \frac{dx}{x} = e^{-.5} \ln x \Big|_{0}^{.5} = e^{.5}(\ln .5 - \ln 0)$$

$$= \infty, \qquad \text{Q.E.D.}$$

We can use the recursion relation

$$\Gamma(\alpha) = (\alpha - 1)\Gamma(\alpha - 1)$$

in the alternative form

$$\Gamma(\alpha - 1) = \frac{1}{\alpha - 1}\Gamma(\alpha)$$

to define the gamma function for negative arguments, or at least for negative nonintegers. For example,

$$\Gamma(-2.5) = \frac{\Gamma(-1.5)}{-2.5}$$

$$= \frac{1}{(-2.5)} \frac{\Gamma(-.5)}{(-1.5)}$$

$$= \frac{1}{(-2.5)} \frac{1}{(-1.5)} \frac{\Gamma(.5)}{(-.5)}$$

$$= \frac{-8\sqrt{\pi}}{15}$$

2A.2 THE BETA FUNCTION

Consider the function defined by

$$B(m,n) = \int_0^1 t^{m-1}(1-t)^{n-1}\, dt \tag{2A.16}$$

This function is a constant whose value is determined by m and n. An alternative form is obtained by the following substitution (with appropriate changes of integration limits):

$$t = \sin^2 \theta$$
$$1 - t = \cos^2 \theta$$
$$dt = 2\sin\theta\cos\theta\, d\theta$$

$$B(m,n) = 2\int_0^{\pi/2} \sin^{2m-1}\theta\cos^{2n-1}\theta\, d\theta \tag{2A.17}$$

To evaluate this we use a trick based on the gamma function, as defined by (2A.5):

$$\Gamma(m)\Gamma(n) = \left(2\int_0^\infty z^{2m-1}e^{-z^2}\, dz\right)\left(2\int_0^\infty w^{2n-1}e^{-w^2}\, dw\right)$$

$$= 4\int_0^\infty\int_0^\infty z^{2m-1}w^{2n-1}e^{-(z^2+w^2)}\, dz\, dw$$

The substitution

$$z = r\sin\theta$$
$$w = r\cos\theta$$

results in

$$z^2 + w^2 = r^2$$
$$dz\, dw \rightarrow r\, dr\, d\theta$$

so that

$$\Gamma(m)\,\Gamma(n) = 4\int_0^{\pi/2}\int_0^\infty r^{2m-1}(\sin\theta)^{2m-1}r^{2n-1}(\cos\theta)^{2n-1}e^{-r^2}r\,dr\,d\theta$$

$$= \left[2\int_0^\infty r^{2(m+n)-1}e^{-r^2}dr\right]\left[2\int_0^{\pi/2}\sin^{2m-1}\theta\,\cos^{2n-1}\theta\,d\theta\right]$$

$$= \Gamma(m+n)\,B(m,n)$$

Thus

$$B(m,n) = \frac{\Gamma(m)\Gamma(n)}{\Gamma(m+n)} \tag{2A.18}$$

Notice that $B(m,n) = B(n,m)$; that is, the beta function is symmetric in its arguments, even though the function in the integrand is not the same in the two cases.

2A.3   THE BETA FAMILY OF RANDOM VARIABLES

Let X be a random variable with density function

$$f(x) = K\,x^{m-1}(1-x)^{n-1} \qquad \text{for } 0 \le x \le 1$$
$$= 0 \qquad\qquad\qquad\qquad \text{elsewhere} \tag{2A.19}$$

Typical curves are shown in Figure 2.43. It can easily be proved that the normalizing constant must be $K = 1/B(m,n)$. The moments of the beta family are

$$\mu_r' = E(X^r) = \int_0^1 x^r f(x)\,dx = \int_0^1 \frac{x^{r+m-1}(1-x)^{n-1}}{B(m,n)}\,dx$$

$$= \frac{B(r+m,n)}{B(m,n)} = \frac{\Gamma(r+m)\Gamma(n)}{\Gamma(m+n+r)}\frac{\Gamma(m+n)}{\Gamma(m)\Gamma(n)} = \frac{\Gamma(m+n)\Gamma(m+n)}{\Gamma(m+n+r)\Gamma(m)} \tag{2A.20}$$

In particular

$$\mu = E(X) \quad = \frac{\Gamma(m+1)\Gamma(m+n)}{\Gamma(m+n+1)\Gamma(m)} = \frac{m\Gamma(m)\Gamma(m+n)}{(m+n)\Gamma(m+n)\Gamma(m)} = \frac{m}{m+n} \tag{2A.21}$$

$$\mu_2' = E(X^2) \quad = \frac{m(m+1)}{(m+n)(m+n+1)} \tag{2A.22}$$

$$\sigma^2 = \mu_2' - \mu^2 = \frac{mn}{(m+n)^2(m+n+1)} \tag{2A.23}$$

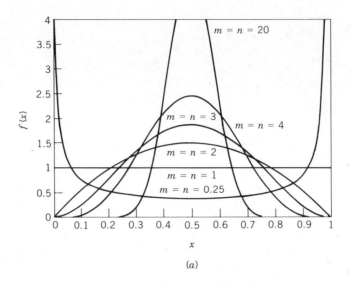

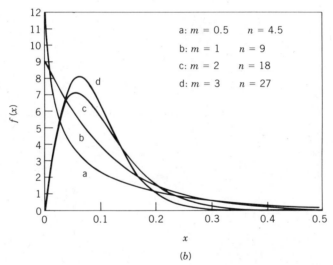

Figure 2.43 Selected beta densities. (*a*) Densities with mean
.5. (*b*) Densities with mean .1.

The mode of the beta distribution is found in the usual way using calculus methods.
Setting

$$y = K x^{m-1}(1-x)^{n-1}$$

then

$$y' = K x^{m-2}(1-x)^{n-2}[-x(n+m-2) + m - 1]$$

and equating this to zero yields the critical point, or mode, at

$$x_0 = \frac{m-1}{m+n-2}$$

We define the *beta function of the second kind* by substituting into (2A.16) as follows:

$$t = \frac{w}{1+w}$$

$$1-t = \frac{1}{1+w}$$

$$dt = \frac{dw}{(1+w)^2}$$

Then $w = t/(1-t)$, so when $t = 0$ we have $w = 0$ and when $t = 1$ we have $w = \infty$. Thus

$$B(m,n) = \int_0^\infty \frac{w^{m-1}}{(1+w)^{m+n}}\, dw$$

We say that $W \sim be_2(m,n)$, or that W is a beta random variable of the second kind, if its density is

$$f(w) = \frac{1}{B(m,n)}\frac{w^{m-1}}{(1+w)^{m+n}} \qquad \text{for } 0 \le w < \infty \qquad (2A.24)$$

A sample density curve is shown in Figure 2.44. The mean and variance are obtained from

$$E(W^r) = \int_0^\infty w^r f(w)\, dw$$

$$E(W^r) = \frac{1}{B(m,n)}\int_0^\infty \frac{w^r w^{m-1}}{(1+w)^{m+n}}\, dw$$

$$= \frac{1}{B(m,n)}\int_0^\infty \frac{w^{r+m-1}}{(1+w)^{m+n}}\, dw$$

$$= \frac{B(m+r,n-r)}{B(m,n)}$$

$$= \frac{\Gamma(m+r)\Gamma(n-r)}{\Gamma(m+n)}\frac{\Gamma(m+n)}{\Gamma(m)\Gamma(n)} = \frac{\Gamma(m+r)\Gamma(n-r)}{\Gamma(m)\Gamma(n)} \qquad (2A.25)$$

For $r = 1$, we get

$$E(W) = \frac{m\Gamma(m)\Gamma(n-1)}{\Gamma(m)(n-1)\Gamma(n-1)} = \frac{m}{n-1} \qquad (2A.26)$$

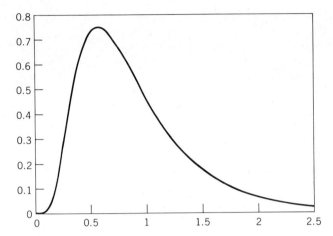

Figure 2.44 The typical beta density of the second kind with $m = 5, n = 7$.

For $r = 2$ we get

$$E(W^2) = \frac{\Gamma(m + 2)\Gamma(n - 2)}{\Gamma(m)\Gamma(n)} = \frac{(m + 1)m\Gamma(m)\Gamma(n - 2)}{\Gamma(m)(n - 1)(n - 2)\Gamma(n - 2)} = \frac{m(m + 1)}{(n - 1)(n - 2)}$$

$$(2A.27)$$

Hence the variance is

$$\sigma^2 = E(W^2) - E^2(W) = \frac{m(m + n - 1)}{(n - 1)^2(n - 2)} \qquad (2A.28)$$

The most important example of such a random variable is known as Snedecor's F, where $m = \nu_1/2$ and $n = \nu_2/2$ and $W = \nu_1 F / \nu_2$.

EXERCISES 2A PROBLEMS IN "INSTANT INTEGRATION"

All these integrations can be performed by reference to the definitions in this appendix, using (in some cases) simple substitutions of variables. Problems 6 and 7 require you to recall the concept of even and odd functions. Except when indicated, no actual integrations are to be performed; this is an exercise in recognition of standard forms.

1. $\displaystyle\int_0^\infty x^5 e^{-x}\, dx$

2. $\displaystyle\int_0^\infty z^{11/2} e^{-x}\, dx$

3. $\displaystyle\int_0^\infty y^7 e^{-3y}\, dy$

4. $\displaystyle\int_0^\infty q^{5/2} e^{-q/4}\, dq$

5. $\displaystyle\int_0^\infty x^2 e^{-x^2/2}\, dx$

6. $\displaystyle\int_{-\infty}^\infty e^{-x^2/2}\, dx$

7. $\displaystyle\int_{-\infty}^\infty y^3 e^{-y^2/2}\, dy$

8. $\displaystyle\int_0^\infty m^4 e^{-m/3}\, dm$

9. $\displaystyle\int_0^\infty v^4 e^{-v^2}\, dv$

10. $\displaystyle\int_{-\infty}^\infty v e^{-v^2/2}\, dv$

11. $\displaystyle\int_0^1 w^3 (1-w)^2\, dw$ (Verify by integration.)

12. $\displaystyle\int_0^1 y^{3/2}(1-y)^{5/2}\, dy$

13. $\displaystyle\int_0^{\pi/2} \cos^4\theta \, \sin^5\theta\, d\theta$

14. $\displaystyle\int_0^4 t^5 (4-t)^3\, dt$

15. $\displaystyle\int_0^\infty \frac{z^7}{(1+z)^{13}}\, dz$

16. $\displaystyle\int_0^\infty \frac{t^{19}}{(1+t)^{20}}\, dt$

Laplace Solution of the Poisson Equation

We present here the solution of the Poisson difference–differential equations of Section 2.3. There we derived the equation set

$$P_n'(t) = -\lambda P_n(t) + \lambda P_{n-1}(t), \qquad n = 1, 2, 3, \ldots \qquad (2.46)$$

$$P_n'(t) = \lambda P_0(t) \qquad (2.47)$$

We denote by s the variable in the Laplace transform space, and by $Z_n = Z_n(s)$ the Laplace transform of $P_n(t)$, so that

$$\mathscr{L}[P_n(t)] = Z_n(s)$$

Taking transforms of (2.46) and (2.47), and recalling that $P_0(0) = 1$, and $P_n(0) = 0$, we have

$$sZ_n - P_n(0) = -\lambda Z_n + \lambda Z_{n-1} \qquad (2B.1)$$

$$sZ_0 - 1 = \lambda Z_0 \qquad (2B.2)$$

which can be written as

$$(s + \lambda)Z_n = \lambda Z_{n-1} \qquad (2B.3)$$

$$(s + \lambda)Z_0 = 1 \qquad (2B.4)$$

Equation (2B.4), written as

$$Z_0 = \frac{1}{s + \lambda}$$

is then operated on with the inverse Laplace operator, yielding

$$P_0(t) = \mathscr{L}^{-1}[Z_0] = \mathscr{L}^{-1}[1/(s + \lambda)] = e^{-\lambda t}$$

Equation (2B.3) is used recursively. For $n = 1, 2, 3, \ldots$, it leads to

$$Z_1 = \frac{Z_0 \lambda}{s + \lambda} = \frac{\lambda}{(s + \lambda)^2} \qquad (2B.5)$$

$$Z_2 = \frac{Z_1 \lambda}{s + \lambda} = \frac{\lambda^2}{(s + \lambda)^3}$$

$$\vdots$$

$$Z_n = \frac{Z_{n-1} \lambda}{s + \lambda} = \frac{\lambda^n}{(s + \lambda)^{n+1}} \qquad (2B.6)$$

Taking the inverse transform, we get

$$P_n(t) = \lambda^n \mathscr{L}^{-1} \left[1/(s_n + \lambda)^{n+1} \right]$$

$$= \lambda^n \frac{e^{-\lambda t} t^n}{n!}$$

$$= \frac{e^{-\lambda t}(\lambda t)^n}{n!}, \qquad n = 0, 1, 2, \ldots \qquad (2B.7)$$

as stated in Equation 2.49.

APPENDIX 2C

Dummy Variables of Integration

We know that

$$\int_1^4 x^2 \, dx = \left. \frac{x^3}{3} \right|_1^4 = 21 \text{ (a constant)}$$

This could also be written as

$$\int_1^4 y^2 \, dy = \int_1^4 q^2 \, dq = \int_1^4 G^2 \, dG = 21$$

The variables of integration x, y, q, and G are merely "place holders." The symbol used is irrelevant, so they are called *dummy* variables of integration.

Now consider the situation in which we want to find the area A between the limits a and b of the function $f(x) = x^2(1 + x^3)^4$. We can write it as

$$A = \int_a^b x^2(1 + x^3)^4 \, dx = \left. \frac{(1 + x^3)^5}{15} \right|_a^b$$

$$= \frac{(1 + b^3)^5 - (1 + a^3)^5}{15}, \qquad \text{(a constant)}$$

Suppose that the problem is changed so that the upper limit of integration is not yet decided on and can be considered as a variable. The area could then be designated as $A(x)$, as shown in Figure 2.45. It is tempting to write it as

$$A(x) = \int_a^x x^2(1 + x^3)^4 \, dx$$

But this is *wrong*! This can be demonstrated as follows. If we employ the following substitution for carrying out the integration,

$$v = 1 + x^3$$

then

$$x = (v - 1)^{1/3}$$

and

$$dv = 3x^2 dx$$

and the limits of integration must be changed (to agree with the change of variable). At the lower limit, where $x = a$, we get $v = 1 + a^3$. We may feel that no change is indicated at the upper limit, which would give us

$$A(x) = \int_{1+a^3}^{x} v^4 \frac{dv}{3} = \frac{v^5}{15}\bigg|_{1+a^3}^{x} = \frac{x^5 - (1 + a^3)^5}{15}$$

which is obviously wrong.

We may feel, correctly, that we should have done something about the upper limit of integration. Recalling that $x = (v - 1)^{1/3}$ we may write

$$A(x) = \int_{1+a^3}^{(v-1)^{1/3}} v^4 \frac{dv}{3} = \frac{v^5}{15}\bigg|_{1+a^3}^{(v-1)^{1/3}} = \frac{(v - 1)^{5/3} - (1 + a^3)^5}{15}$$

$$= \frac{x^5 - (1 + a^3)^5}{15}$$

which is still wrong.

Conclusion. Our problem is that we are using the same letter as a dummy variable of integration and also as a limit of integration. Let's do it the right way, using different symbols for these two variables. Thus we must write

$$A(x) = \int_{a}^{x} t^2(1 + t^3)^4 dt$$

Now let $1 + t^3 = v$ so that $3t^2 dt = dv$ and $t^2 dt = dv/3$. The change of variable requires

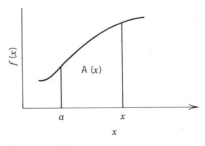

Figure 2.45

a corresponding change of integration limits as follows. If $t = a$ then $v = 1 + a^3$, and if $t = x$ then $v = 1 + x^3$, whereupon the integral becomes

$$A(x) = \int_{1+a}^{1+x^3} v^4 \frac{dv}{3}$$

$$= \frac{v^5}{15}\bigg|_{1+a^3}^{1+x^3} = \frac{(1 + x^3)^5 - (1 + a^3)^5}{15}$$

which is the correct answer.

Moral.   Never use the same letter to represent both an integration limit and a dummy variable of integration, especially when there is a need to integrate by substitution of variable.

APPENDIX 2D

Differentiating an Integral

We develop here the rule for differentiation of a function when the independent variable appears in both the integration limits and the integrand function of a definite integral. Let

$$Q(w) = \int_{a(w)}^{b(w)} H(w, \theta)\, d\theta$$

then

$$Q + \Delta Q = \int_{a+\Delta a}^{b+\Delta b} H(w + \Delta w, \theta)\, d\theta$$

and

$$\Delta Q = \int_{a+\Delta a}^{b+\Delta b} H(w + \Delta w, \theta)\, d\theta - \int_{a}^{b} H(w, \theta)\, d\theta$$

$$= \int_{a}^{b} H(w + \Delta w, \theta)\, d\theta + \int_{b}^{b+\Delta b} H(w + \Delta w, \theta)\, d\theta -$$

$$\int_{a}^{a+\Delta a} H(w + \Delta w, \theta)\, d\theta - \int_{a}^{b} H(w, \theta)\, d\theta$$

$$= \int_{a}^{b} [H(w + \Delta w, \theta) - H(w, \theta)]\, d\theta + \int_{b}^{b+\Delta b} H(w + \Delta w, \theta)\, d\theta -$$

$$\int_{a}^{a+\Delta a} H(w + \Delta w, \theta)\, d\theta$$

Dividing by Δw and applying the mean value theorem to the last two integrals yields

$$\frac{\Delta Q}{\Delta w} = \int_a^b \frac{H(w + \Delta w, \theta) - H(w, \theta)}{\Delta w} d\theta + \frac{\Delta b}{\Delta w} H(w + \Delta w, \theta^*) - \frac{\Delta a}{\Delta w} H(w + \Delta w, \theta^{**})$$

where θ^* and θ^{**} are intermediate values defined by

$$b \le \theta^* \le b + \Delta b, \qquad a \le \theta^{**} \le a + \Delta a$$

Taking the limit as $\Delta w \to 0$ yields

$$\frac{dQ}{dw} = \int_a^b \frac{\partial H(w, \theta)}{\partial w} d\theta + \frac{db}{dw} H(w, b) - \frac{da}{dw} H(w, a)$$

If a and b are constants rather than functions of w, the formula reduces to

$$\frac{dQ}{dw} = \int_a^b \frac{\partial H(w, \theta)}{\partial w} d\theta$$

If a is a constant, $b(w) = w$, and H is a function of θ alone, then

$$\frac{dQ}{dw} = H(w)$$

The most familiar application of this in statistical work is the situation in which $Q(w)$ is the CDF function $F(\cdot)$ and the integrand function is the density function $f(\cdot)$. In short, if

$$F(x) = \int_0^x f(v) \, dv$$

then $F'(x) = f(x)$, as is intuitively obvious.

APPENDIX 2E

Proof of Theorem 2.1

I learned this proof from my colleague James Higgins and consider it rather elegant. By definition we have

$$\mu = \int_0^\infty t f(t) \, dt$$

The integral can be split into two pieces, and using a new dummy variable of integration, we have for any t

$$\int_0^\infty s f(s) \, ds = \int_0^t s f(s) \, ds + \int_t^\infty s f(s) \, ds$$

In the second integral on the right we observe that $s > t$ over the range of integration, so we can write

$$E(T) \geq \int_0^t sf(s)\,ds + t\int_t^\infty f(s)\,ds$$

$$\geq \int_0^t sf(s)\,ds + tP(T \geq t)$$

$$\geq \int_0^t sf(s)\,ds + tR(t)$$

Rearranging gives

$$E(T) - \int_0^t sf(s)\,ds \geq tR(t)$$

and taking the limit of both sides as $t \to \infty$ yields $0 \geq \lim_{t\to\infty} tR(t)$. On the other hand, $tR(t) \geq 0$ always, so $0 \geq \lim_{t\to\infty} tR(t) \geq 0$, which is the desired result. It depends only on the existence of the mean, which must then be the equivalence of well-behavedness mentioned in Section 2.2.

APPENDIX 2F

Computer Programs for Section 2.9

Two programs are given. The first uses SAS simplified plotting to construct plots of t_i versus the hazard function $h(t_i)$. The printout for only the exponential program is given. To revise it for the other distributions (gamma, normal, Weibull), replace the indicated exponential block on the first page of the program with the appropriate block. Then make a global editing change from the word "exponential" to the appropriate name (gamma, normal, Weibull). Figures 2.28 through 2.31 were generated using these programs.

The program given here is an abbreviated version for five samples (NS = 5). If more samples are desired, multiple runs of the program must be made starting with a different seed each time. If fewer than five samples are desired, simply change NS to the desired number. There will be an error message from SAS but the program will run. The problem is that we cannot run a DO-loop between the FORTRAN and SAS portions of the program, or between PROC statements in SAS. Readers who want to accomplish a large number of samples in a single run should eliminate the SAS portion of the programs and utilize the IMSL routine USPLOD, or something equivalent, to get plots of t_i versus $h(t_i)$.

The second program uses the IMSL routine USPRP to plot t_i against the appropriately transformed value of $\hat{F}(t_i)$. The JCL supplied is for the National Advanced System 6630 using IBM's VM/SP. As given, the program generates samples of all the population types under study and tests them against all six of the available models of USPRP. Changing some DO-loops appropriately will make the program more restricted if that is desired. The second program was used to generate Figures 2.34 through 2.39.

PROGRAM 1

This program or its variants can be used to generate random samples from the exponential, normal, Weibull, or gamma distributions, then compute the empirical hazard function and plot it. The program requires the IMSL library and SAS.

```
//*++ REGION 600K
//*++ PRINT
//*++ VMMSG
//*++ SERVICE  *  TIME ,10
//  EXEC      FORTVCLG
      INTEGER NR, NS
C          NR IS THE SAMPLE SIZE.
C          NS IS THE NUMBER OF SAMPLES (PLOTS) TO BE GENERATED.
C   REAL XM, R(100), H(100), X(100), Q(100)
C          XM IS THE MEAN OF THE EXPONENTIAL DISTRIBUTION AND IS THE
C             RECIPROCAL OF THE FAILURE RATE LAMBDA.
C           R(.) IS THE OUTPUT VECTOR OF RANDOM EXPONENTIAL VALUES
C              FROM THE IMSL GENERATOR.  AFTER SORTING, IT IS ALSO
C              THE VECTOR OF ORDERED FAILURE TIMES.
C           H(.) IS THE VECTOR OF HAZARD FUNCTION VALUES CORRESPONDING
C              TO THE TIMES GIVEN BY R(.).
      DOUBLE PRECISION DSEED
      DSEED = 123457.D0
C***********EXPONENTIAL BLOCK*********************************
C  REPLACE THIS BLOCK OF CODE FOR THE OTHER DISTRIBUTIONS      *
      XM = 100.                                                *
      NR =  20                                                 *
      NS =   5                                                 *
      DO 200 M=1,NS                                            *
      CALL GGEXN (DSEED, XM, NR, R)                            *
C          THIS IS THE INVOCATION OF THE RANDOM NUMBER GENERATOR.*
C          XM IS THE MEAN OF THE DISTRIBUTION.                 *
C          NR IS THE SAMPLE SIZE.                              *
C          R IS THE OUTPUT VECTOR.                             *
C                                                              *
C***************************************************************
      CALL VSRTA(R,NR)
C          THIS IS THE IMSL SORT PROCEDURE WHICH ORDERS THE FAILURE TIMES.
```

```
      H(1)=  1./(R(1)* NR)
C        NOW THE HAZARD FUNCTION VALUES ARE CALCULATED.
C        THE VALUE FOR H(1) MUST BE DONE SEPARATELY BECAUSE FORTRAN
C        DOESN'T LIKE ZERO AS AN INDEX.  ALL THE OTHER H VALUES ARE
C        CALCULATED INSIDE THE DO_100  DO LOOP.
      DO 100 I = 2,NR
      H(I)=  1./((NR-I+1)*(R(I)-R(I-1)))
 100  CONTINUE
      DO 101 I=1,NR
C        NOW THE DATA FOR A SAMPLE IS WRITTEN TO DISK (UNIT 10).
      WRITE(10,99) I, R(I), H(I),  NR
  99  FORMAT (5X,I5,5X,F10.3,5X,F10.5,5X,I5)
 101  CONTINUE
 200  CONTINUE
      STOP
      END
C        THE NEXT FEW JCL CARDS (AFTER THE /* WHICH TERMINATES THIS
C        FORTRAN PROGRAM) ARE THE JCL NEEDED TO READ THE FORTRAN
C        GENERATED DATA SETS OFF DISK AND ALLOW THEM TO BE USED
C        AS DATA IN THE S.A.S. PROGRAM WHICH FOLLOWS.
/*
//LKED.SYSLIB   DD
//             DD
//             DD
//     DD DSN=SYS1.IMSL.SPFLIB,DISP=SHR
//GO.FT10F001 DD DSN=&&TEMP,DISP=(NEW,PASS),
//    UNIT=SYSDA,SPACE=(100,(80,10)),
//    DCB=(RECFM=FB,LRECL=80,BLKSIZE=6160)
//GO.SYSIN   DD*
/*
// EXEC  SAS
//DATASET  DD DSN=&&TEMP,DISP=(OLD,DELETE)
OPTIONS LS=80;
DATA BIGLIST;
INFILE DATASET;
INPUT I 6-10   T 16-25 H 31-40   NR 46-50
TITLE  SAMPLE OF SIZE TWENTY:  EXPONENTIAL CASE WITH MEAN 100. ;
DATA SAMPLE1 SAMPLE2 SAMPLE3 SAMPLE4 SAMPLE5;
SET BIGLIST;
IF            _N_ <= NR THEN OUTPUT SAMPLE1;
IF _N_ > NR AND _N_ <=2*NR THEN OUTPUT SAMPLE2;
IF _N_ > 2*NR AND _N_  <=3*NR THEN OUTPUT SAMPLE3;
IF _N_ > 3*NR AND _N_  <= 4*NR THEN OUTPUT SAMPLE4;
IF _N_ > 4*NR AND _N_  <= 5*NR THEN OUTPUT SAMPLE5;
PROC PLOT DATA=SAMPLE1;
TITLE3 PLOT OF EXPONENTIAL SAMPLE1. HAZARD FUNCTION H AGAINST TIME T.;
PLOT H*T='*';
PROC PLOT DATA=SAMPLE2;
TITLE3 PLOT OF EXPONENTIAL SAMPLE2. HAZARD FUNCTION H AGAINST TIME T.;
PLOT H*T='*';
```

```
PROC PLOT DATA =SAMPLE3;
TITLE3 PLOT OF EXPONENTIAL SAMPLE3. HAZARD FUNCTION H AGAINST TIME T.;
PLOT H*T='*';
PROC PLOT DATA=SAMPLE4;
TITLE3 PLOT OF EXPONENTIAL SAMPLE4. HAZARD FUNCTION H AGAINST TIME T.;
PLOT H*T = '*';
PROC PLOT DATA=SAMPLE5;
TITLE3 PLOT OF EXPONENTIAL SAMPLE5. HAZARD FUNCTION H AGAINST TIME T.;
PLOT H*T='*';
/*
C
C**********************NORMAL BLOCK******************************
      MEAN = 10.0
      NR =20
      NS= 5
      DO 200 M=1,NS
      CALL GGNML (DSEED,NR,Q)
      DO 10  I = 1,NR
      R(I) = MEAN + Q(I)
  10 CONTINUE
C***************************************************************
C**********************GAMMA BLOCK******************************
      NR =20
      NS= 5
      A= 6.0
      DO 200 M=1,NS
      CALL GGAMR (DSEED, A, NR, WK, R)
C          WK IS A WORK VECTOR OF LENGTH 2*NR IF A .LT. 1.0.
C          IF A .GE. 1.0 THEN WK IS NOT USED AND MAY BE DIMENSIONED
C          AS WK(1).
C***************************************************************
C***********************WEIBULL BLOCK**************************
      NR =20
      NS= 5
      THETA = 1.3
      BETA = 6.0
      DO 200 M = 1,NS
      CALL GGWIB (DSEED,BETA,NR,R)
      DO 3 I = 1,NR
      R(I) = THETA * R(I)
   3  CONTINUE
C***************************************************************
```

PROGRAM II

This program generates a random sample from one of seven distributions and then plots linearized ogive curves for six candidate model distributions using the IMSL routine USPRP.

```
//*++ PRINT
//*++ VMMSG    IQL
//*++ SERVICE *
// EXEC  FORTQCLG
      INTEGER     NR, NS, CODE, IDIST, IOPT, IER
      REAL        MEAN, A, B, WK, R(50), Z(50), T(50), WK2(100), X(50)
      REAL        ALPHA, BETA, MEAN, STDEV, CONST, COEF, MODE, SCALE
      DOUBLE PRECISION    DSEED
C
      NR   = 20
C  NR IS THE SAMPLE SIZE
      NS   = 1
C  NS IS THE NUMBER OF SAMPLES
C
      DO 200 M = 1,NS
      DSEED = 123457.D0
C
C  HERE IS WHERE THE SIMULATED DISTRIBUTION IS CHOSEN
      DO 99 MM = 1,1
       CODE = MM
      IF (CODE.EQ.1) GO TO 1
      IF (CODE.EQ.2) GO TO 2
      IF (CODE.EQ.3) GO TO 3
      IF (CODE.EQ.4) GO TO 4
      IF (CODE.EQ.5) GO TO 5
      IF (CODE.EQ.6) GO TO 6
      IF (CODE.EQ.7) GO TO 7
C CODE=1 GENERATES THE EXPONENTIAL DISTRIBUTION
C CODE=2 GENERATES THE WEIBULL DISTRIBUTION
C CODE=3 GENERATES THE GAMMA DISTRIBUTION
C CODE=4 GENERATES THE NORMAL DISTRIBUTION
C CODE=5 GENERATES THE HALF-NORMAL DISTRIBUTION
C CODE=6 GENERATES THE LOG-NORMAL DISTRIBUTION
C CODE=7 GENERATES THE EXTREME VALUE DISTRIBUTION
C
C  NEGATIVE EXPONENTIAL CASE
1     MEAN = 100.
      CALL GGEXN (DSEED,MEAN,NR,X)
      GO TO 600
C
C  WEIBULL CASE
2     ALPHA = 10.0
      BETA = 6.0
      CONST = -1.0/BETA
      COEF = ALPHA ** CONST
      CALL GGWIB (DSEED,BETA,NR,R)
      DO 30 I=1,NR
      X(I) = COEF * R(I)
```

```
30    CONTINUE
      GO TO 600
C
C  GAMMA CASE
3     A = 6.0
      CALL GGAMR (DSEED, A, NR, WK, X)
      GO TO 600
C
C  NORMAL CASE
4     MEAN = 10.
      STDEV = 1.
      CALL GGNML (DSEED, NR, R)
      DO 40 I =1, NR
      X(I) = MEAN + STDEV * R(I)
40    CONTINUE
      GO TO 600
C
C HALF—NORMAL CASE
5     CONTINUE
      CALL GGNML (DSEED, NR,R)
      DO 50 I=1,NR
      X(I) = ABS (R(I))
50    CONTINUE
      GO TO 600
C
C  LOG—NORMAL CASE
6     CONTINUE
      CALL GGNML (DSEED, NR, R)
      DO 60 I=1,NR
      X(I) = EXP(R(I))
60    CONTINUE
      GO TO 600
C
C  EXTREME VALUE CASE
7     CONTINUE
      MODE=10.
      SCALE = 1.0
      CALL GGUBS (DSEED, NR, R)
      DO 70 I=1, NR
      X(I)= MODE — SCALE*ALOG(ALOG(1.0/R(I)))
70    CONTINUE
      GO TO 600
C
600   CONTINUE
      DO 700 L=1,6
      CALL USPRP (X,NR,1,NR,L,1,WK2,IER)
700   CONTINUE
      DO 17 I=1,NR
```

```
      WRITE (6,101) I,X(I)
101   FORMAT(' ',20X,I2,5X,F10.5)
17    CONTINUE
99    CONTINUE
200   CONTINUE
      STOP
      END
/*
//LKED.SYSLIB  DD
//            DD
//            DD
//            DD  DSN=SYS1.IMSL.SPFLIB,DISP=SHR
//GO.SYSIN    DD   *
/*
```

CHAPTER 3

Multicomponent Systems

3.1 SERIES AND PARALLEL SYSTEMS

In Chapter 2 we dealt with some theoretical considerations involving the statistical failure law that may apply to individual components. Here we recognize that, in fact, most equipment is a combination of several or many components. For most units, then, it may be necessary or convenient to compute the equipment system reliability from the individual component reliabilities. For example, a newly proposed design using familiar components could be evaluated without actually being built. In the work that follows, the symbol R_i will stand for the reliability of the ith component in the system. Sometimes the time dependence will be explicitly indicated as $R_i = R_i(t)$, and sometimes R_i will be considered as a constant. It will be considered a constant if either (1) the component reliability is (relatively) independent of time, or (2) the discussion is concerned with a mission of some stated length T_0.

The simplest configuration is the *series* system, in which *all* components must function correctly for the system to function correctly. Schematically it is represented as in Figure 3.1, the symbol C_i indicating the ith component. Notice that, in this representation, there is no closure of the graph as there is in the somewhat analogous case of a series electrical system. In the electrical system we know that the circuit must be closed for the electrons to "go 'round and 'round." Here there is no such circularity implied. There may be an implication that the movement or energy or action is required to go from left to right, and if there is no defect in the path, the

111

Figure 3.1   A series system.

system will function adequately. Sometimes the electrical analogy is quite apt, as for a flashlight which consists of

C_1: a battery,
C_2: another battery,
C_3: a light bulb, and
C_4: a switch.

At other times the electrical analogy is totally inappropriate; the system task might be the preparation of a business letter. The components then could be

C_1: the supervisor who dictates the letter into a dictaphone (or tape recorder),
C_2: the dictaphone (tape recorder),
C_3: the typist who transcribes the letter, and
C_4: the typewriter.

The components C_2 and C_4 may be subdivided, if necessary, into many subsidiary components.

The system reliability for series systems has the form

$$
\begin{aligned}
R_{sys} &= P(\text{system works}) \\
&= P(\text{all components work}) \\
&= P(C_1 \text{ works}) P(C_2 \text{ works} \mid C_1 \text{ works}) \ldots \\
&\quad P(C_n \text{ works} \mid C_1, \ldots, C_{n-1} \text{ work})
\end{aligned}
\tag{3.1}
$$

Equation 3.1 allows for the possible dependence of one component on another— usually a realistic assumption. It is convenient, however, to assume independence between the various components; this can sometimes appropriately be done, especially if the components that are likely to fail together are regarded as a single unit. When the simplifying assumption of independence can be made, Equation 3.1 becomes

$$
\begin{aligned}
R = R_{sys} &= P(C_1 \text{ works}) P(C_2 \text{ works}) \ldots P(C_n \text{ works}) \\
&= R_1 R_2 \ldots R_n
\end{aligned}
\tag{3.2}
$$

The consequence of Equation 3.2 is that a series system is much less reliable than any of its individual elements. For example, a system of ten equally reliable components with $R_i = .99$ has reliability of only .90. For twenty such components the system reliability is only .82.

Example 3.1

How good must individual and identical components of a six-component series system be for the system to have a reliability of at least .95?

Solution: Set $(R_i)^6 \geq .95$; then solving for R_i yields

$$R_i \geq (.95)^{1/6} = .9915$$

Another simple configuration of components is the *parallel* system, represented graphically in Figure 3.2. The assumption is that the system performs satisfactorily if any one or more of the components performs satisfactorily; independence is again assumed between the components and the components are assumed not to impede or interfere with one another. Such a system is of course "overdesigned," with built-in redundancy, but the added cost is presumably a worthwhile trade-off for increased reliability. For $n = 2$ the system reliability is

$$R = R_{sys} = P(\text{either } C_1 \text{ or } C_2 \text{ works})$$
$$= R_1 + R_2 - R_1R_2 \qquad (3.3)$$

As may be seen by analogy with the Venn diagram in Figure 3.3, the corresponding system reliability for $n = 3$ is

$$R = P(\text{at least one of } C_1, C_2, C_3 \text{ works})$$
$$= R_1 + R_2 + R_3 - R_1R_2 - R_1R_3 - R_2R_3 + R_1R_2R_3 \qquad (3.4)$$

For $n = 4$ the system reliability is

$$R = R_1 + R_2 + R_3 + R_4 - R_1R_2 - R_1R_3 - R_1R_4 - R_2R_4 - R_3R_4 - R_2R_3$$
$$+ R_1R_2R_3 + R_1R_2R_4 + R_1R_3R_4 + R_2R_3R_4 - R_1R_2R_3R_4 \qquad (3.5)$$

It is apparent that with increasing n the expression for system reliability grows increasingly difficult to write in the given form. It thus becomes expedient to use the form which depends on complementary events. Thus for $n = 2$ the system reliability is the probability that both items do not fail simultaneously, that is,

$$R = 1 - P(\text{both fail})$$
$$= 1 - P(C_1 \text{ fails}) P(C_2 \text{ fails})$$
$$R = 1 - (1 - R_1)(1 - R_2) \qquad (3.6)$$

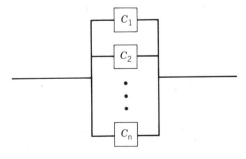

Figure 3.2 A parallel system.

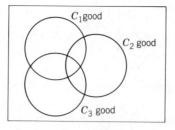

Figure 3.3   Venn diagram for
three parallel components.

which is equivalent to Equation 3.3. For $n = 4$ we could write

$$R = 1 - P(\text{all fail})$$
$$= 1 - P(C_1 \text{ fails}) \ldots P(C_4 \text{ fails})$$
$$= 1 - (1 - R_1)(1 - R_2)(1 - R_3)(1 - R_4)$$

which is equivalent to Equation 3.5. Of course, the expression for general n is

$$R_{\text{sys}} = 1 - \prod_{i=1}^{n} (1 - R_i) \tag{3.7}$$

which is certainly easier to use than the alternative "direct" probability expression.

Example 3.2

A housewife needs a box of a certain type of food for making a special meal. It is not carried at every store. She has no car but she has two children with bicycles, and there are two grocery stores within biking distance that may carry the item, Store A with probability .5 and Store B with probability .7. Assuming the children to be perfectly reliable, what is the probability that she can make the special meal if she sends one child to each store? (She may end up with an extra box of the ingredient, but that is the price she must pay for redundancy.) Her situation is that of a parallel procurement system with $R_1 = .5$ and $R_2 = .7$. Then

$$R_{\text{sys}} = .5 + .7 - (.5)(.7) = .85$$

or

$$R_{\text{sys}} = 1 - (1 - .5)(1 - .7) = .85$$

Notice that the parallel system in Example 3.2 is more reliable than any of its individual components. This is a general characteristic of parallel systems and is their *raison d'être*.

Example 3.3

How low may the reliability of five equivalent parallel components be if the system must have reliability of at least .95?

Solution: $R = 1 - (1 - R_i)^5$ is set $\geq$ .95 so that $R_i \geq 1 - (.05)^{1/5} = .45$.

Obviously components can be related in a manner that is a combination of series and parallel. Figure 3.4 shows two such possibilities. The terms "series–parallel" and "parallel–series" have not been standardized in the literature—Smith (1976) uses one nomenclature and Lloyd and Lipow (1962) reverse it.

Both configurations shown in Figure 3.4 may be regarded as different ways of incorporating redundancy into a more basic system that originally consisted of C_1 and C_2 in series. The question is whether the components C_1 and C_2 should be protected by redundant elements individually (as in Figure 3.4*b*) or as a pair (as in Figure 3.4*a*). The author has posed the question to successive "generations" of reliability students, asking for an intuitive response before any analytical attack on the problem. The response has been about equally divided, with a slight favoring of configuration *b*. That configuration *b* is, in fact, the more reliable of the two can be seen by considering the difference of the two system reliabilities $R_b - R_a$, which has the form

$$
\begin{aligned}
R_b - R_a &= (R_1 + R_3 - R_1R_3)(R_2 + R_4 - R_2R_4) - (R_1R_2 + R_3R_4 - R_1R_2R_3R_4) \\
&= R_1R_4 + R_2R_3 - R_1R_2R_4 - R_2R_3R_4 \\
&\quad - R_1R_2R_3 - R_1R_3R_4 + 2R_1R_2R_3R_4 \qquad (3.8) \\
&= R_1R_4(1 - R_2 - R_3 + R_2R_3) \\
&\quad + R_2R_3(1 - R_4 - R_1 + R_1R_4) \\
&= R_1R_4(1 - R_2)(1 - R_3) + R_2R_3(1 - R_1)(1 - R_4) \qquad (3.9)
\end{aligned}
$$

Since every factor and hence every term in Equation 3.9 is positive, it is apparent that $R_b - R_a > 0$ and that configuration *b* is more reliable than configuration *a*.

The pure parallel system is sometimes replaced by a generalization called the *r*-out-of-*n* system, in which the system functions if any *r* of the *n* components are

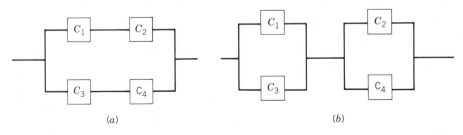

(a)   (b)

Figure 3.4   Mixed series and parallel configurations.

functioning adequately. Obviously, ordinary parallel is merely 1-out-of-n functioning, whereas series is equivalent to n-out-of-n functioning. For $r = 2$ and $n = 4$ the expression for R has the form

$$
\begin{aligned}
R = {} & R_1 R_2 (1 - R_3)(1 - R_4) + R_1 (1 - R_2) R_3 (1 - R_4) \\
& + R_1 (1 - R_2)(1 - R_3) R_4 + (1 - R_1) R_2 R_3 (1 - R_4) \\
& + (1 - R_1) R_2 (1 - R_3) R_4 + (1 - R_1)(1 - R_2) R_3 R_4 \\
& + R_1 R_2 R_3 (1 - R_4) + R_1 R_2 (1 - R_3) R_4 \\
& + R_1 (1 - R_2) R_3 R_4 + (1 - R_1) R_2 R_3 R_4 \\
& + R_1 R_2 R_3 R_4
\end{aligned}
\tag{3.10}
$$

The system reliability of the general r-out-of-n system is

$$
R = \sum_{k=r}^{n} \sum_{j=1}^{\binom{n}{k}} \prod_{i=1}^{n} R_i^{V_{i,j}} (1 - R_i)^{\bar{V}_{i,j}}
\tag{3.11}
$$

where

$$
\sum_{i=1}^{n} V_{i,j} = k
$$

and $V_{i,j}$ and $\bar{V}_{i,j}$ are complementary indicator functions for which $V_{i,j} = 1$ if the ith component functions and $V_{i,j} = 0$ if the ith component fails. For identical or equivalent components where $R_i = p$, say, Equation 3.11 assumes the cumulative binomial form

$$
R = \sum_{k=r}^{n} \binom{n}{k} p^k (1 - p)^{n-k}
\tag{3.12}
$$

EXERCISES 3.1

1. Find the reliability of a system having five identical components each having a reliability of .9 if the components are (a) connected in series, (b) connected in parallel.

2. The reliability of three components are respectively .9, .88, and .95. Find the system reliability if the components are (a) connected in series, (b) connected in parallel.

3. Five components having reliabilities of .92, .94, .89, .95, and .97 are connected in series. What is the system reliability?

4. Find the reliability for the given system.

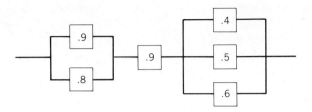

5. By inspection, we know without any calculation that the system reliability in Exercise 4 cannot exceed what value?

6. If every component in Exercise 4 had reliability .9, what would the system reliability be?

7. One hundred equally reliable components are connected in series, resulting in a system with a reliability of .9. What is the reliability of each component?

8. Two identical units are connected in parallel to create a system having a reliability of .95. What is the reliability of each component?

9. Three components are connected in parallel. What is the system reliability if each component has a reliability of .9?

10. A system consists of six identical items connected in parallel. What must the reliability of each component be if the reliability of the system is to be at least .99?

11. If a system must have a reliability of .9, what is the maximum number of components that may be put in series, if each component has a reliability of .95?

12. A system has twenty equally good components in parallel and has a system reliability of .8. How many components should be added to the system to improve the system reliability by 20 percent?

13. A system consists of seven components connected as shown. Find the overall reliability of the system.

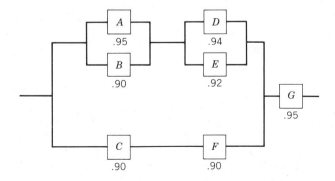

14. A system has two components in parallel. Each component has a reliability of .8. What is the system reliability? How many components must be added to the system to improve the system reliability by 3 percent?

15. A system consists of three components; the reliability of the system is .9. The reliabilities of two components are .92 and .87. What must the reliability of the third component be if the components are connected in series? Connected in parallel?

16. A system consists of components connected as shown in the following figures. Find the value of R if the reliability of system is given as .92. Assume that all components have equal reliability.

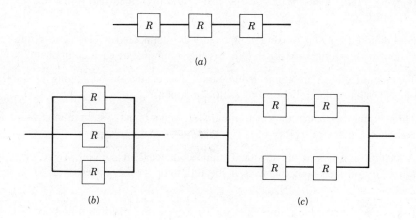

17. Determine the reliability of the system shown in the following figure.

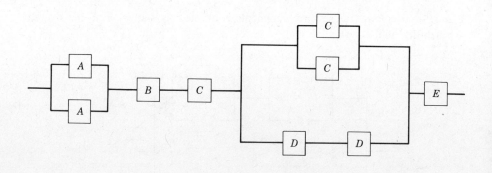

$$A = .9, \quad B = .87, \quad C = .92, \quad D = .95, \quad E = .85$$

18. A system is made of three components with reliabilities as shown in the following figure. The system functions properly if at least two of the three components function properly. Find the reliability of the system.

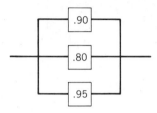

19. Determine the reliability for the system shown when at least two out of three of the elements 3 through 5 and three out of four of the elements 7 through 10 are needed for successful operation. Assume all components have equal reliability.

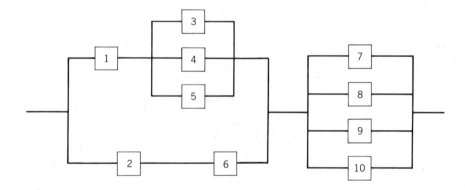

20. Determine the reliability of the system shown if at least two of the three elements (A, B, C) must function properly for the system to function properly. Each element has a reliability of .90.

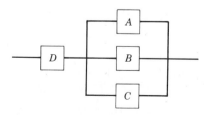

21. Find the reliabilities of the systems shown. Each element has a reliability of .90.

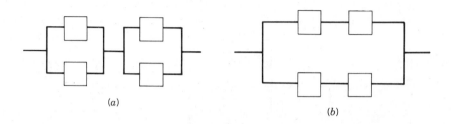

(a)   (b)

3.2 NON-SERIES–PARALLEL SYSTEMS

There are equipment configurations that do not fall into the types mentioned in Section 3.1 (which can be decomposed into building blocks of series and parallel components). These are sometimes called "complex" systems. Such a system is depicted graphically in Figure 3.5. The role of component $\mathcal{C}$ is difficult to define except to say that its presence seems to increase system reliability; the system obviously functions adequately without it.

There are three standard methods for arriving at the reliability of the system. They are enumeration, path tracing, and conditioning on a key element (sometimes incorrectly referred to as an application of Bayes' theorem).

Method I: Enumeration   It is convenient to use the following notation:

$$A = \text{the event that component } \mathcal{A} \text{ is good}$$
$$\overline{A} = \text{the event that component } \mathcal{A} \text{ is failed}$$
$$a = P(A)$$
$$1 - a = P(\overline{A})$$

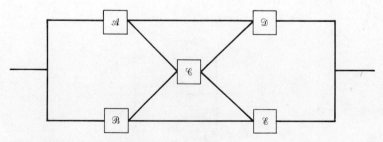

Figure 3.5 A non-series–parallel system.

We can use similar notation for all the other components. Since each of the components can be in either of two states, good or failed, there are 2^5 possible configurations for the system, as listed in Table 3.1. Each configuration is compared with Figure 3.5 to see whether the system can function in that configuration. A check mark beside the configuration in Table 3.1 indicates a satisfactory condition. The system reliability can then be written explicitly as

$$R = abcde + (1 - a)bcde + a(1 - b)cde + ab(1 - c)de + abc(1 - d)e$$
$$+ abcd(1 - e) + (1 - a)b(1 - c)de + (1 - a)bc(1 - d)e$$
$$+ (1 - a)bcd(1 - e) + a(1 - b)(1 - c)de + a(1 - b)c(1 - d)e$$
$$+ a(1 - b)cd(1 - e) + ab(1 - c)(1 - d)e + ab(1 - c)d(1 - e)$$
$$+ a(1 - b)(1 - c)d(1 - e) + (1 - a)b(1 - c)(1 - d)e \qquad (3.13)$$

After onerous algebraic manipulation Equation 3.13 reduces to

$$R = ad + be + ace + bcd - acde - abcd - abde - abce - bcde + 2abcde$$
$$(3.14)$$

It is evident that the construction of the list in Table 3.1 is a tedious operation, becoming nearly impossible to do accurately for $n > 5$. The alternative, if the enumeration method is to be used, is to construct, or invoke, a computer routine for making the table. We begin with the listing of the integers from 0 to $2^n - 1$ in binary form, expressing the leading zeros in all n positions. Thus for the $n = 5$ example

decimal 0 is expressed as 00000
decimal 1 is expressed as 00001

$\vdots$

decimal 31 is expressed as 11111

TABLE 3.1 The Possible States for the System in Figure 3.5

$ABCDE$	✓	$\overline{A}\overline{B}CDE$		$\overline{A}\overline{B}\overline{C}DE$		$\overline{A}\overline{B}\overline{C}\overline{D}E$	‘
		$\overline{A}B\overline{C}DE$	✓	$\overline{A}\overline{B}C\overline{D}E$		$\overline{A}\overline{B}\overline{C}D\overline{E}$	
$\overline{A}BCDE$	✓	$\overline{A}BC\overline{D}E$	✓	$\overline{A}\overline{B}CD\overline{E}$	✓	$\overline{A}\overline{B}C\overline{D}\overline{E}$	
$A\overline{B}CDE$	✓	$\overline{A}BCD\overline{E}$	✓	$\overline{A}B\overline{C}\overline{D}E$		$\overline{A}B\overline{C}\overline{D}\overline{E}$	
$AB\overline{C}DE$	✓	$A\overline{B}\overline{C}DE$	✓	$\overline{A}B\overline{C}D\overline{E}$		$A\overline{B}\overline{C}\overline{D}\overline{E}$	
$ABC\overline{D}E$	✓	$A\overline{B}C\overline{D}E$	✓	$\overline{A}BC\overline{D}\overline{E}$			
$ABCD\overline{E}$	✓	$A\overline{B}CD\overline{E}$	✓	$A\overline{B}\overline{C}\overline{D}E$		$\overline{A}\overline{B}\overline{C}\overline{D}\overline{E}$	
		$AB\overline{C}\overline{D}E$	✓	$A\overline{B}\overline{C}D\overline{E}$	✓		
		$AB\overline{C}D\overline{E}$	✓	$A\overline{B}C\overline{D}\overline{E}$			
		$ABC\overline{D}\overline{E}$		$AB\overline{C}\overline{D}\overline{E}$			

TABLE 3.2 The Binary-to-Literal Correspondences for Table 3.1 (Where *ABC* is Short for *A* ∩ *B* ∩ *C*, etc.)

Decimal Form	Binary Form	Literal Form
0	00000	$\overline{ABCDE}$
1	00001	$\overline{ABCD}E$
2	00010	$\overline{ABC}D\overline{E}$
3	00011	$\overline{ABC}DE$
4	00100	$\overline{AB}C\overline{DE}$
5	00101	$\overline{AB}C\overline{D}E$
6	00110	$\overline{AB}CD\overline{E}$
7	00111	$\overline{AB}CDE$
.	.	.
.	.	.
.	.	.
29	11101	$ABC\overline{D}E$
30	11110	$ABCD\overline{E}$
31	11111	$ABCDE$

A one-to-one correspondence is established between the components $\{\mathcal{A}, \mathcal{B}, \mathcal{C}, \mathcal{D}, \mathcal{E}\}$ of the diagram and the position $i (i = 1, \ldots, 5)$ of the binary number expression, with a binary zero indicating that the component is bad and a binary one indicating that the component is good. For example, the binary number 01011 stands for $\overline{A}B\overline{C}DE$ and 11011 stands for $AB\overline{C}DE$. Because it is necessary to examine each binary number with reference to the graph, it is probably advisable for psychological reasons to construct a computer subroutine that will convert the binary number into the corresponding literal displayed in Table 3.1. Such a correspondence is partially shown in Table 3.2. As an exercise, the student is urged to complete the table.

Method II: Path Tracing Examining Figure 3.5, we see that the system will function if the following components are good.

$$
\left.
\begin{array}{l}
\mathcal{A} \text{ and } \mathcal{D} \\
\mathcal{A} \text{ and } \mathcal{C} \text{ and } \mathcal{E} \\
\mathcal{B} \text{ and } \mathcal{C} \text{ and } \mathcal{D} \\
\mathcal{B} \text{ and } \mathcal{E}
\end{array}
\right\}
\tag{3.15}
$$

Of course, one could also list the combinations

$\mathcal{A}$ and $\mathcal{C}$ and $\mathcal{D}$

$\mathcal{B}$ and $\mathcal{C}$ and $\mathcal{E}$

but since these are subsets of the first and last events in event set 3.15, they need not be listed. This can be seen as follows. Let G stand for the event

{both $\mathcal{A}$ and $\mathcal{D}$ good}

Then it can be seen that

$$P[(\mathcal{A} \text{ and } \mathcal{D} \text{ good}) \text{ or } (\mathcal{A} \text{ and } \mathcal{C} \text{ and } \mathcal{D} \text{ good})]$$
$$= P(G \text{ or } GC) = P(G) + P(GC) - P(GGC)$$
$$= P(G) + P(GC) - P(GC) = P(G) = P(\mathcal{A} \text{ and } \mathcal{D} \text{ good}) \qquad (3.16)$$

Thus, although it is not incorrect to list subset events such as ACD, along with the "main" event AD, it is foolish because of the added complication. The student is advised to be alert to the presence of subset events and to avoid listing them.

Using the nonredundant list in equation set 3.15, we can write the system reliability as

$$R = P(AD \text{ or } ACE \text{ or } BCD \text{ or } BE)$$
$$= P(AD) + P(ACE) + P(BCD) + P(BE)$$
$$\quad - P(ACDE) - P(ABCD) - P(ABDE) - P(ABCDE)$$
$$\quad - P(ABCE) - P(BCDE)$$
$$\quad + P(ABCDE) + P(ABCDE) + P(ABCDE)$$
$$\quad + P(ABCDE) - P(ABCDE) \qquad (3.17)$$

$$= P(AD) + P(ACE) + P(BCD) + P(BE)$$
$$\quad - P(ACDE) - P(ABCD) - P(ABDE) - P(ABCE)$$
$$\quad - P(BCDE) + 2P(ABCDE) \qquad (3.18)$$

$$= ad + be + ace + bcd - acde - abcd$$
$$\quad - abde - abce - bcde + 2abcde \qquad (3.19)$$

which of course agrees with Equation 3.14.

Method III: The Key Element Method In this technique we choose a "key" element to form the basis for writing some conditional probability statements. It is probably most fruitful to choose a complicating element—that is, one that prevents the system from being decomposable into simple series and parallel subsets, but this is not necessary. The technique may be used when no such complicating element is present. In our example the component $\mathcal{C}$ is chosen as the key element. Then, using the decomposition rules of probability, we have

$$R = P(\text{system good} \mid C)P(C) + P(\text{system good} \mid \overline{C})P(\overline{C})$$
$$= P(AD \text{ or } BE \text{ or } AE \text{ or } BD)P(C) + P(AD \text{ or } BE)P(\overline{C}) \qquad (3.20)$$

$$
\begin{aligned}
&= c(ad + be + ae + bd - abde - ade - abd - abe - bde - abde + abde \\
&\quad + abde + abde + abde - abde) + (1 - c)(ad + be - abde) \\
&= c(ad + be + ae + bd - ade - abd + abde - abe - bde) \\
&\quad + (1 - c)(ad + be - abde) \\
&= ad + be + ace + bcd - acde - abcd - abce - bcde - abde + 2abcde
\end{aligned}
$$
$$(3.21)$$

which is the same as (3.14) and (3.16).

In the event that all components are equally reliable, so that $a = b = c = d = e = r$, the system reliability reduces to

$$
R = 2r^2 + 2r^3 - 5r^4 + 2r^5 \tag{3.22}
$$

This is an interesting expression because we see that the highest power in the polynomial is five, the number of components in the system, and that the lowest power is two, the minimum number of components that must be functional for system functioning. The author has seen this result frequently enough that she formerly believed it to be an absolute rule which could serve as a rough algebraic check on the expression for R, until she devised the simple examination question shown in Exercise 3.2.1a.

It is important to convert complex configurations to series and parallel equivalents whenever possible, since these equivalents are so much easier to analyze. Sometimes the delta–star methods discussed in Section 3.4 can be helpful.

It is easy to conceive a system whose non-series–parallel structure cannot be reduced and whose number of components renders unattractive or infeasible the exact methods just presented. In such a system approximation methods may suffice. We will develop the procedure along the lines laid out in Hillier and Lieberman (1986, Chapter 21). We use the configuration of Figure 3.6, where it will be more convenient to indicate the components as $\{C1, \ldots, C5\}$.

We have seen that it is tedious to compute exact reliability for a complex system with only five components; it is fearsome to contemplate the work involved for even moderately large systems. Crude bounds are obtainable by computing the reliabilities of the systems that would be formed by using all the system components in series for the lower bound, and by using all the components in parallel for the upper bound. Such a set of bounds is crude because it fails to consider the actual structure of the system. Better bounds can be obtained by applying the concepts of minimal paths

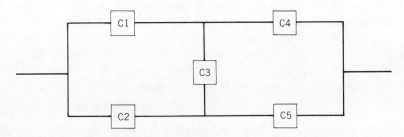

Figure 3.6 The single-bridge network.

and minimal cuts. The results will be presented first as a "cookbook" method, then justified formally later in this section. First, we need a pair of new definitions.

A *minimal path* is a minimal set of components which by functioning ensures the system operation. The enumeration of the various minimal paths for our example was carried out in Equation 3.17. A *minimal cut* is a minimal set of components which by failing guarantee the failure of the system. For the example of Figure 3.6 the minimal paths and cuts are

Minimal Paths	Minimal Cuts
C1, C4	C1, C2
C2, C5	C4, C5
C1, C3, C5	C1, C3, C5
C2, C3, C4	C2, C3, C4

Notice that a set of components can constitute both a minimal path and a minimal cut. Based on the paths and cuts, two auxiliary systems are constructed. Auxiliary network N1, shown in Figure 3.7a, is composed of the parallel configuration of all the minimal path elements in series. Its reliability R_{N1} is an upper bound for the original system; the value is

$$R_{N1} = 1 - (1 - r_1r_4)(1 - r_2r_5)(1 - r_1r_3r_5)(1 - r_2r_3r_4) \qquad (3.23)$$

Auxiliary network N2, shown in Figure 3.7b, is composed of the series

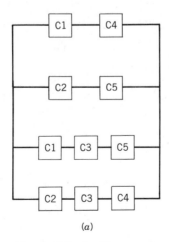

(a)

Figure 3.7 Auxiliary configurations for calculating upper and lower bounds. (a) Network N_1.

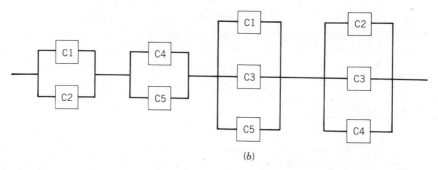

(b)

Figure 3.7 continued (b) Network N_2.

configuration of all the minimal cut elements in parallel. Its reliability R_{N2} is a lower bound for the original system; the value is

$$R_{N2} = [1 - (1 - r_1)(1 - r_2)][1 - (1 - r_4)(1 - r_5)]$$
$$\cdot [1 - (1 - r_1)(1 - r_3)(1 - r_5)][1 - (1 - r_2)(1 - r_3)(1 - r_4)] \quad (3.24)$$

The intuitive justification of the two bounds is that the complex system of Figure 3.6 cannot possibly be any more reliable than a system composed of the parallel union of all its possible good paths, and *mutatis mutandis* for the lower bound.

Example 3.4

Suppose all components in Figure 3.6 are identical (or have equivalent reliability) equal to .9. We have seen that the exact system reliability from (3.22) is $R = .97848$. The crude lower bound $(.9)^5 = .590$ and the crude upper bound is $1 - (1 - .9)^5 = .99999$. The improved upper bound from (3.23) is

$$R_U = 1 - (1 - .9^2)^2(1 - .9^3)^2 = .99735$$

and the improved lower bound from (3.24) is

$$R_L = [1 - (.1)^2]^2[1 - (.1)^3]^2 = .97814$$

We proceed now with the development of the results just stated; it is somewhat lengthy and may be omitted by readers with no taste for mathematical details. We define for each of the components an indicator function or binary variable:

$$X_i = \begin{cases} 1 \text{ if } C_i \text{ is good} \\ 0 \text{ if } C_i \text{ is not good} \end{cases}$$

In general, for a system of n components the performance of the whole system is described by the corresponding system indicator function $\Phi(X_1, \ldots, X_n)$ called the *structure function*, where

$$\Phi(X_1, \ldots, X_n) = \begin{cases} 1 \text{ if the system performs satisfactorily} \\ 0 \text{ if the system fails} \end{cases}$$

For a series system the structure function clearly has the form

$$\Phi(X_1, \ldots, X_n) = X_1 X_2 \ldots X_n = \min\{X_1, \ldots, X_n\} \tag{3.25}$$

For a parallel system it is

$$\Phi(X_1, \ldots, X_n) = 1 - (1 - X_1)(1 - X_2)\cdots(1 - X_n) \tag{3.26}$$
$$= \max\{X_1, \ldots, X_n\}$$

For the r-out-of-n system the form is

$$\Phi(X_1, \ldots, X_n) = 1 \qquad \text{if } \sum_{i=1}^{n} X_i \geq r$$

$$= 0 \qquad \text{if } \sum_{i=1}^{n} X_i < r \tag{3.27}$$

Simple examples can be worked out on an individual basis. Consider a three-component system which has C_1 and C_2 in parallel, in series with C_3. The structure function is

$$\Phi(X_1, X_2, X_3) = X_3[1 - (1 - X_1)(1 - X_2)]$$
$$= X_3(X_1 + X_2 - X_1 X_2)$$
$$= \min\{X_3, \max\{X_1, X_2\}\}$$
$$= X_3 \max\{X_1, X_2\} \tag{3.28}$$

The relationship between the indicator functions and the reliability is given by

$$r_i = P(C_i \text{ good}) = P(X_i = 1) = E(X_i) \tag{3.29}$$

and

$$R = P(\text{system good}) = P(\Phi(X_1, \ldots, X_n) = 1)$$
$$= 1 \cdot P(\Phi = 1) + 0 \cdot P(\Phi = 0)$$
$$= E[\Phi(X_1, \ldots, X_n)] \tag{3.30}$$

For a series system of independently functioning components,

$$R = P(X_1, \ldots, X_n = 1) = P(X_1 = 1, X_2 = 1, \ldots, X_n = 1)$$
$$= P(X_1 = 1)P(X_2 = 1)\ldots P(X_n = 1)$$
$$= r_1 r_2 \cdots r_n$$

a familiar result. Similarly, for independent parallel components,

$$R = P(\max\{X_1,\ldots,X_n\} = 1)$$
$$= P(\text{at least one } X_i \text{ equals } 1)$$
$$= 1 - P(\text{all } X_i \text{ are zero})$$
$$= 1 - P(X_1 = 0, X_2 = 0,\ldots,X_n = 0)$$
$$= 1 - P(X_1 = 0)P(X_2 = 0)\ldots P(X = 0)$$
$$= 1 - (1 - r_1)(1 - r_2)\ldots(1 - r_n)$$

as we have seen before. The events of interest, and their associated indicator functions, are related to the list of minimal paths and minimal cuts previously described. They are shown in Table 3.3. The minimal path calculation requires the evaluation of

$$R = P(E_1 \text{ or } E_2 \text{ or } E_3 \text{ or } E_4)$$
$$= P[X_1X_4 = 1 \cup X_2X_5 = 1 \cup X_1X_3X_5 = 1 \cup X_2X_3X_4 = 1]$$

which is a reformulation of the previous path-tracing example. A more convenient method than the one used there is to recognize that the paths play a "parallel" role with each other; it takes only one good path to ensure system functioning. Thus we can write the structure function Φ as

$$\Phi = \max\{X_1X_4, X_2X_5, X_1X_3X_5, X_2X_3X_4\}$$

or alternatively as

$$\Phi = 1 - (1 - X_1X_4)(1 - X_2X_5)(1 - X_1X_3X_5)(1 - X_2X_3X_4) \qquad (3.31)$$

Using the idempotent property of the X_i (the fact that $X_i^2 = X_i$) and the independence of X_i and X_j, we can reduce Φ by successive algebraic steps to

$$\Phi = X_1X_4 + X_2X_5 + X_1X_3X_5 + X_2X_3X_4 - X_1X_3X_4X_5 - X_1X_2X_3X_4$$
$$-X_1X_2X_3X_5 - X_2X_3X_4X_5 - X_1X_2X_4X_5 + 2X_1X_2X_3X_4X_5 \qquad (3.32)$$

Taking expected values and recognizing that

$$R = E(\Phi)$$

and

$$E(X_iX_j) = r_ir_j$$

TABLE 3.3   Events and Indicator Functions for Figure 3.6

System Operational	System Failed
E_1: $X_1X_4 = 1$	E_1^*: $X_1 = X_2 = 0$
E_2: $X_2X_5 = 1$	E_2^*: $X_4 = X_5 = 0$
E_3: $X_1X_3X_5 = 1$	E_3^*: $X_1 = X_3 = X_5 = 0$
E_4: $X_2X_3X_4 = 1$	E_4^*: $X_2 = X_3 = X_4 = 0$

we achieve the result obtained in Equation 3.14 by the three methods used earlier.

Using the minimal cuts, we see that system failure occurs if any one of the events $E_i^*(i = 1, 2, 3, 4)$ occurs. That is,

$$1 - R = P(E_1^* \cup E_2^* \cup E_3^* \cup E_4^*)$$

This is the same type of "parallel" enumeration just analyzed, so we will not pursue it further.

A second way to make use of the cut sets is to recognize that cuts act as a series system, since failure of all the components in any cut guarantees system failure. Thus from Equation 3.25

$$\Phi = \min \{\Phi_1^*, \Phi_2^*, \Phi_3^*, \Phi_4^*\} = \Phi_1^* \Phi_2^* \Phi_3^* \Phi_4^* \tag{3.33}$$

where $\Phi_i^* = \Phi(E_i^*)$, with $i = 1, 2, 3, 4$. The form of Φ_i^* must be such that it vanishes if and only if all the corresponding cut components fail. Thus

$$\Phi_1^* = 1 - (1 - X_1)(1 - X_2) \tag{3.34a}$$

$$\Phi_2^* = 1 - (1 - X_4)(1 - X_5) \tag{3.34b}$$

$$\Phi_3^* = 1 - (1 - X_1)(1 - X_3)(1 - X_5) \tag{3.34c}$$

$$\Phi_4^* = 1 - (1 - X_2)(1 - X_3)(1 - X_4) \tag{3.34d}$$

Readers who care to may verify that the expansion of Equation 3.33 using (3.34a–d) does indeed lead to the result given by (3.32).

A system is said to be *coherent* if improving the individual components improves, or at least does not degrade, the system. (Can you think of any system for which this would not be true?) A system is coherent if its structure function is monotone increasing in the component indicator functions. Thus, if $Y_i \geq X_i$, coherency requires that

$$\Phi(Y_1, \ldots, Y_n) \geq \Phi(X_1, \ldots, X_n)$$

If the inequality is strict for all components, we say that all components are *relevant*; certainly a component C1, say, for which

$$\Phi(1, \ldots) = \Phi(0, \ldots)$$

can be considered irrelevant to the functioning of the system.

The binary variables Y_1 and Y_2 are said to be *associated* if their covariance is nonnegative, that is, if $\text{cov}(Y_1, Y_2) \geq 0$. Two such variables might be associated, or positively correlated, because they represent one of the following situations (Barlow and Proschan, 1975, page 29).

1. Minimal path structures of a coherent system with components in common.

2. Components subjected to the same set of stresses.

3. Structures in which components share the load, so that failure of one component increases the load on each of the remaining components.

If Y_1 and Y_2 are associated, so are $Z_1 = 1 - Y_1$ and $Z_2 = 1 - Y_2$ because of the well-known statistical property that

$$
\begin{aligned}
\operatorname{cov}(a_1 + b_1 Y_1, a_2 + b_2 Y_2) \\
= E[(a_1 + b_1 Y_1)(a_2 + b_2 Y_2)] - (a_1 + b_1 \mu_1)(a_2 + b_2 \mu_2) \\
= b_1 b_2 E(Y_1 Y_2) - b_1 b_2 \mu_1 \mu_2 \\
= b_1 b_2 \operatorname{cov}(Y_1, Y_2)
\end{aligned} \tag{3.35}
$$

Hence, in particular, if $a_1 = a_2 = 1$, $b_1 = b_2 = -1$, then

$$
\begin{aligned}
\operatorname{cov}(Z_1, Z_2) &= \operatorname{cov}(1 - Y_1, 1 - Y_2) \\
&= \operatorname{cov}(Y_1, Y_2)
\end{aligned} \tag{3.36}
$$

We need the following important result for associated variables:

$$
P(Y_1 = 0, Y_2 = 0, \ldots, Y_r = 0) \ge P(Y_1 = 0)P(Y_2 = 0) \cdots P(Y_r = 0) \tag{3.37}
$$

This is easily obtained for $r = 2$ by starting with the complementary variables Z_1 and Z_2 which are also associated, so that

$$
\begin{aligned}
&\operatorname{cov}(Z_1, Z_2) \ge 0 \\
&E[(Z_1 - E(Z_1))][Z_2 - E(Z_2)] \ge 0 \\
&E(Z_1 Z_2) \ge E(Z_1)E(Z_2) \\
&0 \cdot P(Z_1 Z_2 = 0) + 1 \cdot P(Z_1 Z_2 = 1) \\
&\quad \ge [0 \cdot P(Z_1 = 0) + 1 \cdot P(Z_1 = 1)][0 \cdot P(Z_2 = 0) + 1 \cdot P(Z_2 = 1)] \\
&P(Z_1 Z_2 = 1) \ge P(Z_1 = 1)P(Z_2 = 1)
\end{aligned}
$$

But $Z_1 Z_2 = 1$ if and only if $Z_1 = 1$ and $Z_2 = 1$. Rewriting in terms of the Y_i gives our needed result,

$$
P(Y_1 = 0, Y_2 = 0) \ge P(Y_1 = 0)P(Y_2 = 0) \tag{3.38}
$$

which can be generalized to Equation 3.37.

Now consider the situation in which the Y_i are the minimal cut or path structure functions,

$$
Y_i = \prod_{j \in J_i} X_j
$$

with J_i the set of subscripts, or components, for the ith minimal cut or path. Clearly the Y_i are associated so that Equation 3.37 holds. The system fails if all the paths fail, so that in our example we see that

$$
R = 1 - P(\text{all paths fail})
$$

If we let $Y_1 = X_1 X_4$, $Y_2 = X_2 X_5$, $Y_3 = X_1 X_3 X_5$, $Y_4 = X_2 X_3 X_4$, then

$$
\begin{aligned}
P(\text{all paths fail}) &= P(X_1 X_4 = 0, \ X_2 X_5 = 0, \ X_1 X_3 X_5 = 0, \ X_2 X_3 X_4 = 0) \\
&\ge P(X_1 X_4 = 0)P(X_2 X_5 = 0)P(X_1 X_3 X_5 = 0)P(X_2 X_3 X_4 = 0) \\
&\ge (1 - r_1 r_4)(1 - r_2 r_5)(1 - r_1 r_3 r_5)(1 - r_2 r_3 r_4)
\end{aligned} \tag{3.39}
$$

Hence the system reliability has the upper bound R_U given by

$$R_U = 1 - (1 - r_1 r_4)(1 - r_2 r_5)(1 - r_1 r_3 r_5)(1 - r_2 r_3 r_4) \qquad (3.40)$$

On the other hand, the system functions if at least one of the components in each cut operates, so that

$R = \ P(\text{at least one of } C_1, C_2 \text{ operates, \textbf{and}}$

$\qquad$ at least one of C_4, C_5 operates, **and**

$\qquad$ at least one of C_1, C_3, C_5 operates, **and**

$\qquad$ at least one of C_2, C_3, C_4 operates)

$\qquad = \ P[(1 - X_1)(1 - X_2) = 0, (1 - X_4)(1 - X_5) = 0,$

$\qquad\quad (1 - X_1)(1 - X_3)(1 - X_5) = 0, (1 - X_2)(1 - X_3)(1 - X_4) = 0]$

$\qquad \geq \ P[(1 - X_1)(1 - X_2) = 0] \cdot P[(1 - X_4)(1 - X_5) = 0]$

$\qquad\quad \cdot P[(1 - X_1)(1 - X_3)(1 - X_5) = 0] \cdot P[(1 - X_2)(1 - X_3)(1 - X_4) = 0]$

Now $P[(1 - X_1)(1 - X_2) = 0]$ is the same form as the probability that two components in parallel do not both fail, so it can be written as

$$r_1 + r_2 - r_1 r_2 = 1 - (1 - r_1)(1 - r_2)$$

and similarly for the other factors. Thus a lower bound R_L is

$$R_L = [1 - (1 - r_1)(1 - r_2)][1 - (1 - r_4)(1 - r_5)]$$
$$\cdot [1 - (1 - r_1)(1 - r_3)(1 - r_5)][1 - (1 - r_2)(1 - r_3)(1 - r_4)] \qquad (3.41)$$

EXERCISES 3.2

1. Use the three methods to find the reliability expressions for the following configurations. Is there one method that always seems superior (less work)? Derive the general formula and then use it to compute the numerical value of the system reliability when $a = .9$, $b = .8$, $c = .75$, $d = .6$, $e = .5$.

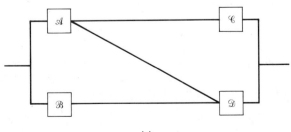

(a)

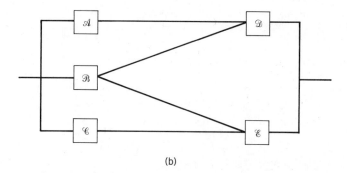

(b)

2. Use any or all of the three methods of this section to show that the single-bridge network in Figure 3.6 is isomorphic to the network of Figure 3.5. They are isomorphic despite the fact that there are differences in the paths over which energy may travel in the two networks. Can you find the differences?

3. Use the three techniques of this section on the following figure to find system reliability. Assume "energy flow" is from left to right.

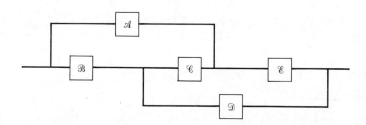

4. Use the method of minimal paths and cuts to determine upper and lower bounds for the networks of Exercises 1 and 3.

3.3 THE TIME-DEPENDENT CASE

The previous sections dealt with systems in which component reliability (and hence system reliability) was considered constant. We now consider systems in which $R_i = R_i(t)$, and—in particular—systems with CFR components in which

$$R_i(t) = e^{-\lambda_i t} \tag{3.42}$$

The series system then has reliability

$$R(t) = \prod_{i=1}^{n} e^{-\lambda_i t}$$

$$= e^{-\lambda_T t} \qquad (\lambda_T = \lambda_1 + \cdots + \lambda_n) \qquad (3.43)$$

so its behavior is like that of a single component whose CFR is the sum of all the component failure rates, and whose mean time to failure is

$$\text{MTTF} = \frac{1}{\lambda_T} = \frac{1}{\lambda_1 + \cdots + \lambda_n} \qquad (3.44)$$

The pure parallel system has reliability

$$R(t) = 1 - \prod_{i=1}^{n} (1 - e^{-\lambda_i t}) \qquad (3.45)$$

which, for identical components with failure rate λ, is

$$R(t) = 1 - (1 - e^{-\lambda t})^n \qquad (3.46)$$

Figure 3.8 shows the behavior of (3.46) for selected values of n, where time is measured in standard units, meaning that λ (and hence also mean life) has been set equal to unity so that t is expressed in multiples of mean life.

Equations 3.42 through 3.46 involve three elements—failure rate(s), mission length, and reliability. If any two of these are given, the third is easily computable. Thus, if a CFR component must have reliability at least .99 for a mission of 500 hours we can solve

$$e^{-500\lambda} \geq .99$$

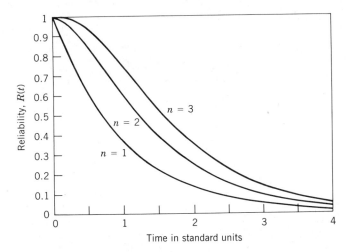

Figure 3.8 System reliability for n identical components in parallel ($n = 1, 2, 3$).

to find that λ may not exceed .0000201 failures per hour, or the mean life must be at least 49,750 hours. Similarly, to determine the mission length which will give reliability at least .9 for a CFR component with MTTF = 15000 hours (so that λ = 1/15000 f/hr), we solve

$$e^{-t/15000} \geq .9$$

to obtain $t \leq 1580$ hours.

It is of interest to find out whether the parallel system, like the series system, is CFR. Recalling that $f(t) = -R'(t)$ and that $h(t) = f(t)/R(t)$, we see that for the parallel system the failure rate is

$$h(t) = \frac{-R'(t)}{R(t)} \tag{3.47}$$

$$= n\lambda B_n(t) \tag{3.48}$$

where

$$B_n(t) = \frac{e^{-\lambda t}(1 - e^{-\lambda t})^{n-1}}{1 - (1 - e^{-\lambda t})^n} \tag{3.49}$$

It is proved in Grosh (1982) that $B_n(t)$ is less than unity and monotone-increasing, so the parallel system has a hazard function which is less than that of the corresponding series system (naturally!) and is IFR. $B_n(0) = 0$ and the factor $B_n(t)$ approaches $1/n$ as $t \to \infty$. Thus the parallel system after a long time has a failure rate that converges to that of a single component. Students who wish to develop a feel for the rate of convergence are encouraged to use an electronic spreadsheet with the DATA-FILL and COPY commands to compute $B_n(t)$ for increasing t.

To obtain the mean time to failure of the parallel system, we use the relationship derived in Chapter 2,

$$\text{MTTF} = \int_0^\infty R(t)\, dt$$

which for equal CFR components becomes

$$\text{MTTF} = \int_0^\infty [1 - (1 - e^{-\lambda t})^n]\, dt \tag{3.50}$$

In order to perform the integration, a trick is necessary. Recall that a finite geometric series can be summed as

$$1 + q + q^2 + \cdots + q^{n-1} = \frac{1 - q^n}{1 - q} \tag{3.51}$$

so that

$$1 - q^n = (1 - q) \sum_{k=0}^{n-1} q^k \qquad (3.52)$$

Now let

$$q = 1 - e^{-\lambda t} \qquad (3.53)$$

Then

$$\text{MTTF} = \int_0^\infty e^{-\lambda t} \sum_{k=0}^{n-1} (1 - e^{-\lambda t})^k \, dt \qquad (3.54)$$

and interchanging summation and integration gives

$$\text{MTTF} = \sum_{k=0}^{n-1} \int_0^\infty (1 - e^{-\lambda t})^k e^{-\lambda t} \, dt \qquad (3.55)$$

Upon substitution this yields

$$\text{MTTF} = \frac{1}{\lambda} \sum_{k=0}^{n-1} \int_0^1 q^k \, dq \qquad (3.56)$$

$$= \frac{1}{\lambda} \sum_{k=0}^{n-1} \frac{1}{k+1} \qquad (3.57)$$

$$= \frac{1}{\lambda}(1 + 1/2 + 1/3 + \cdots + 1/n) \qquad (3.58)$$

$$= \frac{1}{\lambda} M(n) \text{ (say)} \qquad (3.59)$$

where $M(n)$ is the sum in parentheses.

Inspection of this equation shows that each component added to the system increases expected life, but at a slower and slower rate. Because $M(n)$ is the nth partial sum of the divergent harmonic series, it is apparent that adding sufficient extra units will extend the mean system life to any desired figure.

We see from Table 3.4 that for a system to have double the mean life of a single component, it must consist of four components. To triple the mean life, the system must have eleven components. Theoretically, there is no limit to how much the system mean life can be extended, but the cost of extending life through mere redundancy is usually prohibitive. Redesign should be preferred to excessive redundancy.

Keep in mind that in this section we have been dealing with CFR components. We emphasize that a series configuration of CFR components is CFR, but a parallel configuration is not.

TABLE 3.4 Values of $\frac{1}{n}$ and $M_n = \sum\limits_{k=0}^{n-1} \frac{1}{k+1}$ for $n = 1, \ldots, 32$

n	$\frac{1}{n}$	M_n	n	$\frac{1}{n}$	M_n
1	1	1	17	.0588	3.4295
2	.5	1.5	18	.0556	3.4951
3	.3333	1.8333	19	.0526	3.5477
4	.25	2.0833	20	.05	3.5877
5	.2	2.2833	21	.0476	3.6453
6	.1667	2.4500	22	.0435	3.7343
7	.1429	2.5928	23	.0435	3.7343
8	.1250	2.7178	24	.0417	3.7759
9	.1111	2.8289	25	.04	3.8159
10	.1	2.9289	26	.0385	3.8544
11	.0909	3.0198	27	.0370	3.8914
12	.0833	3.1032	28	.0357	3.9271
13	.0769	3.1801	29	.0345	3.9616
14	.0714	3.2515	30	.0333	3.9950
15	.0667	3.3182	31	.0323	4.0272
16	.0625	3.3807	32	.0313	4.0585

EXERCISES 3.3

1. On the same sheet of graph paper plot the curves of $R(t)$ for parallel systems with $n = 1, 2, \ldots, 6$ for $\lambda = 1$. Do a few calculations with a pocket calculator to get a feel for it; then do the rest with an electronic spreadsheet.

2. Write the general formula for $h(t)$ for unspecified n. Then, for $n = 2$ decide whether the system is IFR, DFR, or neither. You may do this analytically or graphically. Can you assume that your conclusion for $n = 2$ is true for larger n? (This problem was written before the results following Equation 3.49 had been developed.)

3. A component has a constant failure rate of 0.001 per hour. Find the following:

(a) The reliability of one such component for an operating time of 300 hours.

(b) The reliability of two components in series for an operating time of 300 hours.

(c) The reliability of two components in parallel for an operating time of 300 hours.

4. Find the MTTF of a system if four components in series have constant failure rates of 1.2, 1.5, 1.4, and 1.7 per 1000 hours respectively.

5. What must the mean life of a CFR component be if it is to have a reliability of .97 for a 400-hour mission?

6. If the mean time to failure of a CFR component is 50,000 hours, what is its reliability for a 100-hour mission?

7. Assuming a component with a constant hazard function, what must the mean time to failure be if we need a reliability of at least .98 for a mission of 200 hours?

8. A system has five units with failure rates of 6, 8, 4.6, 7, and 9 per 10^6 hours, respectively. Find the MTTF and the reliability for 100 hours for the five units in series.

9. Trucks are equipped with tires having a failure rate of 2×10^{-6} per mile. Two types of trucks are used: one type with four tires and the other type with six tires (i.e., two on each end of the rear axle). The same tires are used on both types of the truck. Sketch a reliability block diagram for each type of truck, and calculate the probability that each type will have tire trouble in 10,000 miles of driving and be unable to complete deliveries.

3.4 STAR AND DELTA CONFIGURATIONS

A clever technique that is sometimes useful for simplifying reliability calculations for complex systems is the conversion of one set of three components in the *delta* configuration, to another set of three components in the *star* configuration. The technique depends on the *duality* that exists between the two patterns, as depicted in Figure 3.9. This method was discussed in preliminary form by Ramamoorty (1970), Banerjee and Rajamani (1970, 1972), Singh and Kankam (1976), Singh and Proctor (1976), and Rosenthal (1978). The best (and correct) presentation is that of Gupta and Sharma (1978).

We begin tentatively with the solution outlined by Banerjee and Rajamani (1972), because of its simplicity of form and historical interest. It was later proved to be only an approximation, so the correct solution will be presented later in this section.

The delta and star configurations are equivalent, as far as reliability is concerned, if, given delta components AB, AC, and BC, the corresponding star compo-

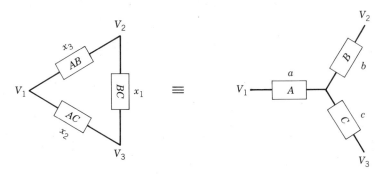

Figure 3.9 The delta-star equivalence.

nents are chosen appropriately, based on the following reasoning. To get from vertex V_1 to vertex V_3 of the delta, one may traverse element AC only, *or* go through AB and then BC. In the star, to get from vertex V_1 to vertex V_3, one must traverse element A and then element C. Thus we can state that

Star: A and C in series

is equivalent to

Delta: (AB in series with BC) parallel with AC.

Hence we can write

$$R_A R_C = R_{AC} + R_{AB} R_{BC} - R_{AB} R_{BC} R_{AC} \qquad (3.60a)$$

and similarly for the other pairs of vertices (V_1 with V_2 and V_2 with V_3)

$$R_B R_C = R_{BC} + R_{AB} R_{AC} - R_{AB} R_{BC} R_{AC} \qquad (3.60b)$$

$$R_A R_B = R_{AB} + R_{AC} R_{BC} - R_{AB} R_{BC} R_{AC} \qquad (3.60c)$$

It is convenient to simplify notation by letting

$$x_1 = R_{BC}, \qquad x_2 = R_{AC}, \qquad x_3 = R_{AB},$$

$$a = R_A, \qquad b = R_B, \qquad c = R_C$$

and this notation is indicated on the graphs in Figure 3.9. Then equation set 3.60a–c becomes, after some reordering,

$$bc = x_1 + x_2 x_3 - x_1 x_2 x_3 = E_1 \text{ (say)} \qquad (3.61a)$$

$$ac = x_2 + x_1 x_3 - x_1 x_2 x_3 = E_2 \text{ (say)} \qquad (3.61b)$$

$$ab = x_3 + x_1 x_2 - x_1 x_2 x_3 = E_3 \text{ (say)} \qquad (3.61c)$$

This is a set of three quadratic equations in a, b, and c. The solution technique involves the trick of multiplying all three equations together to get

$$a^2 b^2 c^2 = E_1 E_2 E_3$$

so that

$$abc = \sqrt{E_1 E_2 E_3} = E_4 \text{ (say)} \qquad (3.62)$$

and then dividing Equation 3.62 in turn by each equation of set (3.61a–c). The result is

$$a = E_4/E_1 \qquad (3.63a)$$

$$b = E_4/E_2 \qquad (3.63b)$$

$$c = E_4/E_3 \qquad (3.63c)$$

Example 3.5

Let the delta component reliabilities be $x_1 = .7$, $x_2 = .8$, $x_3 = .9$. Then

$$E_1 = .7 + .72 - .504 = .916$$

$$E_2 = .8 + .63 - .504 = .926$$

$$E_3 = .9 + .56 - .504 = .956$$

$$E_4 = \sqrt{(.916)(.926)(.956)} = .900497$$

and hence the star component reliabilities are

$$a = .983\ 075\ 115$$

$$b = .972\ 458\ 753$$

$$c = .941\ 942\ 265$$

(It is obviously ridiculous to carry so many significant figures in the answer, but these will be needed later for solving the inverse problem.)

In the event that a star is given and the corresponding delta is wanted, equation set 3.61a–c must be solved for the x's—a much more difficult procedure because it involves cubic equations. The form of solution was indicated by Banerjee and Rajamani (1972), but it is more convenient to use the Gauss–Seidel iteration technique[1] since a solution of any desired degree of accuracy can be obtained using a pocket calculator or microcomputer BASIC program in much less time than would be required to figure out the coefficients of the cubic equation. The procedure is to solve each member of equation set 3.61a–c for one of the unknowns. Specifically, we have

$$x_1 = \frac{bc - x_2 x_3}{1 - x_2 x_3} \tag{3.64a}$$

$$x_2 = \frac{ac - x_1 x_3}{1 - x_1 x_3} \tag{3.64b}$$

$$x_3 = \frac{ab - x_1 x_2}{1 - x_1 x_2} \tag{3.64c}$$

Trial values $x_2^{(0)}$ and $x_3^{(0)}$ are inserted in to Equation 3.64a, and the corresponding solution $x_1^{(1)}$, along with the trial value $x_3^{(0)}$, is inserted into Equation 3.64b to arrive at an improved value $x_2^{(1)}$, and so on. The cyclic iteration process is repeated as long as desired. Ten iterations are usually adequate for five-digit accuracy, no matter what starting values are tried; these can be carried out in less than a minute on a previously programmed pocket calculator. To demonstrate the solution technique,

---

[1] See Hamming, 1962, page 7.

TABLE 3.5   Solutions to the Star–Delta Problem of Example 3.5

					Iteration Number						
	0	1	2	3	4	5	6	7	8	9	10
x_1	*	.8687	.7149	.6886	.6918	.6957	.6979	.6990	.6995	.6998	.6999
x_2	.6	.8454	.8166	.8074	.8036	.8017	.8008	.8004	.8002	.8001	.8000
x_3	.6	.8343	.8943	.9009	.9009	.9005	.9003	.9001	.9001	.9000	.9000

					Iteration Number						
	0	1	2	3	4	5	6	7	8	9	10
x_1	*	.9152	.7493	.6780	.6823	.6905	.6953	.6977	.6989	.6995	.6997
x_2	.1	.9185	.8383	.8160	.8078	.8038	.8019	.8009	.8004	.8002	.8001
x_3	.1	.7240	.8817	.9015	.9020	.9011	.9006	.9003	.9001	.9001	.9000

	Iteration Number	
	0	1
x_1	*	.700000000
x_2	.8	.800000000
x_3	.9	.900000000

we use the data of Example 3.5, starting with the a, b, c as given values, and with the correct solution already known to be (.7, .8, .9). Table 3.5 shows the results of three iteration series, one with reasonable starting values (*, .6, .6), one with poor starting values (*, .1, .1), and one with the true solution's starting values — since one criterion of a good solution technique is its ability to "recognize the truth when it sees it." The method is very satisfactory on that score.

The same problem was solved using the Newton–Raphson technique, with poor results. The coefficient matrix is

$$\begin{bmatrix} 1 - x_2 x_3 & x_3(1 - x_1) & x_2(1 - x_1) \\ x_3(1 - x_2) & 1 - x_1 x_3 & x_1(1 - x_2) \\ x_2(1 - x_3) & x_1(1 - x_3) & 1 - x_1 x_2 \end{bmatrix} \tag{3.65}$$

Starting values of (.9, .9, .9) led to the following sequences of updated solutions (!):

$$\begin{bmatrix} x_1^{(0)} \\ x_2^{(0)} \\ x_3^{(0)} \end{bmatrix} = \begin{bmatrix} .9 \\ .9 \\ .9 \end{bmatrix} \rightarrow \begin{bmatrix} 1.1930 \\ .4035 \\ .9247 \end{bmatrix} \rightarrow \begin{bmatrix} .9625 \\ .0027 \\ 1.0099 \end{bmatrix} \rightarrow \begin{bmatrix} 2.4504 \\ -6.3764 \\ .9079 \end{bmatrix}$$

At this point the solution procedure was abandoned as hopeless. Only when the starting values are nearly correct does the Newton–Raphson method yield good results, probably because the matrix is ill conditioned; in the previous sequence the determinant values were .0037, −.0011, −.0305, and −.0066 respectively.

The greatest power of the delta–star duality is its usefulness in transforming a complex system such as the single bridge of Figure 3.10a. (Students will recognize this as the simple network of Exercise 3.2.2.) In Figure 3.10b the network has been relabeled in recognition of the delta formation of elements 1, 2, and 3, and in Figure 3.10c the delta is replaced by the corresponding star. As pointed out by Rosenthal (1978), the transformation of Equations 3.61 is not entirely correct, since it does not yield correct results when the star and delta are imbedded in a system with other elements. Example 3.6 illustrates the difficulty.

Example 3.6

In Figure 3.10 let the five component reliabilities be

$$r_1 = .90 \qquad r_4 = .85$$
$$r_2 = .95 \qquad r_5 = .93$$
$$r_3 = .99$$

Be careful with the notation here! Components C1, C2, C3 in Figure 3.10 correspond to AB, AC, BC respectively in Figure 3.9, so that r_3 must be used as x_1 and r_1 as x_3 in (3.61). The star elements have reliabilities

$$a = .995022$$
$$b = .999024$$
$$c = .999526$$

and the system reliability is then calculated as .984450. But if we use the exact formula given by Equation 3.14 or 3.19, we get a system reliability of .984433. The error is trivial in this particular problem.

Rosenthal derived the exact formula for the error in the event that all components have the same reliability. The exact reliability of the single-bridge network, from Equation 3.14, is

$$R_s = 2r^2 + 2r^3 - 5r^4 + 2r^5 \qquad (3.66)$$

Expressing the equation sets 3.61, 3.62, and 3.63 in terms of $x_1 = x_2 = x_3 = r$, say, we have

$$E_1 = E_2 = E_3 = r + r^2 - r^3$$
$$E_4 = (r + r^2 - r^3)^{3/2}$$
$$a = b = c = (r + r^2 - r^3)^{1/2} = r_s \text{ (say)} \qquad (3.67)$$

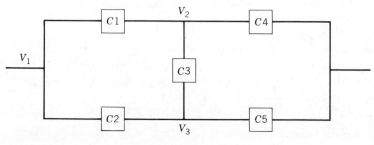

(a) the original bridge network.

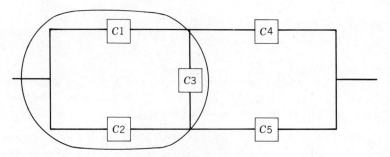

(b) recognizing and labeling the delta.

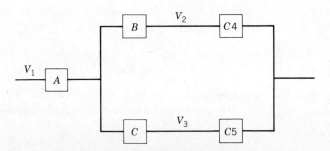

(c) replacing the delta by its dual star.

Figure 3.10 Reduction of a single-bridge network. (a) The original bridge network. (b) Recognizing and labeling the delta. (c) Replacing the delta by its dual star.

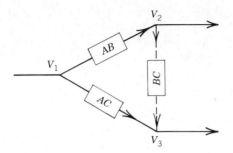

(a) energy flows through BC from V_2 to V_3 (event H_1)

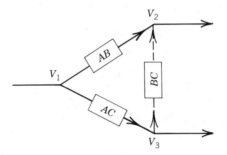

(b) energy flows through BC from V_3 to V_2 (event H_2)

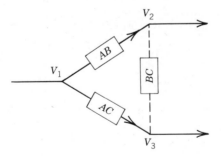

(c) no energy flows through BC (event H_3)

Figure 3.11 Depiction of the three cases of "energy flow." (a) Event H_1: flow from V_2 to V_3 through BC. (b) Event H_2: flow from V_3 to V_2 through BC. (c) Event H_3: no flow through BC.

The system reliability, upon substitution, would then be

$$r_s[2r\,r_s - r^2 r_s^2] = (r + r^2 - r^3)^{1/2}[2r(r + r^2 - r^3)^{1/2} - r^2(r + r^2 - r^3)] \quad (3.68)$$

It is quite apparent that Equations 3.66 and 3.68 are not equal. What is the reason for the discrepancy? It appears that the symmetric formulation of the equation set 3.60 is at fault, since it does not take into consideration the direction in which "energy is flowing" through the system; that is, it ignores which nodes of the delta and star are "input" nodes and which are "output" nodes. Gupta and Sharma (1978) have resolved the problem. First of all, directionality must be recognized. Here we replace the Gupta–Sharma formulation, which some students find difficult to understand, by an equivalent description. We assume that node V_1 is an input node and that V_2 and V_3 are output nodes. There are three mutually exclusive events to consider.

H_1: "Energy" flow through BC is from V_2 to V_3.

H_2: "Energy" flow through BC is from V_3 to V_2.

H_3: No "energy" flows through BC at all.

These three events can be depicted graphically as in Figure 3.11. In the three cases, the corresponding behavior of the system can be represented as in Figure 3.12. The configuration of Figure 3.12a gives rise to the equation

$$ac = x_2 + x_1 x_3 - x_1 x_2 x_3$$

and of Figure 3.12b to

$$ab = x_3 + x_1 x_2 - x_1 x_2 x_3$$

We have met these equations before as Equations 3.10b and 3.10c. Now Figure 3.12c can be interpreted as follows. The truncated delta on the left will function as long as either AB or AC is good. Thus we can state

$$R(\text{delta}) = x_2 + x_3 - x_2 x_3 \quad (3.69)$$

This expression looks like that for a parallel configuration, and indeed in the Gupta and Sharma formulation it is diagrammed as such. But this expression is the logical structure rather than a reflection of any physical structure. Similarly, the star in the right-hand side of Figure 3.12c will function if A functions *and* B or C functions. This functioning can be written as

$$R(\text{star}) = a(b + c - bc) \quad (3.70)$$

This expression looks like that for an element A in series with a parallel pair B and C, and it is so depicted by Gupta and Sharma.

Equating (3.69) with (3.70) we have

$$a(b + c - bc) = x_2 + x_3 - x_2 x_3 \quad (3.71)$$

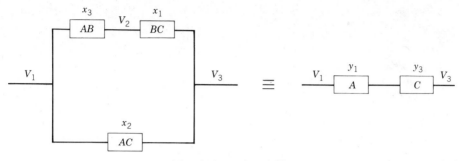

(a) equivalence of event H_1.

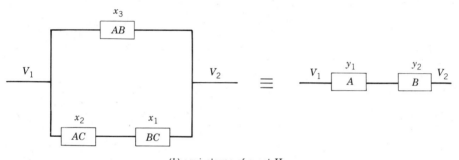

(b) equivalence of event H_2.

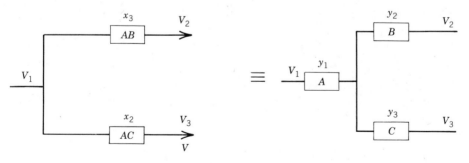

(c) equivalence of event H_3.

Figure 3.12 Equivalence event graphs. (a) Equivalence of event H_1. (b) Equivalence of event H_2. (c) Equivalence of event H_3.

which, with Equations 3.61b and 3.61c, makes up the needed transformation set. We reproduce it here:

$$a(b + c - bc) = x_2 + x_3 - x_2x_3 \tag{3.72a}$$

$$ac = x_2 + x_1x_3 - x_1x_2x_3 = E_2 \tag{3.72b}$$

$$ab = x_3 + x_1x_2 - x_1x_2x_3 = E_3 \tag{3.72c}$$

If a more symmetrical form is desired for the first equation, it can be obtained by subtracting Equation 3.72a from the sum of Equations 3.72b and 3.72c. This gives

$$abc = x_1x_2 + x_1x_3 + x_2x_3 - 2x_1x_2x_3 = F_1 \text{ (say)} \tag{3.73}$$

The star component reliabilities are thus easily obtained as

$$a = E_2E_3/F_1 \tag{3.74a}$$

$$b = F_1/E_2 \tag{3.74b}$$

$$c = F_1/E_3 \tag{3.74c}$$

In the event that $x_1 = x_2 = x_3 = r$, then

$$E_2 = E_3 = r + r^2 - r^3$$
$$F_1 = 3r^2 - 2r^3$$

so that

$$a = \frac{(r + r^2 - r^3)^2}{3r^2 + 2r^3} = \frac{(1 + r - r^2)^2}{3 - 2r} \tag{3.75}$$

whereas

$$b = c = \frac{3r^2 - 2r^3}{r + r^2 - r^3}$$
$$= \frac{r(3 - 2r)}{1 + r - r^2} = r_s \text{ (say)} \tag{3.76}$$

The bridge then transforms into the series–parallel system of Figure 3.10c, with a system reliability given by

$$R = a(2r\,r_s - r^2r_s^2)$$

which reduces readily to the correct value given by Equation 3.66.

Example 3.7

For the delta of Example 3.5, the exact Equations 3.72b–c and 3.73 lead to

$$F_1 = .902$$
$$E_2 = .926$$
$$E_3 = .956$$

so that the exact star reliabilities are

$$a = .981\ 436\ 807$$
$$b = .974\ 082\ 073$$
$$c = .943\ 514\ 644$$

If a star is given and the delta is wanted, the Gauss–Seidel technique cannot be recommended as enthusiastically as before. The former relationship 3.64a must now be replaced by

$$x_1 = \frac{abc - x_2 x_3}{x_2 + x_3 - 2x_2 x_3} \qquad (3.77)$$

In the sample problem used earlier, the method does not converge at all for some "reasonable" starting values like (*, .5, .5), although it does—*very* slowly—for the poor starting values (*, .1, .1). Good starting values (*, .75, .75) require twice as many iterations for convergence as were needed in the approximate case. Students are encouraged to experiment, just for practice.

So far we have been concentrating on the mechanical and algebraic details of the equivalence between the delta and the star. It should be emphasized that the equivalence is a conceptual one, essentially a mathematical fiction; there may in fact be no physical components *A*, *B*, *C* that can be connected in a star to do exactly the same work or energy transformation carried out by the delta elements *AB*, *AC*,

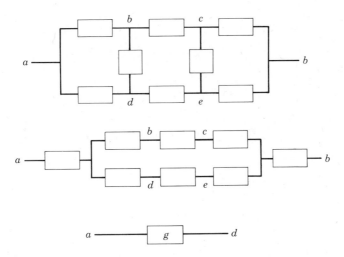

Figure 3.13 Reduction of a double-bridge network. From S. K. Banerjee and K. Rajamani (1972), "Parametric Representation of Probability in Two Dimensions—A New Approach to System Reliability Evaluations," *IEEE Trans. Reliability*, Vol. R-21, (February), pp. 56–60. Copyright © 1972 by IEEE.

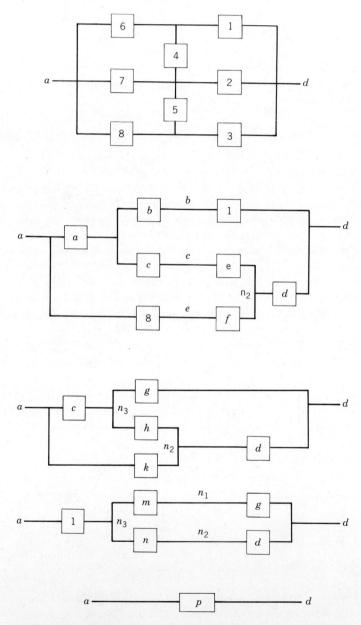

Figure 3.14 Reduction of a different kind of double-bridge network. From Banerjee and Rajamani, op cit.

BC. Nevertheless, the concept is a fruitful one for computing reliabilities for complex systems that would be frightening to tackle using only the methods of Section 3.2. Figures 3.13 and 3.14 show two such networks from Banerjee and Rajamani (1972). Others are shown in the literature survey by Hwang, Tillman, and Lee (1981).

EXERCISES 3.4

1. Verify that a complex system does not always lend itself to the techniques of this section, by trying these techniques on the network of Exercise 3.2.1b.

2. Rework Example 3.6 for $x_1 = .6$, $x_2 = .7$, $x_3 = .5$. Obtain both approximate and exact solutions. Then use these solution values for a, b, c to solve the inverse problem in the next exercise.

3. Demonstrate to yourself that the fastest convergence of the exact inverse problem (using Equation 3.77 in place of 3.64a) is obtained when the approximate solution from Equations 3.64a–c is carried out for several iterations and the results from Exercise 2 are used as starting values for the exact formulation.

3.5 RELIABILITY ALLOCATION

The reliability engineer who is faced with complex equipment will no doubt wonder whether it is possible to get more reliability for the money. In this section we present three techniques that may be of interest.

The first technique is the AGREE (Advisory Group on Reliability of Electronic Equipment) method of allocation of system failure rates. It is described in pages 52–57 of the AGREE report (1957), where it is discussed in the context of aircraft electronic equipment. The method applies to any unit that can be decomposed into a series of independent subsystems. The analysis is carried out for k subsystems, each consisting of independent standard modules, one for each vacuum tube. Let

R = the desired reliability for a mission

m_i = required mean life of the ith subsystem

t_i = time from takeoff until the ith subsystem is no longer needed (i.e., the mission time for this subsystem)

w_i = probability that the total mission will fail if the ith subsystem fails

= $\dfrac{\text{number of mission failures owing to } i\text{th subsystem failures}}{\text{number of } i\text{th subsystem failures}}$

= importance index

n_i = number of modules (tubes) in the ith subsystem

$N = \sum_{i=1}^{k} n_i$ = total number of modules in aircraft

With the application óf these definitions, the required mean lives for the subsystems are approximately

$$m_i = \frac{Nw_i t_i}{-n_i \ln R}, \qquad \text{for } i = 1, \ldots, k \tag{3.78}$$

The derivation of this formula follows. Notice first that the total system under consideration is not truly a series system unless all the w_i equal unity and all t_i are equal. Recognize that the reliability of the ith subsystem for its part of the mission is based on the CFR assumption

$$R_i(t_i) = e^{-t_i/m_i} \tag{3.79}$$

so that the probability that the ith subsystem will fail is $1 - R_i(t_i)$. The compound or joint probability that the subsystem i will fail *and* will cause mission failure is

$$w_i[1 - R_i(t_i)] \tag{3.80}$$

and thus

$$1 - w_i[1 - R_i(t_i)] \tag{3.81}$$

is the probability that the ith subsystem does not cause mission failure. The total system reliability is therefore

$$R(t) = \prod_{i=1}^{k} [1 - w_i(1 - e^{-t_i/m_i})] \tag{3.82}$$

The goal is to choose the m_i values so as to maximize (3.82). To simplify the work we invoke the well known first-order series approximation

$$e^{-t/m} \sim 1 - t/m \tag{3.83}$$

for small t/m. It has already been noted in previous sections that for equipment to have decent reliability, the length of a mission must be much shorter than the equipment's mean life, so the requirement of small t/m is not a problem. The quantity $1 - t/m$ is an underestimate of $e^{-t/m}$, since the first neglected term in the series of alternating signs is positive. When the approximation 3.83 is used in the right-hand side of (3.82), it becomes

$$R(t) = \prod_{i=1}^{k} \left(1 - \frac{w_i t_i}{m_i}\right) \tag{3.84}$$

Invoking (3.83) again in the opposite direction gives

$$R(t) = \prod_{i=1}^{k} \exp\left(-\frac{w_i t_i}{m_i}\right) = \exp\left(-\sum_{i=1}^{k} \frac{w_i t_i}{m_i}\right) \tag{3.85}$$

Now recall that the ith subsystem consists of n_i modules. If T_i is the mean life of each module (tube) in the ith subsystem, its failure rate is $1/T_i$. Since there are n_i modules in the subsystem, the subsystem failure rate is n_i/T_i and thus the mean life is

$$m_i = \frac{T_i}{n_i} \tag{3.86}$$

We can now write

$$\sum_{i=1}^{k} \frac{w_i t_i}{m_i} = \sum_{i=1}^{k} \frac{w_i t_i n_i}{T_i} \tag{3.87}$$

$$= \sum_{i=1}^{k} \sum_{j=1}^{n_i} \frac{w_i t_i}{T_i} \tag{3.88}$$

It can be seen that the quantity n_i has been written in the artificial form $\sum_{j=1}^{n_i} 1$, in order to convert (3.87) into a double sum with $\sum_{i=1}^{k} n_i = N$ terms. The purpose is to create a sum with as many terms as there are modules in the whole system. For each module to make the same contribution to system reliability, we ask that the m_i be chosen so that all the summands are equal:

$$\frac{w_1 t_1}{T_1} = \frac{w_2 t_2}{T_2} = \cdots = \frac{w_k t_k}{T_k} \tag{3.89}$$

We can now write

$$R(t) = \left[\exp\left(-\frac{w_i t_i}{T_i} \right) \right]^N \tag{3.90}$$

or

$$\frac{1}{N} \ln R = \frac{-w_i t_i}{T_i}$$

$$= \frac{-w_i t_i}{n_i m_i} \tag{3.91}$$

and thus the m_i should be chosen so that, as given in (3.78), we have

$$m_i = \frac{N w_i t_i}{n_i (-\ln R)} \tag{3.92}$$

Notice the intuitive appeal of (3.92). It says that m_i should increase proportionately with both the importance number w_i and length t_i of the subsystem mission. It decreases with the relative module count n_i/N. (Does this seem natural?) Using the approximation

$$-\ln R \sim 1 - R$$

which is a system unreliability, and inverting (3.92), we get

$$\frac{1}{m_i} = \frac{(1-R)n_i}{Nw_i t_i} \tag{3.93}$$

so that equipment failure rate $1/m_i$ is a fraction of the system's allowable unreliability.

Example 3.8

Consider a system with four equipment subsystems. In 125 previous 50-hour missions there have been nine mission failures, whose causes are allocated to the subsystems as shown.

Subsystem	t_i hour	Number of Times This Subsystem Has Failed in the Past	Number of Subsystem Failures That Caused System Failure	n_i
1	0.1	10	1	4
2	2	7	2	10
3	10	6	4	5
4	50	10	2	2
			9	21

With what mean lives should the components be designed if the mission reliability must be $R = .9$?

Solution: The weights w_i are 1/10, 2/7, 4/6, and 2/10 and $N = 21$.
The solution values are

$$m_1 = (21/4)\frac{(1/10)(0.1)}{(-\ln.9)} = 0.50 \text{ hour}$$

$$m_2 = (21/10)\frac{(2/7)(2)}{(-\ln.9)} = 11.39 \text{ hours}$$

$$m_3 = (21/5)\frac{(4/6)(10)}{(-\ln.9)} = 265.75 \text{ hours}$$

$$m_4 = (21/2)\frac{(2/10)(50)}{(-\ln.9)} = 996.58 \text{ hours}$$

The AGREE document suggests that it may not be feasible to insist that all modules contribute equally to system reliability, especially those of little importance: "This is especially true if they are also complex. Including such equipments would distort the allocation." Any unit for which

$$w_i < 1 - R^{n_i/N} \qquad (3.94)$$

should be eliminated from the allocation and the module count, and should have its mean life determined in some other way. In case of doubt, the more exact formula

$$m_i = \frac{-t_i}{\ln[1 - 1/w_i(1 - R^{n_i/N})]} \qquad (3.95)$$

can be used instead of (3.92). The error of using (3.92) in place of (3.95) should be the order of magnitude of

$$\frac{1}{2t_i} \frac{n_i^2}{N^2} \left(\frac{1}{w_i} - 1 \right) (1 - R)^2$$

for the failure rate $1/m_i$.

Suppose that an entire unit is duplicated in parallel. Then the required unit mean life is calculated as for a single component and "distributed" to the parallel elements. Thus if α is the required individual mean life, we would set

$$
\begin{aligned}
e^{-t/m} &\equiv 1 - (1 - e^{-t/\alpha})^2 \\
&\simeq 1 - (t_i/\alpha)^2 && (3.96) \\
&= e^{-(t_i/\alpha)^2} && (3.97)
\end{aligned}
$$

Then the individual mean life α should be chosen as

$$\alpha = (mt)^{1/k} \qquad (3.98)$$

for two units in parallel. For three in parallel (3.96) is replaced by

$$e^{-t/m} = 1 - (1 - e^{-t/\alpha})^3$$

$$1 - \frac{t}{m} = 1 - \frac{t^3}{\alpha^3}$$

from which it is seen that

$$\alpha \simeq (mt^2)^{1/3}$$

In general, for k units in parallel,

$$\alpha \simeq (mt^{k-1})^{1/k}$$

According to von Alven (1964, page 193), these redundancy formulae are poor approximations, at best.

In the parallel system we must ask how the module count is affected. The total module count $N = \sum n_i$ of the previous development has a series implication about it. According to the AGREE document (page 56),

> The solution to [the parallel] problem . . . is not as clear. Obviously the module count used as index of complexity of the parallel combination should be less than

that of a single member. If the two members were in series, we would consider the combination to be precisely twice as complex as one member. Considering parallel connection to be dual to series connection, we therefore use *half the tube count of a single member* as the tube count for the parallel combination.

Two different equipments with module counts $n_1 + n_2$ performing essentially parallel functions should be treated as a combination with tube count $1/4(n_1 n_2)$.

The formula above should then be replaced with

$$r_1 = (1/n_1)\sqrt{mtn_1n_2}$$
$$r_2 = (1/n_2)\sqrt{mtn_1n_2}$$

The AGREE document finishes its section on allocation with the following comments.

The module-count method of accounting for relative difficulty of attainment leaves much to be desired. Progress here should follow progress in prediction. Adaptability to derating, air conditioning, relaxed tolerances, etc., should all be considered. The objective is to make proper allowance for relative difficulty of attainment inherent in the function of the equipments, which may be quite different from relative reliability exhibited by past designs.

The module-count method may be unfair in that it discriminates against manufacturers developing a successor equipment to one which has already had its tube count reduced as compared to other equipments performing the same functions. For this reason it would not be wise to make a permanent practice of using module counts in the way they are used here.

A much simpler method, which may be used for simple series configurations, is proposed by von Alven (1964, page 189). It is based on the assumption that a certain maximum failure rate λ^* can be tolerated, and that the current component failure rates of the series elements are $\lambda_i (i = 1, \ldots, n)$. If the current system failure rate is too high, that is, if

$$\lambda_s = \lambda_1 + \cdots + \lambda_n \geq \lambda^* \tag{3.99}$$

then the individual failure rates must be decreased. There are an infinite number of ways to convert (3.99) into an equality. The easiest way (as a mathematical conception) is to decrease each λ_i by the same ratio λ^*/λ_s so that the new component failure rates are

$$\hat{\lambda}_i = \lambda_i \frac{\lambda^*}{\lambda_s}$$
$$= \lambda^* \frac{\lambda_i}{\lambda_s}$$
$$= \lambda^* w_i \tag{3.100}$$

and the w_i are called the relative weights.

If system performance is stated in terms of minimum acceptable reliability $R^*(t)$, the resulting allocation formula is

$$R_j(t) \geq [R^*(t)]^{w_j} \qquad (3.101)$$

but the details are essentially the same.

Example 3.9

Assume that a five component series system has failure rates $\lambda_i = \{.003, .002, .0015, .001, .005\}$ failures per hour. The maximum allowable is $\lambda^* = .01$ (a mean life of 100 hours). We see that $\lambda_1 + \cdots + \lambda_5 = .0125$, so that the weights are the ratios $w_i = \{.24, .16, .12, .08, .40\}$. The desired component failure rates must then be no worse than $\hat{\lambda}_i = \{.0024, .0016, .0012, .0008, .0040\}$.

The final reliability allocation method to be presented here is one suggested by Albert (1958) and reported by Lloyd and Lipow (1962). It is valid only for a series configuration, and it makes certain assumptions about the cost of improving elements, which can be stated as follows. Let x and y be the initial and terminal reliability of a component, and let $G(x,y)$ be the effort (measured in some appropriate units) of raising component reliability from x to a higher value y. The assumptions on $G(x,y)$ are the following.

1. $G(x,y) \geq 0$ for $y \geq x$.
2. $G(x,y)$ is nondecreasing in y for fixed x and nonincreasing in x for fixed y. That is,

$$G(x,y) \leq G(x,y + \Delta y) \text{ and}$$

$$G(x,y) \geq G(x + \Delta x,y)$$

3. $G(x,y) + G(y,z) = G(x,z)$ for $x \leq y \leq z$.
4. $G(0,y)$ has a derivative $h(y)$ such that $y h(y)$ is strictly increasing in $0 < y < 1$.

The first three assumptions are self-explanatory and the first two are simple, sensible, and nonrestrictive. Assumption 3 is rather restrictive and a good case might be made for inequality rather than equality. Assumption 4 is needed for theoretical reasons and has to do with the convexity of $G(0,y)$.

Each component in the system would normally have its own effort function, but Albert assumed as a first approximation that there was a common effort function for all components. Under these assumptions, Albert's work provided the following rule, which will be stated more formally later. *Leave the best components in the system alone and spend the effort budget to bring the worst components up to a common level.* The technique is best illustrated with an example.

Example 3.10

Consider a system with five components in series, whose reliabilities are { .89, .75, .85, .90, .80} respectively. The components are renumbered { C1, . . . ,C5} so that $R_1 \le R_2 \le \ldots \le R_n \le R_{n+1} = 1$. The perfect $(n + 1)$st dummy component is needed for the mathematical formulation given later but not for this example. Thus we write $R_1 = .75$, $R_2 = .80$, $R_3 = .85$, $R_4 = .89$, $R_5 = .90$. Assume now that the required system reliability is $R^* = .50$, which is greater than the current system reliability of .409. Instead of improving all components in some equal ratio, we ask "Is the best component C5 good enough?" That is, if all were as good as the best, would the system pass inspection? The answer is "yes" since if all were as good as the best, the system reliability would be $(.90)^5 = .59$, which exceeds the required .50. Another way to answer the same question is to form the test criterion $r_5 = (.50)^{1/5} = .871$. This is the value that five equally reliable components would need to have for the system to meet exactly the required .50 value. Since $R_5 > r_5$, we again have the answer, "Yes, the best component C5 is good enough. Leave it alone."

Now we ask "Is the next best component C4 good enough? We test this by leaving $R_5 = .90$. Then if C4 and all those below it were as good as $R_4 = .89$, would the system pass inspection? The answer is yes because if all (except C5) were as good as C4, the system reliability would be $(.89)^4(.90) = .565$, which exceeds the required system value .50. Another way to answer the same question is to "factor out" of the required .50 the satisfactory value of $R_5 = .90$ and "distribute" the remaining reliability over the four lower components C1, . . . ,C4. This is equivalent to forming the test criterion $r_4 = (.50/.90)^{1/4} = .863$. This value would be adequate, and since we have $R_4 > r_4$, we can say "Yes, C4 is good enough. Leave it alone."

Now we ask, "Is the third component C3 good enough? If it and all those below it had its value of $R_3 = .85$, would the system pass inspection?" Leaving $R_5 = .90$ and $R_4 = .89$, we compute $(.85)^3(.89)(.90) = .492$, and since this is less than the required .50, we say, "No, C3 is not good enough. It and all those below it must be improved." To what value? The second way of asking the question will supply the answer. We factor out of the desired system reliability of .50 the part that belongs to the satisfactory components C5 and C4, and the residual is distributed over the three remaining components.

The corresponding test value is

$$r_3 = \left[\frac{.50}{(.90)(.89)} \right]^{1/3} = .855$$

Since the actual $R_3 = .85$ is less than this, we see that C3 is not good enough, nor obviously are those components below it in the list. The value r_3 provides us with the common value to which C1, C2, C3 must be raised. Thus

$$(.855)(.855)(.855)(.89)(.90) = .500 \qquad (3.102)$$

There are many other solutions to the requirement that

$$\prod_{i=1}^{5} R_i \geq .5$$

For example, one solution is to leave C2 through C5 alone and raise C1 to the value .92, but Albert demonstrated that the one described minimizes total effort.

Now we state Albert's solution formally. Subject to possible reordering so that $R_1 \leq R_2 \leq \ldots R_n \leq R_{n+1} = 1$, and letting R^* denote the required system reliability, form the series of test values

$$r_j = \left(\frac{R^*}{\displaystyle\prod_{i=j+1}^{n+1} R_i} \right)^{1/j} \qquad (3.103)$$

If $R_j \geq r_j$ the component Cj is good enough. The lowest numbered component, with index k_0 (say) for which $R_j < r_j$, is not good enough, nor are those below it in the list. Their reliabilities should all be raised to the common value

$$R_0^* = r_{k_0} \qquad (3.104)$$

It is evident that this will achieve the desired result since

$$(R_0^*)^{k_0} R_{k_0+1} \ldots R_n = R^* \qquad (3.105)$$

In Example 3.10, we had $r_5 < R_5$, $r_4 < R_4$, but $r_3 > R_3$ so $k_0 = 3$, and R_1, R_2, R_3 were replaced by $R_0^* = r_3$.

Two functions that obey Albert's assumptions for effort functions are

$$G(x,y) = a \ln \frac{y+b}{x+b}, \qquad a,b > 0 \qquad (3.106)$$

and

$$G(x,y) = a(\sqrt{y} - \sqrt{x}) \qquad (3.107)$$

In general $G(x,y)$ will have the form

$$G(x,y) = H(y) - H(x) \qquad (3.108)$$

Students can verify that both (3.106) and (3.107) for $x = 0$ are convex and have $y G'(0,y)$ strictly increasing.

Afzali (1979) investigated numerically a limited number of systems involving Albert's approach (a) when the components were in parallel rather than series, (b) when the components were in series, but the effort functions were not the same,

158   CHAPTER 3 MULTICOMPONENT SYSTEMS

and (c) when the components were in the single-bridge network encountered earlier
in Section 2.4. Albert (1981) outlined a method of solution for the series case with
unequal effort functions. He indicated two solution techniques, one using dynamic
programming and one using Lagrangian multipliers and the Kuhn–Tucker conditions.
Neither method is of sufficient simplicity to be appropriate for inclusion here.

EXERCISES 3.5

1. Test the accuracy of the approximation of (3.85) to the exact value (3.82) for
$t_i/m_i = .5$ $(i = 1, 2, 3)$ and $w_1 = .1$, $w_2 = 1$, $w_3 = .4$. Do you consider the
approximation to be good? If not, why not?

2. Redo Exercise 1 using $t_i/m_i = .1$ $(i = 1, 2, 3)$.

3. Using the AGREE method, solve the problem displayed in the following table
by finding m_i and $R_i(t)$ for $i = 1, \ldots, 4$. The required system reliability is
$R = .9$.

Equipment	n_i	w_i	t_i
RCVR	20	.7	4
XMTR	30	.5	4
RADAR	200	.8	4
IFF	50	.2	4
	300		

4. Repeat Exercise 3 with $t_1 = 1$, $t_2 = 4$, $t_3 = 5$, $t_4 = 4$. Find m_i and $R_i(t_i)$.

5. (a) Use Albert's method to make a reliability reallocation for three series components with reliabilities $\{.6, .7, .8\}$ when the required system reliability is
.5.

(b) Use the second method (Equation 3.100) on this problem. Assume $t = 100$
hours.

6. (a) Use Albert's method to make a reliability reallocation for four series components with reliabilities $\{.6, .7, .8, .9\}$ if the system requirement is $R^* = .4$.

(b) Use the second method (Equation 3.100) on this problem. Assume $t = 100$
hours.

7. (a) Use Albert's method to make a reliability reallocation for a series system
whose reliabilities are $\{.9, .9, .9, .9, .95, .95, .96, .96, .96, .99, .99, .99\}$
if the system requirement is $R^* = .65$.

(b) Noting your sequence of r_i values in this problem, what do you conclude
is the possibility that one of a string of equal components could be good
enough when the others are not?

8. Verify that (3.106) and (3.107) satisfy Albert's effort function assumptions. Plot the curves for $G(0, y)$ for some convenient values of a, b, c.

3.6 TWO MODES OF FAILURE

In our discussions up to now, we have considered components that have only two possible states, good and failed. It is sometimes important when working with electrical components, especially switches, to be aware that there may be two modes of failures, open circuit and short circuit. If the manufacturer wishes to increase the reliability of systems that incorporate such units, how should redundancy be introduced? Consider the switches in Figure 3.15a. If S_1 fails open, S_2 still serves, and vice versa. However, if S_1 (or S_2) fails short, the other switch becomes "a dead letter" and the whole system must be considered failed. Thus we can say that the system fails when either of the events E_1 and E_2 occurs:

E_1: Both switches fail open.

E_2: At least one switch fails short.

Let

$$R = \text{system reliability}$$
$$Q = 1 - R = \text{system unreliability}$$
$$r_i = \text{reliability of } i\text{th element}$$
$$q_{oi} = \text{"open" unreliability of } i\text{th element}$$
$$q_{si} = \text{"short" unreliability of } i\text{th element}$$

Then $r_i + q_{oi} + q_{si} = 1$. We can write

$$Q = P(E_1 \text{ or } E_2) = P(E_1) + P(E_2) \tag{3.109}$$

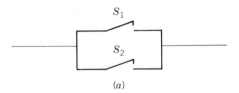

(a)

(b)

Figure 3.15 Systems consisting of two switches. (a) Two switches in parallel. (b) Two switches in series.

since E_1 and E_2 are mutually exclusive events. This equation becomes

$$Q = q_{o1}q_{o2} + (q_{s1} + q_{s2} - q_{s1}q_{s2}) \tag{3.110}$$

Therefore

$$\begin{aligned} R &= 1 - q_{o1}q_{o2} - q_{s1} - q_{s2} + q_{s1}q_{s2} \\ &= (1 - q_{s1})(1 - q_{s2}) - q_{o1}q_{o2} \end{aligned} \tag{3.111}$$

In general, for n switches in parallel, we have

$$R = \prod_{i=1}^{n} (1 - q_{si}) - \prod_{i=1}^{n} q_{oi} \tag{3.112}$$

If all elements are identical, then (3.112) reduces to

$$R^{(n)} = (1 - q_s)^n - q_o^n \tag{3.113}$$

Notice that if $q_o = 0$, the system behaves like a series system, wherein the fewer components the better. If $q_s = 0$, the system behaves like a parallel system, wherein the more components the better. We may conclude that for nonzero q_o and q_s, there is some optimal number of switches. To find the optimum number, we use the standard discrete optimization technique of finding an n_o for which it is true that

$$R^{(n_o-1)} < R^{(n_o)} \tag{3.114}$$

and

$$R^{(n_o)} \geq R^{(n_o+1)} \tag{3.115}$$

The equality is used in (3.115) so that in the event of a tie, the smaller value of n will be used. From (3.114) we have

$$(1 - q_s)^{n-1} - q_o^{n-1} < (1 - q_s)^n - q_o^n \tag{3.116}$$

Transposing some terms,

$$(1 - q_s)^{n-1} - (1 - q_s)^n < q_o^{n-1} - q_o^n$$

and factoring,

$$(1 - q_s)^{n-1}[1 - (1 - q_s)] < q_o^{n-1}(1 - q_o)$$

yields

$$\left(\frac{1 - q_s}{q_o} \right)^{n-1} < \frac{1 - q_o}{q_s}$$

so that

$$(n - 1) \ln\left(\frac{1 - q_s}{q_o} \right) < \ln\left(\frac{1 - q_o}{q_s} \right)$$

from which we obtain the solution

$$n < 1 + \frac{\ln(1 - q_o) - \ln q_s}{\ln(1 - q_s) - \ln q_o} \tag{3.117}$$

Starting from (3.115) and proceeding similarly, we arrive at

$$n \geq \frac{\ln(1 - q_o) - \ln q_s}{\ln(1 - q_s) - \ln q_o} \tag{3.118}$$

Thus n_o is the integer that is at least as great as

$$A(n) = \frac{\ln(1 - q_o) - \ln q_s}{\ln(1 - q_s) - \ln q_o} = \frac{\ln\left(\frac{1 - q_o}{q_s}\right)}{\ln\left(\frac{1 - q_s}{q_o}\right)} \tag{3.119}$$

Example 3.11

For the following cases, the optimum is as given.

(a) $q_o = .1$, $q_s = .1$ $A(n) = 1.00$ $n_o = 1$ $R = .8$
(b) $q_o = .1$, $q_s = .2$ $A(n) = .72$ $n_o = 1$ $R = .7$
(c) $q_o = .2$, $q_s = .1$ $A(n) = 1.38$ $n_o = 2$ $R = .77$
(d) $q_o = .05$, $q_s = .2$ $A(n) = .56$ $n_o = 1$ $R = .75$
(e) $q_o = .2$, $q_s = .2$ $A(n) = 1.78$ $n_o = 2$ $R = .8625$
(f) $q_o = .01$, $q_s = .2$ $A(n) = .36$ $n_o = 1$ $R = .78$
(g) $q_o = .2$, $q_s = .01$ $A(n) = 2.74$ $n_o = 3$ $R = .9623$

If $q_o \leq q_s$ as in cases (a), (b), (d), and (f), the system acts more like a series system than a parallel system, so that $n_o = 1$. This can be seen from the algebra of (3.119). For any component it is obvious that

$$q_s + q_o \leq 1 \tag{3.120}$$

This equation can be multiplied by the positive quantity $q_s - q_o$, giving

$$q_s^2 - q_o^2 \leq q_s - q_o \tag{3.121}$$

where transposing some terms gives

$$q_o - q_o^2 \leq q_s - q_s^2 \tag{3.122}$$

which factors into

$$q_o(1 - q_o) \leq q_s(1 - q_s) \tag{3.123}$$

yielding

$$\frac{1 - q_o}{q_s} \le \frac{1 - q_s}{q_o} \tag{3.124}$$

so that

$$\ln\left(\frac{1 - q_o}{q_s}\right) \le \ln\left(\frac{1 - q_s}{q_o}\right) \tag{3.125}$$

and thus $A(n) \le 1$. Conversely, if $q_o > q_s$, it is obvious that the system behaves more like parallel than series, and that $n_o > 1$.

A similar study can be made of switches in series, as in Figure 3.15b. Let the events E_3 and E_4 be defined as

E_3: Both switches fail short.

E_4: At least one switch fails open.

Then

$$Q = P(E_3 \text{ or } E_4) \tag{3.126}$$

The analysis proceeds exactly as in the previous configuration, *mutatis mutandis*, leading to

$$R^{(n)} = (1 - q_o)^n - q_s^n \tag{3.127}$$

where q_o and q_s are interchanged from the parallel case.

EXERCISES 3.6

1. Master the details for the discrete optimization procedure used in deriving Equation 3.119.

2. Use the same technique to derive the formula for the series configuration of Figure 3.15b.

3. Find the optimum number of components when $q_o = .01$ and $q_s = .10$ for (a) series operation and (b) parallel operation.

4. Do the same for $q_o = .05$, $q_s = .001$.

5. Pretend you do not have formulae available for finding the optimum n, and find it by trial and error in the previous problems. Set $n = 1, 2, \ldots$, and calculate $R^{(n)}$ from (3.113) or (3.127).

6. Consider the four-component series–parallel configurations of Figure 3.4. Suppose that each element performs the same function, and "success" is defined

as output from at least one element. Find the reliability in the two config-urations. Then try to generalize the result to the system with n identical series blocks, each containing m identical elements in parallel.

7. Some authors use an approximation technique to find the optimum number of components, by treating Equation 3.113 as if it were continuous in the variable n, and differentiating to find the maximum. Carry out the procedure and verify that the approximation is very good in nearly every case.

CHAPTER 4

Standby Redundancy

In the previous chapter we considered the simplest form of redundancy, ordinary parallel, in which all n elements are set to function concurrently and the system works as long as even a single component is still alive. In this section we consider what is often the more common and more sensible situation: a single unit is put in service, to be replaced by a spare as soon as failure occurs.

One question of interest is, "If there are two components of differing quality available, which should be selected as primary and which as secondary?" It is regrettably necessary, for mathematical reasons, to assume that replacement of the failed unit is made instantaneously. The approach used here, patterned after Bazovsky (1961), is to find the density function of the system's time to failure. Then we find the reliability from the density function.

Let the probability density functions be defined as

$f_s(t)$ = pdf for system life

$f_1(t)$ = pdf for first (primary) unit life

$f_2(t)$ = pdf for second (standby) unit life

We work with the p.e. (probability element) $f_s(t)dt$, which is the probability that the system fails in the incremental interval dt after living for time t. This event corresponds to the intersection of two simpler events:

E_1: The primary fails some time before t.

E_2: The secondary functions for the balance of the interval $(0,t)$ and fails in dt.

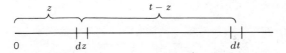

Figure 4.1 Service intervals for primary and standby units.

Let z stand for the length of life of the primary unit, and $t - z$ for the length of life of the standby unit. Now we can restate E_1 and E_2 as depicted in Figure 4.1.

E_1: The primary lives for period z and fails in dz.

E_2: The standby lives for period $t - z$ and fails in dt.

It can be seen that

$$P(E_1) = f_1(z)\, dz \tag{4.1}$$

and

$$P(E_2) = f_2(t - z)\, dt \tag{4.2}$$

and the system failure in dt is related to their product

$$f_1(z)\, dz\, f_2(t - z)\, dt \tag{4.3}$$

Only one thing is missing in (4.3). As written, it depends explicitly on z, the time of the primary's failure, which is of no interest in determining the system's failure law. Therefore it is necessary to "average out" z by integration. This is done by forming the integral

$$f_s(t)dt = \int_{z=0}^{t} f_1(z)\, dz\, f_2(t - z)\, dt \tag{4.4}$$

so that the system failure pdf is (dividing by dt)

$$f_s(t) = \int_{z=0}^{t} f_1(z) f_2(t - z) dz \tag{4.5}$$

Experienced statistics students will recognize (4.5) as a convolution, which is only natural since

$$\text{system life} = \text{primary's life} + \text{secondary's life}$$

or

$$T = T_1 + T_2$$

It is apparent that with a third component (secondary standby) the system failure pdf is the double convolution arising from the situation depicted in Figure 4.2.

$$f_s(t) = \int_{z=0}^{t} \int_{w=0}^{z} f_1(w) f_2(z - w) f_3(t - z)\, dw\, dz \tag{4.6}$$

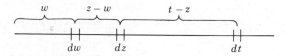

Figure 4.2 Service intervals for primary and two standby units.

Only in the special case of CFR units are (4.5) and (4.6) easy to evaluate. In this case (4.5) becomes

$$f_s(t) = \int_0^t \lambda_1 e^{-\lambda_1 z} \lambda_2 e^{-\lambda_2(t-z)} \, dz \tag{4.7}$$

$$= \lambda_1 \lambda_2 e^{-\lambda_2 t} \int_0^t e^{-(\lambda_1 - \lambda_2)z} \, dz \tag{4.8}$$

$$= \frac{\lambda_1 \lambda_2 e^{-\lambda_2 t}}{\lambda_1 - \lambda_2} [1 - e^{-(\lambda_1 - \lambda_2)t}] \tag{4.9}$$

$$= \frac{\lambda_1 \lambda_2}{\lambda_1 - \lambda_2} e^{-\lambda_2 t} + \frac{\lambda_1 \lambda_2}{\lambda_2 - \lambda_1} e^{-\lambda_1 t} \tag{4.10}$$

Integrating over (t, ∞), we obtain by inspection the reliability function

$$R_s(t) = \frac{\lambda_1 e^{-\lambda_2 t}}{\lambda_1 - \lambda_2} + \frac{\lambda_2 e^{-\lambda_1 t}}{\lambda_2 - \lambda_1} \tag{4.11}$$

The form of (4.11) is symmetric in the subscripts 1 and 2, and thus it is clear that it does not matter which of two components is chosen as primary and which as standby. The mean life of the system, obtained by integrating (4.11) over $(0, \infty)$, is

$$\text{MTTF} = \frac{\lambda_1}{\lambda_1 - \lambda_2} \frac{1}{\lambda_2} - \frac{\lambda_2}{\lambda_1 - \lambda_2} \frac{1}{\lambda_1} \tag{4.12}$$

$$= \frac{1}{\lambda_1 - \lambda_2} \left[\frac{\lambda_1}{\lambda_2} - \frac{\lambda_2}{\lambda_1} \right] = \frac{\lambda_1^2 - \lambda_2^2}{\lambda_1 \lambda_2 (\lambda_1 - \lambda_2)}$$

$$= \frac{1}{\lambda_1} + \frac{1}{\lambda_2} \tag{4.13}$$

as is to be expected, since

$$T = T_1 + T_2$$

implies that

$$E(T) = E(T_1) + E(T_2)$$

The solution of (4.6) for CFR units proceeds in a very similar manner but is considerably more tedious. It is convenient to use Laplace transforms to obtain the answer; this is carried out in Appendix 4A. The result is

$$R_s(t) = \frac{\lambda_2\lambda_3 e^{-\lambda_1 t}}{(\lambda_2 - \lambda_1)(\lambda_3 - \lambda_1)} + \frac{\lambda_1\lambda_3 e^{-\lambda_2 t}}{(\lambda_1 - \lambda_2)(\lambda_3 - \lambda_2)} + \frac{\lambda_1\lambda_2 e^{-\lambda_3 t}}{(\lambda_1 - \lambda_3)(\lambda_2 - \lambda_3)}$$

(4.14)

Integration of (4.14) over $(0, \infty)$ yields

$$\text{MTTF}_s = \frac{\lambda_2\lambda_3}{\lambda_1(\lambda_2 - \lambda_1)(\lambda_3 - \lambda_1)} + \frac{\lambda_1\lambda_3}{\lambda_2(\lambda_1 - \lambda_2)(\lambda_3 - \lambda_2)} + \frac{\lambda_1\lambda_2}{\lambda_3(\lambda_1 - \lambda_3)(\lambda_2 - \lambda_3)}$$

(4.15)

The student who delights in manipulative algebra may verify that this expression reduces, as it should, to

$$\text{MTTF} = \frac{1}{\lambda_1} + \frac{1}{\lambda_2} + \frac{1}{\lambda_3}$$

It is apparent that the general form for the reliability of n components, consisting of one primary and $n - 1$ standby units, is

$$R_s(t) = \sum_{i=1}^{n} e^{-\lambda_i t} \prod_{\substack{j=1 \\ j \neq i}}^{n} \frac{\lambda_j}{\lambda_j - \lambda_i}$$

(4.16)

Returning to the $n = 2$ case of (4.11), we find it instructive to write $R_s(t)$ in a different form by adding zero to the second term in the form

$$\frac{\lambda_1 - \lambda_1}{\lambda_2 - \lambda_1} e^{-\lambda_1 t}$$

(4.17)

so that

$$R_s(t) = \frac{\lambda_2 + \lambda_1 - \lambda_1}{\lambda_2 - \lambda_1} e^{-\lambda_1 t} + \frac{\lambda_1}{\lambda_1 - \lambda_2} e^{-\lambda_2 t}$$

$$= e^{-\lambda_1 t} - \frac{\lambda_1}{\lambda_1 - \lambda_2} e^{-\lambda_1 t} + \frac{\lambda_1}{\lambda_1 - \lambda_2} e^{-\lambda_2 t}$$

(4.18)

$$= e^{-\lambda_1 t} + \frac{\lambda_1}{\lambda_1 - \lambda_2}\left(e^{-\lambda_2 t} - e^{-\lambda_1 t}\right)$$

(4.19)

or in the more symmetric form

$$R_s(t) = e^{-\lambda_1 t} + \lambda_1\left(\frac{e^{-\lambda_1 t}}{\lambda_2 - \lambda_1} + \frac{e^{-\lambda_2 t}}{\lambda_1 - \lambda_2}\right)$$

$$= e^{-\lambda_1 t} + \text{AR}_{(2)}$$

(4.20)

It is apparent that the first term is the reliability of the primary component and that the second term $\text{AR}_{(2)}$ is the reliability added by the presence of the standby component.

It can be proved that for $n = 3$ the system reliability has the same form as (4.19) but with the added term

$$\text{AR}_{(3)} = \lambda_1 \lambda_2 \left[\frac{e^{-\lambda_1 t}}{(\lambda_2 - \lambda_1)(\lambda_3 - \lambda_1)} + \frac{e^{-\lambda_2 t}}{(\lambda_1 - \lambda_2)(\lambda_3 - \lambda_2)} + \frac{e^{-\lambda_3 t}}{(\lambda_1 - \lambda_3)(\lambda_2 - \lambda_3)} \right]$$

$$(4.21)$$

The student is invited to construct the term $\text{AR}_{(4)}$ that represents the added reliability of a third standby.

So far, it has been assumed that switchover from the primary component is perfect and instantaneous. If, in fact, the ith sensing-and-switching device has reliability only r_{Si} (say), then (4.19) requires the insertion of the factor r_{Si} with the second term. The general case would be

$$R_s(t) = e^{-\lambda_1 t} + r_{S1} \{ \text{AR}_{(2)} + r_{S2}[\text{AR}_{(3)} \ldots \ldots] \} \qquad (4.22)$$

When switching is perfect and components are identical, we are led to a special case. The second term of (4.19) degenerates, when $\lambda_1 = \lambda_2 = \lambda$, to zero divided by zero. This can be handled using l'Hôpital's rule. Set $\lambda_1 = \lambda$ (a constant) and let the variable λ_2 approach λ. The result is

$$\lim_{\lambda_2 \to \lambda} \lambda \frac{e^{-\lambda_2 t} - e^{-\lambda t}}{\lambda - \lambda_2} = \lim_{\lambda_2 \to \lambda} \lambda \frac{-t e^{-\lambda_2 t}}{-1}$$

$$= \lambda t e^{-\lambda t} \qquad (4.23)$$

and hence for identical components and $n = 2$

$$R_s(t) = e^{-\lambda t} + \lambda t e^{-\lambda t}$$

Repeated use of l'Hôpital's rule for $n = 3$ gives

$$R_s(t) = e^{-\lambda} + \lambda t e^{-\lambda t} + \frac{(\lambda t)^2 e^{-\lambda t}}{2!} \qquad (4.24)$$

and for general n it is

$$R_s(t) = e^{-\lambda t} + \lambda t e^{-\lambda t} + \cdots + \frac{(\lambda t)^{n-1} e^{-\lambda t}}{(n-1)!} \qquad (4.25)$$

These results should not be surprising, since for identical components a renewal process is set up with identical negative exponential interfailure times; of course it is a Poisson process. The mean life for an n-element system under these conditions is n/λ.

Formula 4.25 makes possible an easy derivation for the failure law of a unit that fails only after k shocks if the shocks arrive in Poisson manner with intensity λ. Let T be the lifetime of the unit, $f^*(t)$ its unknown density, and $F^*(t)$ its unknown

CDF. Then the unit will survive for time t only if there have been strictly *fewer* than k shocks by time t. Thus

$$P(T > t) = 1 - F^*(t)$$

$$= \sum_{m=0}^{k-1} \frac{(\lambda t)^m e^{-\lambda t}}{m!} \tag{4.26}$$

This gives

$$F^*(t) = \sum_{m=k}^{\infty} \frac{(\lambda t)^m e^{-\lambda t}}{m!} \tag{4.27}$$

But we have seen in Section 2.6 that the right-hand tail of a Poisson distribution equals the left-hand tail of a suitably chosen gamma distribution. Letting $\lambda = 1/\beta$, we see that in the notation of Section 2.6

$$F^*(t) = F(t; k, 1/\lambda) = F(t; k, \beta) \tag{4.28}$$

and thus

$$T \sim G(k, \beta) = G(k, 1/\lambda)$$

We have hereby demonstrated, without the use of moment generating functions, the useful result that the sum of k identical exponential waiting time random variables is a gamma random variable. Symbolically we can write it as

$$\text{NGEX}(\lambda) + \ldots + \text{NGEX}(\lambda) = G(k, 1/\lambda)$$

or

$$G(1, 1/\lambda) + \ldots + G(1, 1/\lambda) = G(k, 1/\lambda)$$

It has been assumed in the foregoing that the standby elements were "cold," or under no stress. We consider now the system in which the standby element is subject to possible failure while in the standby mode, through heat, vibration, or other system stress. We describe this state as *hot* or *idling* standby. Such a standby component has its failure law described by

$f_2(t)$ while in active service.

$f_2^*(t)$ while on hot standby.

The system fails in the incremental interval dt if E_3 or E_4 occurs, where

E_3: Primary lives for time z and fails in dz; secondary has not failed while on standby, takes over, and lives for time $t - z$, finally failing in dt.

E_4: Primary fails in dt but secondary cannot take over because it has failed while on standby.

Notice that the events E_3 and E_4 have a compound structure not present in the simple events E_1 and E_2 which were used at the beginning of this section. The probability element for system failure is

$$
\begin{aligned}
f_s(t)dt &= P(E_3) + P(E_4) \\
&= \int_0^t f_1(z)dz\, R_2^*(z)\, f_2(t-z)\, dt \\
&\quad + f_1(t)dt[1 - R_2^*(t)]
\end{aligned}
\tag{4.29}
$$

For CFR components this element becomes, on dividing out dt,

$$
\begin{aligned}
f_s(t) &= \int_0^t \lambda_1 e^{-\lambda_1 z}dz\, e^{-\lambda_2^* z}\, \lambda_2 e^{-\lambda_2(t-z)} \\
&\quad + \lambda_1 e^{-\lambda_1 t}(1 - e^{-\lambda_2^* t})
\end{aligned}
\tag{4.30}
$$

$$
\begin{aligned}
&= \lambda_1 \lambda_2 e^{-\lambda_2 t}\int_0^t e^{-(\lambda_1 + \lambda_2^* - \lambda_2)z}dz \\
&\quad + \lambda_1 e^{-\lambda_1 t} - \lambda_1 e^{-(\lambda_1 + \lambda_2^*)t}
\end{aligned}
\tag{4.31}
$$

$$
\begin{aligned}
&= \frac{\lambda_1 \lambda_2}{\lambda_1 + \lambda_2^* - \lambda_2}\left[e^{-\lambda_2 t} - e^{-(\lambda_1 + \lambda_2^*)t}\right] \\
&\quad + \lambda_1 e^{-\lambda_1 t} - \lambda_1 e^{-(\lambda_1 + \lambda_2^*)t}
\end{aligned}
\tag{4.32}
$$

The system reliability is found by integrating over (t, ∞) and rearranging terms to get

$$
R_s(t) = e^{-\lambda_1 t} + \frac{\lambda_1}{\lambda_1 + \lambda_2^* - \lambda_2}\left[e^{-\lambda_2 t} - e^{-(\lambda_1 + \lambda_2^*)t}\right]
\tag{4.33}
$$

The mean life of the system is

$$
\begin{aligned}
\text{MTTF} &= \frac{1}{\lambda_1} + \frac{\lambda_1}{\lambda_1 + \lambda_2^* - \lambda_2}\left(\frac{1}{\lambda_2} - \frac{1}{\lambda_1 + \lambda_2^*}\right) \\
&= \frac{1}{\lambda_1} + \frac{\lambda_1}{\lambda_2(\lambda_1 + \lambda_2^*)}
\end{aligned}
\tag{4.34}
$$

The added mean life attributable to the hot standby element is thus reduced by the factor $\lambda_1/(\lambda_1 + \lambda_2^*)$ compared with the added mean life attributable to a cold standby element. Of course, when $\lambda_2^* = 0$, (4.33) and (4.34) reduce to (4.19) and (4.13) respectively.

Still another situation which is a variation of parallel redundancy Bazovsky (1961, page 125) calls "shared load." The assumption is that two components start functioning together at time $t = 0$ with constant failure rates λ_1 and λ_2. If one component fails, the other carries on alone, but with higher failure rate, say λ_1^* or λ_2^*. The system fails in the incremental interval dt if any of the following three compound events occurs.

E_5: Component 1 fails in interval dz and component 2 carries on alone for time $t - z$ at increased failure rate, finally failing in dt.

E_6: Component 2 fails in interval dz and component 1 carries on alone for time $t - z$ at increased failure rate, finally failing in dt.

E_7: Both components fail in dt.

The system p.e. of failure in dt is thus

$$f_s(t)dt = P(E_5) + P(E_6) + P(E_7)$$

$$= \int_0^t f_1(z)\, dz\, R_2(z)\, f_2^*(t - z)dt$$

$$+ \int_0^t f_2(z)\, dz\, R_1(z)\, f_1^*(t - z)dt + f_1(t)\, f_2(t)\, (dt)^2 \qquad (4.35)$$

The last term vanishes because it is a differential of higher order in dt (recall that simultaneous events are assumed to be impossible in a Poisson process). Thus, for CFR components we have

$$f_s(t) = g_1(t) + g_2(t), \quad \text{say} \qquad (4.36)$$

The form for $g_1(t)$ is

$$g_1(t) = (\lambda_1 \lambda_2^*) e^{-\lambda_2^* t} \int_0^t e^{-(\lambda_1 + \lambda_2 - \lambda_2^*) z} dz$$

$$= \frac{\lambda_1 \lambda_2^* e^{-\lambda_2^* t}}{\lambda_1 + \lambda_2 - \lambda_2^*} [1 - e^{-(\lambda_1 + \lambda_2 - \lambda_2^*) t}] \qquad (4.37)$$

The expression for $g_2(t)$ is identical, with subscripts 1 and 2 interchanged. Integrating over (t, ∞) we find

$$R_s(t) = + \frac{\lambda_1 \lambda_2^*}{\lambda_1 + \lambda_2 - \lambda_2^*} \left[\frac{e^{-\lambda_2^* t}}{\lambda_2^*} - \frac{e^{-(\lambda_1 + \lambda_2) t}}{\lambda_1 + \lambda_2} \right]$$

$$+ \frac{\lambda_1^* \lambda_2}{\lambda_1 + \lambda_2 - \lambda_1^*} \left[\frac{e^{-\lambda_1^* t}}{\lambda_1^*} - \frac{e^{-(\lambda_1 + \lambda_2) t}}{\lambda_1 + \lambda_2} \right] \qquad (4.38)$$

This equation can be written in a form similar to the earlier equations by converting the coefficient of the exponential factor from the second term in each bracket like this,

$$\frac{\lambda_1 \lambda_2^*}{\lambda_1 + \lambda_2 - \lambda_2^*} \frac{-1}{\lambda_1 + \lambda_2} = \frac{-\lambda_1}{\lambda_1 + \lambda_2 - \lambda_2^*} + \frac{\lambda_2}{\lambda_1 + \lambda_2}$$

and

$$\frac{\lambda_2 \lambda_1^*}{\lambda_1 + \lambda_2 - \lambda_1^*} \frac{-1}{\lambda_1 + \lambda_2} = \frac{-\lambda_2}{\lambda_1 + \lambda_2 - \lambda_1^*} + \frac{\lambda_2}{\lambda_1 + \lambda_2}$$

The result of these manipulations is

$$R(t) = e^{-(\lambda_1 + \lambda_2)t}\left(1 - \frac{\lambda_1}{\lambda_1 + \lambda_2 - \lambda_2^*} - \frac{\lambda_2}{\lambda_1 + \lambda_2 - \lambda_1^*}\right)$$
$$+ \frac{\lambda_1}{\lambda_1 + \lambda_2 - \lambda_2^*}\, e^{-\lambda_2^* t} + \frac{\lambda_2}{\lambda_1 + \lambda_2 - \lambda_1^*}\, e^{-\lambda_1^* t} \qquad (4.39)$$

If $\lambda_1^* = \lambda_2^* = \lambda_1 + \lambda_2$ then

$$R_s(t) = e^{-(\lambda_1 + \lambda_2)t} + (\lambda_1 + \lambda_2)t\, e^{-(\lambda_1 + \lambda_2)t} \qquad (4.40)$$

If $\lambda_1 = \lambda_2 = \lambda$ and $\lambda_1^* = \lambda_2^* = \lambda^*$ (that is, the components are alike), then

$$R_s(t) = e^{-2\lambda t} + \frac{2\lambda}{\lambda^* - 2\lambda}\left(e^{-2\lambda t} - e^{-\lambda^* t}\right) \qquad (4.41)$$

If $\lambda^* = 2\lambda$, then l'Hôpital's rule gives

$$R_s(t) = e^{-2\lambda t} + 2\lambda t\, e^{-2\lambda t} \qquad (4.42)$$

which is an entirely expected result, since it represents the probability of fewer than two failures in time t of a Poisson process with failure rate 2λ.

EXERCISES 4.1

1. Verify the algebra in the transition from Equation 4.8 to (4.10) and (4.11).

2. Verify Equation 4.15 starting from (4.14).

3. Prove that (4.21) is correct.

4. Derive the expression for $AR_{(4)}$ as suggested following Equation 4.21.

5. Verify the details of obtaining (4.23) from (4.19).

6. Two components A and B can be used in various ways for a ten-hour mission. It has been determined that when acting alone the failure rates are

$\lambda_A = .012$
$\lambda_B = .030$

When working together the failure rates are

$\lambda_A = .010$
$\lambda_B = .015$

When acting on hot standby the failure rates are

$\lambda_A = .003$
$\lambda_B = .002$

Examine the system reliability for these two components under the following assumptions.

(a) Cold standby with perfect switching.

(b) Cold standby with a switch whose probability of operating correctly is .9 (two configurations).

(c) Hot standby with perfect switching (two configurations).

(d) Hot standby with the same switch as in assumption b (two configurations).

(e) Shared load.

Rank the results in order of reliability.

7. Repeat Exercise 6 if the failure rates are $\lambda_A = .15$, $\lambda_B = .10$ for acting alone, $\lambda_A = .10$, $\lambda_B = .05$ for working together, and $\lambda_A = .01$, $\lambda_B = .02$ for hot standby.

The Laplace Transform Solution
for Standby Redundancy

The solution of the convolution equation 4.5 for the total lifetime $T_s = T_1 + T_2$ is easily carried out in the transform domain. We use the notation

$$f_i^*(s) = \mathscr{L}[f_i(t)] = \int_0^\infty e^{-st} f_i(t)\, dt \qquad (4A.1)$$

For the CFR failure law, the densities

$$f_i(t) = \lambda_i e^{-\lambda_i t} \qquad (4A.2)$$

have Laplace transforms

$$f_i^*(s) = \frac{\lambda_i}{s + \lambda_i} \qquad (4A.3)$$

The convolution (indicated by the operator *) defined as

$$f_s(t) = f_1(t) * f_2(t) = \int_0^t f_1(z) f_2(t - z)\, dz \qquad (4A.4)$$

has transform

$$\mathscr{L}[f_s(t)] = f_1^*(s)\, f_2^*(s) \qquad (4A.5)$$

For three components the convolution of Equation (4.6) has transform

$$\mathscr{L}[f_s(t)] = f_1^*(s) f_2^*(s) f_3^*(s) \qquad (4A.6)$$

$$= \frac{\lambda_1}{s + \lambda_1} \frac{\lambda_2}{s + \lambda_2} \frac{\lambda_3}{s + \lambda_3} \qquad (4A.7)$$

$$= \lambda_1 \lambda_2 \lambda_3 \left[\frac{1}{s + \lambda_1} \frac{1}{s + \lambda_2} \frac{1}{s + \lambda_3} \right] \qquad (4A.8)$$

The product in brackets in (4A.8) is decomposed into partial fractions

$$\frac{a}{s + \lambda_1} + \frac{b}{s + \lambda_2} + \frac{c}{s + \lambda_3} \qquad (4A.9)$$

and it is easily proved that

$$a = 1/(\lambda_2 - \lambda_1)(\lambda_3 - \lambda_1)$$
$$b = 1/(\lambda_1 - \lambda_2)(\lambda_3 - \lambda_2)$$
$$c = 1/(\lambda_1 - \lambda_3)(\lambda_2 - \lambda_3) \tag{4A.10}$$

Hence (4A.8) assumes the form

$$\mathcal{L}[f_s(t)] = \frac{\lambda_1\lambda_2\lambda_3}{(s + \lambda_1)(\lambda_2 - \lambda_1)(\lambda_3 - \lambda_1)} + \frac{\lambda_1\lambda_2\lambda_3}{(s + \lambda_2)(\lambda_1 - \lambda_2)(\lambda_3 - \lambda_2)}$$
$$+ \frac{\lambda_1\lambda_2\lambda_3}{(s + \lambda_3)(\lambda_1 - \lambda_3)(\lambda_2 - \lambda_3)} \tag{4A.11}$$

Taking inverse transforms, and recalling that

$$\mathcal{L}^{-1}[1/(\lambda + s)] = e^{-\lambda t} \tag{4A.12}$$

we have

$$f_s(t) = \frac{\lambda_1\lambda_2\lambda_3 e^{-\lambda_1 t}}{(\lambda_2 - \lambda_1)(\lambda_3 - \lambda_1)} + \frac{\lambda_1\lambda_2\lambda_3 e^{-\lambda_2 t}}{(\lambda_1 - \lambda_2)(\lambda_3 - \lambda_2)} + \frac{\lambda_1\lambda_2\lambda e^{-\lambda_3 t}}{(\lambda_1 - \lambda_3)(\lambda_2 - \lambda_3)} \tag{4A.13}$$

Integrating over $(0,t)$ gives

$$R_s(t) = \frac{\lambda_2\lambda_3 e^{-\lambda_1 t}}{(\lambda_2 - \lambda_1)(\lambda_3 - \lambda_1)} + \frac{\lambda_1\lambda_3 e^{-\lambda_2 t}}{(\lambda_1 - \lambda_2)(\lambda_3 - \lambda_2)} + \frac{\lambda_1\lambda_2 e^{-\lambda_3 t}}{(\lambda_1 - \lambda_3)(\lambda_2 - \lambda_3)} \tag{4A.14}$$

as stated in Equation 4.14.

CHAPTER 5

Life Testing

5.1 INTRODUCTION

The material in this chapter is based on the very important series of papers by Epstein and Sobel (1953 through 1960). In order to follow Epstein's notation as closely as possible, we introduce some important changes of usage from those in previous chapters. In this chapter the symbol $\tilde{t}$ denotes the random variable life of a component, whereas T (formerly used for component life) is given a new meaning, as explained later. Most of the literature on life testing is based on the assumption of CFR components, for reasons of mathematical tractability. Therefore the model for component life t is assumed to be

$$f(t) = \lambda e^{-\lambda t} \qquad \text{for } t \geq 0 \qquad (5.1)$$

or in alternate form

$$f(t) = \frac{1}{\theta} e^{-t/\theta} \qquad \text{for } t \geq 0 \qquad (5.2)$$

where of course $\lambda = 1/\theta$, with λ the failure rate and θ the mean life. Both formulations 5.1 and 5.2 are convenient, so we use them interchangeably, as needed. The goal of life testing is to find an estimate of the mean life θ. There are a variety of ways in which the data may be gathered. We will be concerned here with both point estimates and interval estimates, as well as hypothesis testing. The various testing situations can be categorized by the answers to the following questions.

1. Were a predetermined number of units tested?
2. Was the test time predetermined?
3. Were failure times noted?
4. Were failed items replaced?

We consider these questions in order of increasing complexity.

Readers are reminded of the several appendices to this chapter which provide a useful review on some topics needed in the development of the results which follow.

Appendix 5A: Order Statistics

Appendix 5B: Maximum Likelihood Estimators

Appendix 5C: Transformation of Variables

Appendix 5D: Confidence Limits for the Binomial Parameter

Appendix 5E: Occurrence Times of a Poisson Process

5.2 NONREPLACEMENT TEST WITH FAILURE TIMES NOT RECORDED

Nonreplacement, with failure times not recorded, is familiar to us through our study of the binomial distribution where n items are put on test for some predetermined length of time t_0, and at the end of the testing period the number of failures r is noted. A point estimate of reliability is then

$$\hat{R}(t_0) = 1 - \frac{r}{n} = \frac{n-r}{n} \tag{5.3}$$

To obtain an interval estimate of reliability, we use the fact that the binomial parameter has confidence bounds based on the F distribution, as described in Appendix 5D. The bounds given there are on the probability of failure, since the "success" is a failed item. To find reliability bounds, we reverse the roles of success and failure and find that a confidence interval for $R(t_0)$ can be expressed by the relationship

$$P(R_L \leq R(t_0) \leq R_U) = 1 - \alpha \tag{5.4}$$

where the lower confidence bound R_L is

$$R_L = R_L(t_0) = \frac{1}{1 + \dfrac{r+1}{n-r}F_2} \tag{5.5}$$

with

$$F_2 = F_{\alpha/2}(2r+2, \ 2n-2r) \tag{5.6}$$

and the upper confidence bound R_U is

$$R_U = R_U(t_0) = \frac{F_1}{F_1 + \dfrac{r}{n - r + 1}} \qquad (5.7)$$

with

$$F_1 = F_{\alpha/2}(2n - 2r + 2, \ 2r) \qquad (5.8)$$

It is most common to want a lower confidence bound only, so Equation 5.5 is the most-needed formula, with α replacing $\alpha/2$.

The confidence bounds given by (5.5) and (5.7) are *exact* in the sense that no approximations were invoked in the mathematical development. Hence they can be used for any sample size, however small, and deserve to be more widely applied than they are. The better known normal approximation to the binomial parameter requires large sample size and intermediate occurrence probability, to be valid.

The foregoing formulae are called nonparametric, or distribution-free, since no assumption is made about the form of the failure law. They are of limited usefulness since they apply only to identical Bernoulli trials of length t_0. If it can be assumed that the units are CFR, reliability estimates for other mission times are possible. For example, the quantity $(n - r)/n$ of (5.3) will be an estimate of

$$R(t_0) = e^{-t_0/\theta} \qquad (5.9)$$

so that an estimate of θ is

$$\hat{\theta} = \frac{-t_0}{\ln\left[(n - r)/n\right]} \qquad (5.10)$$

This is a point estimate of θ; a confidence interval on θ based on (5.5) and (5.7) is obtained by setting

$$R_L \le R(t_0) \le R_U$$

and solving for θ in $R(t_0) = e^{-t_0/\theta}$. This gives

$$\frac{-t_0}{\ln R_L} \le \theta \le \frac{-t_0}{\ln R_U} \qquad (5.11)$$

For missions of length other than t_0, we can write

$$\hat{R}(t) = e^{-t/\hat{\theta}} \qquad (5.12)$$

$$= (e^{-t_0/\hat{\theta}})^{t/t_0}$$

$$= \left(\frac{n - r}{n}\right)^{t/t_0} \qquad (5.13)$$

A lower confidence bound on general mission reliability from (5.5) will be

$$R(t) \ge (R_L)^{t/t_0} \qquad (5.14)$$

Example 5.1

Suppose that ten identical units are put on test for a week (168 hours). At the end of that time, two units have failed. A point estimate of $R(168)$ is 8/10. Ninety percent confidence bounds on $R(168)$ are

$$R_U = F_1(F_1 + 2/9)^{-1}$$

where, by interpolation,

$$F_1 = F_{.05}(18, 4) = 5.82$$

so that $R_U = .963$. Similarly, from (5.5) and with $F_2 = F_{.05}(6, 16) = 2.74$, we find $R_L = [1 + (3/8)2.74]^{-1} = .492$. A point estimate of θ is

$$\hat{\theta} = -168(\ln 8/10)^{-1} = 752.9 \text{ hours}$$

and a 90 percent confidence interval on θ is 237.5 hours $\le \theta \le$ 4483.4 hours. Such wide confidence intervals are relatively worthless, so we could rework the problem using $F_{.25}(\nu_1, \nu_2)$ instead of $F_{.05}(\nu_1, \nu_2)$. This yields the 50 percent confidence intervals $.645 \le R(168) \le .903$ and 383.1 hours $\le \theta \le$ 1646.5 hours.

EXERCISES 5.2

1. Twelve items are put on test for a period of 50 hours, at the end of which nine are still functioning. The distribution of lifetimes is unknown.
   (a) Make a point estimate and an exact 90 percent interval estimate of $R(50)$.
   (b) Compare the exact confidence interval of part a with the normal approximation usually used.

2. If the equipment tested in the previous problem consists of CFR components, make a point estimate and a 90 percent confidence interval for mean life. Use them to find an estimate and a lower confidence bound for $R(40)$.

3. Verify that Equation 5.10 follows from (5.9).

4. Fifteen items were put on test for a period of 200 hours; at the end of the test only one item is found to have failed. The distribution of lifetimes is unknown.
   (a) Compute a point estimate and an exact 90 percent lower confidence bound for $R(200)$.
   (b) Compare the lower confidence bound of part a with the normal approximation which is commonly used.

5. If the components in the previous problem are assumed to be CFR, compute the point estimate and 90 percent lower confidence bound for $R(100)$.

5.3 STANDARD TESTING, FAILURE TIMES NOTED

The standard procedure in classical statistics is to observe a sample of n items and note the lifetimes of all items. Usually nothing is said about the testing procedure, but it can be either of two extreme cases as well as a mixture. We can assume that there is a single testing station, or "socket," and that the items under test are placed in it sequentially until they fail. If no time is wasted at the test station, that is, if replacement of the failed items is instantaneous, a Poisson renewal process is generated as shown in Figure 5.1. Notice that the notation here is different from that used in Section 2.3. Here the $\{t_i\}$ are not the Poisson arrival instants, as they were there, but are the component lifetimes themselves, since we consider these more fundamental to the present discussion. The Poisson character of the renewal process is not of interest at the moment; what is of interest is that an unbiased maximum likelihood estimator, or MLE, point estimate of θ is found to be

$$\hat{\theta} = \bar{t} = \left(\sum_{i=1}^{n} t_i \right) / n \tag{5.15}$$

If, instead of a single test station, all items can be tested concurrently, the time graph will look like Figure 5.2. Of course the data will be available much sooner in this multiple-socket case than in the single-socket case of Figure 5.1, but formula 5.15 is still valid.

The construction of a confidence interval for θ depends on knowing the distribution of $\hat{\theta}$. Recall that the t_i are realizations (sample observations) of a common

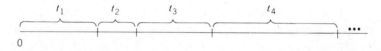

Figure 5.1 Single-socket testing.

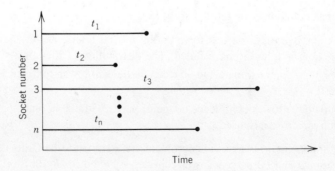

Figure 5.2 Multiple-socket testing.

random variable $\tilde{t}$ for which we can write either $\tilde{t} \sim \text{NGEX}(\lambda)$ or $\tilde{t} \sim G(1, \theta)$. Hence the total test time

$$\tilde{T} = \tilde{t}_1 + \cdots + \tilde{t}_n \tag{5.16}$$

has distribution (see Theorem 2.5 of Section 2.6) of the form

$$\tilde{T} \sim G(n, \theta) \tag{5.17}$$

and hence (see Theorem 5C.1 of Appendix 5C) the sample mean life has distribution given by

$$\tilde{\tilde{t}} \sim G(n, \theta/n)$$

This could be used to construct a confidence interval on θ, if good tables of the standard critical points on the gamma family of distributions were widely available. The only such commonly available table is that for the χ^2 distribution, so a linear transformation is used to convert (5.17). Recall (see Appendix 5C) that if $X \sim G(\alpha, \beta)$, then $CX \sim G(\alpha, C\beta)$. In particular, we can state that

$$\frac{2\tilde{T}}{\theta} \sim G(n, 2) = \chi^2(2n) \tag{5.18}$$

When dealing with confidence statements for the gamma distribution, we are beset with a minor notational problem, since α, the usual confidence level indicator, can be easily confused with the α that is the shape parameter. To obviate the confusion we will frequently use the letter γ to indicate the confidence parameter since it is not needed for any other purpose in this work.

The $100(1 - \gamma)$ percent confidence interval on θ can be constructed by the usual pivot method based on the statement

$$P[\chi^2_{1-\gamma/2}(\nu) \leq \chi^2(\nu) \leq \chi^2_{\gamma/2}(\nu)] = 1 - \gamma \tag{5.19}$$

which in our case becomes

$$P[\chi^2_{1-\gamma/2}(2n) \leq \frac{2T}{\theta} \leq \chi^2_{\gamma/2}(2n)] = 1 - \gamma \tag{5.20}$$

Inverting the three-way inequality and multiplying by $2T$ gives the $100(1 - \gamma)$ percent confidence bounds for θ as the pair (θ_L, θ_U) given by

$$(\theta_L, \theta_U) = \left(\frac{2T}{\chi^2_{\gamma/2}(2n)}, \frac{2T}{\chi^2_{1-\gamma/2}(2n)} \right) \tag{5.21}$$

Usually of more interest is a one-sided lower confidence bound on θ given by

$$\theta_L^* = \frac{2T}{\chi^2_\gamma(2n)} \tag{5.22}$$

Notice that the lower bound on θ comes from the upper critical point on the χ^2 curve, and vice versa. The expressions in (5.15), (5.21), and (5.22) can be used to construct point estimates and confidence bounds for $R(t)$. Thus

$$\hat{R}(t) = e^{-t/\hat{\theta}} \tag{5.23}$$

where $\hat{\theta}$ is based on (5.15). We can also write

$$P[r_1(t) \le R(t) \le r_2(t)] = 1 - \gamma \tag{5.24}$$

where $r_1(t)$ is

$$r_1(t) = e^{-t/\theta_L} = \exp\left[\frac{-t\chi^2_{\gamma/2}(2n)}{2T}\right] \tag{5.25}$$

and similarly for $r_2(t)$.

Example 5.2

Ten items with a presumed negative exponential failure law are tested, yielding lifetimes in hours: {317, 735, 886, 5, 916, 1263, 1020, 586, 636, 830}. The total time on test is $T = 7194$, so $\hat{\theta} = 719.4$. The 95 percent lower confidence bound for θ is

$$\theta_L^* = \frac{2(7194)}{\chi^2_{.05}(20)} = \frac{2(7194)}{31.41} = 458.1 \text{ hours}$$

If a mission of 100 hours is contemplated, we estimate $R(100)$ by

$$\hat{R}(100) = e^{-100/719.4} = .87$$

and a 95 percent lower confidence bound on $R(100)$ is

$$r_1^*(t) = e^{-t/\theta_L^*} = e^{-100/458.1} = .804$$

Note.   The transformation of Equation 5.18 can be used to solve the guarantee problem of Section 2.6. Suppose that the lifetimes of a class of equipment are known to be well described by the $G(\alpha, \beta)$ distribution. A guarantee value t_0 is sought for which there will be, on the average, no more than 5 percent claims. That is, we desire a value t_0 for which $P(t_0 \le \tilde{t}) \le \gamma = .05$. Then by (5.18)

$$P\left(\frac{2t_0}{\beta} \le \frac{2\tilde{t}}{\beta}\right) = \gamma$$

whereupon it follows that

$$\frac{2t_0}{\beta} = \chi^2_{1-\gamma}(2\alpha)$$

and thus that

$$t_0 = \frac{\beta \chi^2_{1-\gamma}(2\alpha)}{2}$$

EXERCISES 5.3

1. A test station of ten sockets is used to investigate the lifetime of certain components which are believed to be adequately described as CFR. Failed components are not replaced but failure times are noted as 26, 10, 42, 17, 20, 50, 32, 12, 16, and 4 hours. Estimate population mean life by a point estimate and a 90 percent confidence interval.

2. Repeat Exercise 1 for the data set {471, 243, 265, 826, 451, 402, 410, 37, 863, 748}.

3. What should the 5 percent guarantee point be for equipment whose life is well described by the $G(5,20)$ distribution?

4. Rework Exercise 3 but assume exponential lifetimes with a mean of 100.

5. Rework Exercise 3 assuming that α and β are unknown but that a sample of size 20 yields $\overline{X} = 48.2$ and $S = 16.3$.
   Hint.   Use method-of-moments estimates for α and β based on $\overline{x}$ and S.

5.4   INDEX-TRUNCATION (TYPE II CENSORING): WITHOUT REPLACEMENT

Whether using single- or multiple-socket testing, as described in Section 5.3, the reliability engineer who is testing good, long-lived product may have a very long wait indeed before the test results are available for analysis. The heavy tail of the negative exponential distribution is the reason. To help students realize the significance of the heavy tail, let us compare the negative exponential curve with the normal curve. They are so different in character that we hardly know what sort of comparison to make.

We present two ways of comparing them. Consider the $N(3,1)$ distribution, chosen with $\sigma = 1$ for simplicity and with $\mu = 3$ so that "essentially all" the distribution will be on the positive side of the origin. The corresponding negative exponential distribution could be said to be NGEX(1/3) which also has mean $\mu = 3$, whereas it has $\sigma = 3$. At $X = 6$, which is three standard deviations above the mean, the normal variable has a right-hand tail area $R_N(6) = .00135$. At $X = 6$ the corresponding negative exponential variable has $R_E(6) = e^{-2} = .135$, one hundred times larger. Perhaps a more valid procedure would be to compare the two distributions at their own $\mu + 3\sigma$ points; this leads to comparing $R_N(6) = .00135$ with $R_E(12) = e^{-4} = .0183$, which is still a full order of magnitude different.

This heaviness of tail means that when a fairly large number of items is put on test, a few of them may last for a very long time and inordinately delay the availability of the test results. We dare not omit the large values in the calculations, since to do so would badly bias the θ estimate. What we need is a method of terminating the testing before all elements have failed, and of correctly interpreting the incomplete information. Such a method, called *Type II censoring*, is presented here, valid when the stopping rule is as follows.

Rule.   Put n items on test at time $t = 0$ in separate testing stations (sockets). Record failure times. Terminate testing when r (a predetermined number) of the items have failed.

This rule is depicted in Figure 5.3. To an observer the failure data become available—not in the order $t_1, t_2, \ldots, t_n$, which corresponds to component number as in Section 5.3—but in order-statistic form. For ease of notation we will use the symbols

$$z_1, z_2, \ldots, z_r$$

for the ordered values that are more often designated as

$$t_{(1)}, t_{(2)}, \ldots, t_{(r)}$$

or sometimes more completely as

$$t_{1n}, t_{2n}, \ldots, t_{rn}$$

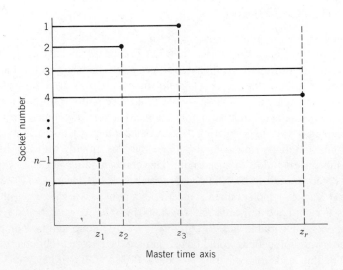

Figure 5.3   Type II (nonreplacement) censoring.

As soon as the failure at time z_r is observed, testing is terminated, and there are $n - r$ unfailed items with an unknown amount of residual life left in them. How will these data be used to construct an estimate of mean life? The process is to form the joint density function of the first r order statistics (see Appendix 5A).

$$g(z_1, \ldots, z_r) = \frac{n!}{(n - r)!} f(z_1) f(z_2) \ldots f(z_r)[1 - F(z_r)]^{n-r} \qquad (5.26)$$

$$= \frac{n!}{(n - r)!} \frac{e^{-z_1/\theta}}{\theta} \frac{e^{-z_2/\theta}}{\theta} \cdots \frac{e^{-z_r/\theta}}{\theta} (e^{-z_r/\theta})^{n-r} \qquad (5.27)$$

$$= \frac{n!}{(n - r)!} \theta^{-r} e^{-[z_1 + \cdots + z_r + (n-r)z_r]/\theta} \qquad (5.28)$$

This expression assumes simpler form if we write

$$T = z_1 + z_2 + \cdots + z_r + (n - r)z_r \qquad (5.29a)$$

$$= z_1 + z_2 + \cdots + z_{r-1} + (n - r + 1)z_r \qquad (5.29b)$$

so that

$$g(z_1, \ldots, z_r) = \frac{n!}{(n - r)!} \theta^{-r} e^{-T/\theta} \qquad (5.30)$$

The variable $\tilde{T}$ is the total lifetime of all items on test; it will be a very important quantity. It must be noted that even though the right-hand side of (5.30) is written in terms of T, it is *not* in fact the density function of $\tilde{T}$, but is still the joint density of the order statistics $\tilde{z}_1, \ldots, \tilde{z}_r$. However, it can be regarded as a likelihood function for θ in the current testing situation, so we can write

$$L(\theta) = \frac{n!}{(n - r)!} \theta^{-r} e^{-T/\theta} \qquad (5.31)$$

and use this for finding a maximum likelihood estimator for θ. The log-likelihood function is

$$\ln L = \ln n! - \ln (n - r)! - r \ln \theta - T/\theta \qquad (5.32)$$

and

$$\frac{d \ln L}{d\theta} = \frac{-r}{\theta} + \frac{T}{\theta^2} \qquad (5.33)$$

when equated to zero, yields the maximum likelihood estimator

$$\hat{\theta} = \frac{T}{r} \qquad (5.34a)$$

$$= \frac{z_1 + \cdots + z_r + (n - r)z_r}{r} \qquad (5.34b)$$

It is interesting to contemplate the structure of (5.34b). We know that if we averaged the available failure times by constructing $(z_1 + \cdots + z_r)/r$, the result would be seriously biased because it would be based on the lifetimes of the poorest units in the collection. So there is a feeling that it should be proper to include the information about the $n - r$ items that did live for time z_r and perhaps could have lived much longer. We feel, however, that dividing total life T by n would still give an underestimate of the mean life, as indeed it does. The proper divisor is the number of failed items r, and we shall see later that (5.34b) is unbiased for θ.

It is important to find the density function for the random variable $\tilde{T}$, and for this a change of variables is necessary; the old variable set $z_1, \ldots, z_r$ is replaced by a new variable set $w_1, \ldots, w_r$ defined as follows. Let

$$\left.\begin{array}{l} w_1 = n z_1 \\[4pt] w_2 = (n - 1)(z_2 - z_1) \\[4pt] w_3 = (n - 2)(z_3 - z_2) \\[4pt] \qquad \cdot \\[2pt] \qquad \cdot \\[2pt] \qquad \cdot \\[4pt] w_{r-1} = (n - r + 2)(z_{r-1} - z_{r-2}) \\[4pt] w_r = (n - r + 1)(z_r - z_{r-1}) \end{array}\right\} \qquad (5.35)$$

Since $z_1 \leq z_2 \leq z_3 \leq \ldots \leq z_r$, the w's are all nonnegative. They represent "partial lives": All n components were put on test and all lived until time z_1, so $n z_1$ is the accumulated group life until z_1. After the first failure, only $n - 1$ components were alive and they lived $z_2 - z_1$ additional hours until the second failure. Hence the accumulated group life over the second interval is $(n - 1)(z_2 - z_1)$. Similar reasoning holds for the rest of the set. Notice that if all the equations of set 5.35 are added together, the right-hand sides sum to

$$z_1 + z_2 + \cdots + z_{r-1} + (n - r + 1)z_r = T$$

and thus

$$T = w_1 + w_2 + \cdots + w_r \qquad (5.36)$$

as could be expected. The inverse transformation must be obtained from (5.35), and this is done in a sequential "domino" manner. The procedure is to solve the first equation in set 5.35 for z_1, giving $z_1 = w_1/n$. Solve the next equation in set 5.35 for z_2 so that

$$z_2 - z_1 = \frac{w_2}{n - 1}$$

$$z_2 = z_1 + \frac{w_2}{n - 1}$$

$$= \frac{w_1}{n} + \frac{w_2}{n - 1}$$

Repetition of this technique leads to the inverse solution set

$$
\left.
\begin{aligned}
z_1 &= \frac{w_1}{n} \\[2mm]
z_2 &= \frac{w_1}{n} + \frac{w_2}{n-1} \\[2mm]
&\;\;\vdots \\[2mm]
z_r &= \frac{w_1}{n} + \frac{w_2}{n-1} + \cdots + \frac{w_r}{n-r+1}
\end{aligned}
\right\}
\tag{5.37}
$$

The Jacobian of the transformation is thus

$$
J =
\begin{vmatrix}
1/n \\
1/n & 1/(n-1) \\
1/n & 1/(n-1) & 1/(n-2) \\
\vdots & \;\;\vdots \\
1/n & 1/(n-1) & 1/(n-2) \ldots 1/(n-r+1)
\end{vmatrix}
$$

$$
= [n(n-1)(n-2)\ldots(n-r+1)]^{-1} = \frac{(n-r)!}{n!}
\tag{5.38}
$$

Now since the new density g^* is developed (see Appendix 5C for information on change of variables) from the relationship

$$
g^*(w_1,\ldots,w_r) = g(z_1,\ldots,z_r)\,|J|
$$

we can write

$$
g^*(w_1,\ldots,w_r) = \frac{n!}{(n-r)!}\,\theta^{-r} e^{-(w_1+\cdots+w_r)/\theta}\frac{(n-r)!}{n!}
\tag{5.39}
$$

$$
= \frac{e^{-w_1/\theta}}{\theta}\,\frac{e^{-w_2/\theta}}{\theta}\cdots\frac{e^{-w_r/\theta}}{\theta}
\tag{5.40}
$$

   Each $\tilde{w}_i$ is a random variable whose lower bound is zero, when z_i occurs "right after" z_{i-1}, and with no upper bound since z_i can occur "a long time" after z_{i-1}. In other words, the domains of the $\tilde{w}_i$ are $0 \le \tilde{w}_i < \infty$. The $\tilde{w}_i$ are also independent, as can be proved by integrating out all but one of the $\tilde{w}_i$ in (5.40). In short, (5.40) is the product of r independent densities of the negative exponential (or gamma) type. Then, since

$$
\tilde{w}_i \sim G(1,\theta)
\tag{5.41}
$$

the total life $\tilde{T} = \tilde{w}_i + \cdots + \tilde{w}_r$ has the distribution

$$
\tilde{T} \sim G(r,\theta)
\tag{5.42}
$$

From this we know that $E(\tilde{T}) = r\theta$ so that the maximum likelihood estimate $\hat{\theta} = T/r$ is also an unbiased estimate of θ.

Most of the results of Section 5.3 can be carried over here, with r replacing n where appropriate. For instance, confidence bounds on θ can be obtained by using equations (5.21) and (5.22) where the χ^2 values have $2r$ degrees of freedom.

Example 5.3

Fifteen items are put on test at time $t = 0$ and testing is terminated as soon as the fifth failure occurs ($r = 5$). The observed failure times are at 17.5, 18.8, 21.0, 31.0, and 42.3 hours. Construct a point estimate and a 50 percent confidence interval for θ, as well as for $R(20)$.

Solution: The total life is

$$T = 17.5 + 18.8 + 21.0 + 31.0 + 42.3 + 10(42.3) = 553.6$$

and $\hat{\theta} = T/r = 553.6/5 = 110.7$. It is interesting to notice that the estimated mean life is greater than any of the observed lifetimes; to the uninitiated this must seem like a very strange sort of average. Since $\chi^2_{.25}(10) = 12.5$ and $\chi^2_{.75}(10) = 6.74$, the 50 percent confidence interval is

$$\frac{2(553.6)}{12.5} \le \theta \le \frac{2(553.6)}{6.74}$$

or

$$P(88.6 \le \theta \le 164.3) = .50$$

The point estimate for $R(20)$ is

$$\hat{R}(20) = e^{-20/110.7} = .835$$

and with 50 percent confidence we can state that

$$e^{-20/88.6} = .80 \le R(20) \le .89 = e^{-20/164.3}$$

It is interesting to compare the expected duration of the truncated test ($r < n$) with that of the complete test ($r = n$). The notation to be used is

ETT_{nn} = expected test time when n items are put on test and all are allowed to fail

ETT_{rn} = expected test time when n items are put on test and only r are allowed to fail

Obviously $\text{ETT}_{rn} = E(z_r)$ could be obtained as the expected value of the rth order statistic, but the calculation is difficult. It is much better to use a trick, letting

$$z_r = z_1 + (z_2 - z_1) + \cdots + (z_{r-1} - z_{r-2}) + (z_r - z_{r-1})$$
$$= \frac{w_1}{n} + \frac{w_2}{n-1} + \frac{w_3}{n-2} + \cdots + \frac{w_r}{n-r+1} \qquad (5.43)$$

But $E(\tilde{w}_1) = E(\tilde{w}_2) = \ldots = E(\tilde{w}_r) = \theta$, so that

$$\text{ETT}_{rn} = E(\tilde{z}_r) = \theta\left(\frac{1}{n} + \frac{1}{n-1} + \cdots + \frac{1}{n-r+1}\right) \tag{5.44}$$

We notice that it is necessary to know θ in order to compute the average length of an experiment performed to estimate θ. Naturally, a guess of some kind will be needed in (5.44). Notice that the formula contains r terms, equal to the number of items that are allowed to fail. Table 5.1 shows values of ETT_{rn} for $\theta = 1$, and these can be used as multipliers for other values of θ.

It is worthwhile to compare the expected test time of the truncated case with the expected test time for nontruncated testing in two different ways. This is done in Tables 5.2 and 5.3. In Table 5.2, we assume that a fixed number of items n is to be tested, and we look at how various r values increase the test time. For instance, for $n = 10$ we see that, if we can settle for $r = 1$, we need wait only 3.4 percent as long as if we waited for all ten items to fail. If we can settle for $r = 5$, we need wait only 22.1 percent as long as if we waited for all ten items to fail. But this knowledge will not be of interest if it has already been decided that, in order to reach a certain level of "validity," the r value must be fixed (at $r = 4$, say). Then we would look at the $r = 4$ column of Table 5.3. There we see that 4-out-of-4 requires more than two lifetimes, that 4-out-of-10 will require only 23 percent as much test time as 4-out-of-4, and so on.

It has been tacitly assumed in the foregoing discussion that the n items have been put on test in n sockets, but it may be that the items have been put into service and that the failure times are data gathered incidentally during a period of use. In either case, it is of interest to inquire about the distribution of interfailure times $\{z_i - z_{i-1}\}$ under the usual CFR assumption. The times $\{z_i\}$ are *not* Poisson arrival times because at the beginning of the observation period there are n items in service so the group failure

TABLE 5.1   Expected Test Time ETT_{rn} for $\theta = 1$

n \ r	1	2	3	4	5	10	15	20	25
1	1.000								
2	0.500	1.500							
3	0.333	0.833	1.833						
4	0.250	0.583	1.083	2.083					
5	0.200	0.450	0.783	1.283	2.283				
10	0.100	0.211	0.336	0.479	0.646	2.929			
15	0.067	0.138	0.215	0.298	0.389	1.035	3.318		
20	0.050	0.103	0.158	0.217	0.280	0.669	1.314	3.598	
25	0.040	0.082	0.125	0.171	0.218	0.489	0.887	1.533	3.816

$$\text{ETT}_{50,50} = 4.499$$
$$\text{ETT}_{100,100} = 5.187$$

TABLE 5.2   ETT$_{rn}$/ETT$_{nn}$, Relative Time Savings for Index-Truncated Test with n Fixed

n \ r	1	2	3	4	5	10	15	20	25
1	1.000								
2	0.333	1.000							
3	0.182	0.455	1.000						
4	0.120	0.280	0.520	1.000					
5	0.088	0.197	0.343	0.562	1.000				
10	0.034	0.072	0.115	0.164	0.221	1.000			
15	0.020	0.042	0.064	0.090	0.117	0.312	1.000		
20	0.014	0.029	0.044	0.060	0.078	0.186	0.365	1.000	
25	0.010	0.021	0.033	0.045	0.057	0.131	0.232	0.402	1.000

rate is $n\lambda$ and hence the first failure time z_1 is distributed NGEX($n\lambda$). After the first failure there are only $n-1$ items in service so the group failure rate is $(n-1)\lambda$ and the interarrival time $(z_2 - z_1)$ is distributed NGEX[$(n-1)\lambda$]. In like manner, after i failures have occurred, there are $n-i$ items still in service and the interfailure time $(z_{i+1} - z_i)$ is distributed NGEX[$(n-i)\lambda$].

TABLE 5.3   ETT$_{rn}$/ETT$_{rr}$, Relative Time Savings for Index-Truncated Tests with r Fixed

n \ r	1	2	3	4	5	10	15	20	25
1	1.000								
2	0.500	1.000							
3	0.300	0.555	1.000						
4	0.250	0.389	0.591	1.000					
5	0.200	0.300	0.427	0.616	1.000				
10	0.100	0.141	0.183	0.230	0.283	1.000			
15	0.067	0.092	0.117	0.143	0.170	0.353	1.000		
20	0.050	0.069	0.086	0.104	0.123	0.228	0.396	1.000	
25	0.040	0.055	0.069	0.082	0.095	0.170	0.267	0.426	1.000

As the number of failures accumulates, the times between failures tend to increase, since they are originating from a stepwise changing Poisson process with ever-decreasing failure rates. Such failure times, when plotted on a time line, do not appear like a sample from a uniform distribution, as with a true Poisson process. Instead, they look "front-end-loaded," as in Figure 5.4.

Example 5.4

A new type of unit is being produced, for which it is hoped that the mean life will be in the neighborhood of 2000 hours. Accelerated testing is possible, whereby the results can be obtained in one-fifth the "real-world" time. Ten units can be spared for the test procedure and there are enough sockets to accommodate them. What should the truncation index r be if management wants the test results in one week's time (168 hours)?

Solution: If θ is really near 2000 hours, we can treat it as though it is 400 hours under accelerated testing. We want to find r (for $n = 10$) so that $\text{ETT}_{r,10} = 168/400 = .420$. Table 5.1 gives $\text{ETT}_{3,10} = .336$ and $\text{ETT}_{4,10} = .479$, so r of 3 or 4 will do.

Comment.   The statement that "r of 3 or 4 will do" is not a very precise answer, but there are many imprecise things in the problem, including the 5:1 ratio of the

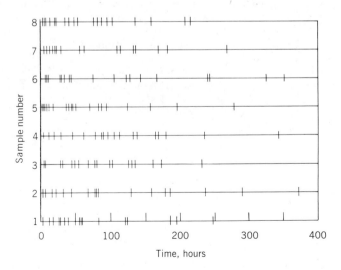

Figure 5.4   The attrition model of nonreplacement multiple-socket testing.

accelerated test, as well as the presumed θ value of 2000 hours. Even so, a rough estimate in a situation like this can be helpful. As the test proceeds, a tentative θ can be used to replace the pure guess.

Readers have probably realized that the nonreplacement sampling treated in this section is rather inefficient. It would be much better to replace each failed item immediately and keep the sockets full at all times; this procedure is considered in the next section. But there may be situations in which nonreplacement sampling is cheaper; if failure times can be recorded automatically, the testing laboratory can be left unattended except for occasional checks to see whether r failures have occurred yet.

EXERCISES 5.4

1. Master the details of proof leading to Equations 5.28, 5.40, and 5.42.

2. A sample of five items is to be subjected to a nonreplacement life test until three fail. The test results in the following failure times: 0.7, 8.8, and 14.5 hours. Assuming CFR components, estimate the MTTF and reliability for a ten-hour mission using

   (a) a point estimate,

   (b) a 95 percent lower confidence bound, and

   (c) a 95 percent two-sided confidence interval.

3. Repeat Exercise 2 on the assumption that 20 items are tested with $r = 10$, resulting in failure times (in hours) of 2.0, 5.9, 9.3, 10.2, 14.3, 15.1, 18.9, 30.0, 35.5, and 40.3. Does the CFR assumption seem reasonable? (Use the technique of Section 2.5.)

4. Verify that equation set 5.37 can be obtained from set 5.35.

5. How much time do you think was saved in the previous two problems by suspending testing at the given r rather than waiting until all items had failed?

6. A batch of ten items was put on test and testing suspended at the fourth failure. The failure times were 380, 506, 1020, and 1500 hours. Assuming CFR components, estimate mean life and reliability for a 5000-hour mission using

   (a) a point estimate.

   (b) a 75 percent lower confidence limit.

   (c) Approximately how long might the life test have been expected to last if testing had continued until all items failed?

7. Repeat parts a and b of Exercise 6 for a 1000-hour mission.

8. Adapt Exercise 6 using the data from Exercise 2.5.3, truncating at $r = 3$ failures. Use mission time ten hours. Compare θ from the full set with θ from the truncated set.

9. Adapt Exercise 6 using the data from Exercise 2.5.4, truncating at $r = 3$ failures. Use mission time 50 hours. Compare θ from the full set with θ from the truncated set.

10. Adapt Exercise 6 using the data from Exercise 2.5.5, truncating at $r = 5$ failures. Use mission time ten hours. Compare θ from the full set with θ from the truncated set.

11. Adapt Exercise 6 using the data from Exercise 2.5.6, truncating at $r = 3$ failures. Use mission time ten hours. Compare θ from the full set with θ from the truncated set.

12. Adapt Exercise 6 using the data from Exercise 2.5.7, truncating at $r = 2$, $r = 3$, and $r = 4$ failures. Use mission time 100 hours. Compare θ from the full set with θ from the truncated sets.

5.5 INDEX TRUNCATION (TYPE II CENSORING): WITH REPLACEMENT

We consider here the situation in which there are n sockets, so that n items are put on test at time $t = 0$. Failed items are replaced immediately so that the test bank is full at all times. This is depicted graphically in Figure 5.5. The assumption, as before, is that testing is terminated at the rth failure, with r a predetermined number. Under the assumption of CFR components, each socket is the scene of a Poisson renewal

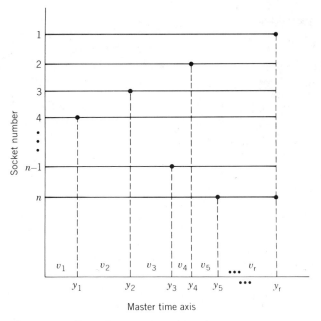

Figure 5.5 Type II censoring with replacement.

process with intensity $\lambda = 1/\theta$. Let $N_i(t)$ be the failure count at the ith socket by time t. On the master time axis, the observed failure count by time t will be

$$N(t) = N_1(t) + \cdots + N_n(t) \tag{5.45}$$

and this is a Poisson variable with intensity $n\lambda = n/\theta$. The points $y_1, y_2, \ldots, y_r$ are Poisson arrival times, and the interarrival times

$$\left.\begin{aligned}
v_1 &= y_1 \\
v_2 &= y_2 - y_1 \\
&\vdots \\
v_r &= y_r - y_{r-1}
\end{aligned}\right\} \tag{5.46}$$

all have the NGEX($n\lambda$) distribution, or alternatively $G(1, \theta/n)$. Now we define

$$\left.\begin{aligned}
w_1^* &= nv_1 = ny_1 \\
w_2^* &= nv_2 = n(y_2 - y_1) \\
&\vdots \\
w_r^* &= nv_r = n(y_r - y_{r-1})
\end{aligned}\right\} \tag{5.47}$$

We see that $w_1^*, \ldots, w_r^*$ are the partial lives lived by all components over the intervals $(0, y_1), (y_1, y_2), \ldots, (y_{r-1}, y_r)$. Furthermore, recalling that $X \sim G(\alpha, \beta)$ implies $CX \sim G(\alpha, C\beta)$, we have the immediate result that

$$w_i^* \sim G(1, \theta) \tag{5.48}$$

just as for the partial lives w_i when there is no replacement. Letting

$$\tilde{T} = \tilde{w}_1^* + \cdots + \tilde{w}_r^*$$

be the total life lived by all components, we know that $\tilde{T} \sim G(r, \theta)$, as before, but note that the formula for T is different here because

$$\begin{aligned}
T &= w_1^* + \cdots + w_r^* \\
&= nv_1 + \cdots + nv_r \\
&= ny_1 + n(y_2 - y_1) + \cdots + n(y_r - y_{r-1}) \\
&= ny_r
\end{aligned} \tag{5.49}$$

The expected test time is

$$\begin{aligned}
\mathrm{ETT}_{rn}^* &= E(\tilde{y}_r) \\
&= E[(\tilde{y}_1) + (\tilde{y}_2 - \tilde{y}_1) + \cdots + (\tilde{y}_r - \tilde{y}_{r-1})] \\
&= E\left(\frac{w_1^*}{n} + \cdots + \frac{\tilde{w}_r^*}{n}\right) \\
&= \frac{r\theta}{n}
\end{aligned} \tag{5.50}$$

TABLE 5.4 Time Saving Ratio ($ETT^*_{rn} \div ETT_{rn}$)

n	r 1	2	3	4	5	10	15	20	25
1	1.000								
2	1.000	.667							
3	1.000	.800	.546						
4	1.000	.858	.692	.480					
5	1.000	.889	.766	.624	.438				
10	1.000	.948	.893	.835	.774	.341			
15	1.000	.966	.930	.895	.857	.644	.301		
20	1.000	.971	.949	.922	.893	.747	.571	.278	
25	1.000	.976	.960	.936	.917	.803	.676	.522	.262

It is worthwhile to compare the factor r/n when there is replacement with the factor $[1/n + 1/(n-1) + \cdots + 1/(n-r+1)]$ when there is no replacement, to see how much time can be saved by keeping the test stations full at all times. This is done for selected values in Table 5.4.

Notice that in the replacement situation, the total number of items that undergo testing is $n + r - 1$, not n as with nonreplacement. It is sometimes awkward to know what to designate as a sample size in these situations. In traditional sampling, the sample size (n or $n + r - 1$ in our cases) is the determiner of how powerful a test is, or how wide the confidence band is. In this life testing situation, the failure count r is the determiner of these things, so it is tempting to speak of r as the sample size. Instead, we will call it the destruction index or the termination number.

Another reason for wanting to treat r like a sample size is that, since in life testing we are talking about destructive testing, no matter how many items are put on test only r items are destroyed. The remaining $n - r$ (or $n - 1$) items can be returned to inventory "as good as new," because of the forgetfulness property of the exponential distribution. The size of r determines the power of the statements we make about θ, that is, how good our information is. The relative size of n to r merely determines how soon that information becomes available. Notice that r can exceed n in replacement testing.

Example 5.5

Management is willing to allow the destruction of only $r = 7$ units for a life test. If testing is done sequentially (only one socket used), the expected test time is 7θ. If testing can be done simultaneously with seven sockets, the expected test time is reduced to $\theta(1 + \cdots + 1/7) = 2.593\,\theta$. (When testing is done with all seven sockets, it must be without replacement, since there are no spares for replacing failed items.) Would there be any advantage in using fewer sockets (say $n = 4$) and doing replacement testing?

With such a limited number of units available, there is no way that we can keep all the sockets full until the seventh failure, so the formula ETT $= r\,\theta/n$ is meaningless here. See Figure 5.6, where four items were placed on test at time $t = 0$, the fifth unit at time $t = y_1$, the sixth at time $t = y_2$, and the seventh at time $t = y_3$. There are no more items available for replacement purposes, so what we have here is "neither fish nor fowl."

This situation emphasizes the fact that if replacement sampling is to be carried out in n sockets, there must be $n + r - 1$ units available. Or, stated in another way, if n' units are available to use, of which only r can be spared for destruction, replacement testing can be carried out only if the number of sockets used is no greater than $n' - r + 1$. If we can persuade management to "lend" us extra units, to be returned "as good as new," then with all n sockets kept full the expected test time is reduced as shown in Table 5.5.

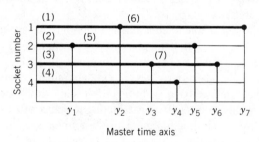

Figure 5.6 A failed attempt at replacement testing ($n = 4$, $r = 7$).

TABLE 5.5   Reduction in Test Time for n Busy Test Sockets

n	$\text{ETT}_{7,n}^{*}$	n	$\text{ETT}_{7,n}^{*}$
1	7.00θ	8	0.88θ
2	3.50θ	9	0.78θ
3	2.33θ	10	0.70θ
4	1.75θ	11	0.64θ
5	1.40θ	12	0.58θ
6	1.17θ	13	0.54θ
7	1.00θ	14	0.50θ

EXERCISES 5.5

1. Master the details of proof in Equations 5.45 through 5.50.

2. Testing is conducted with *replacement* of failed items; it will be terminated at the seventh failure, and $n = 20$ items are to be kept on test at all times. The failure times are 42, 59, 79, 129, 160, 172, and 188 hours. Compute a point estimate and a 95 percent confidence interval for (a) the MTTF and (b) the reliability of a mission of 600 hours.

Comment. The true mean is 600 hours.

3. Repeat Exercise 2 with $r = 5$. The failure times are 4.0, 7.0, 7.6, 9.0, and 9.1. Let the mission length be ten hours.

Comment. The true mean is 40.

4. Repeat Exercise 2 with $r = 11$. The failure times are 18, 81, 129, 138, 154, 187, 227, 297, 300, 324, and 402 hours. Let the mission length be 500 hours.

Comment. The true mean is 1000.

5. Rework Exercise 2 using a 50 percent confidence level.

6. Rework Exercise 3 using a 50 percent confidence level.

7. On the basis of the stated true mean value, how long would you have expected the testing to last in Exercise 2?

8. Rework Exercise 7 based on the data in Exercise 3.

9. Rework Exercise 7 based on the data in Exercise 4.

10. After studying Example 5.5, decide whether it is possible to carry out a replacement test using $n = 5$ sockets and $r = 8$ as termination number. Make a sketch of various possible outcomes (like Figure 5.6) of the test.

5.6 TIME TRUNCATION (TYPE I CENSORING): WITH REPLACEMENT

Time truncation was dealt with in a preliminary way in Section 5.2, in the nonreplacement situation which gives rise to the binomial distribution. With replacement sampling, each socket is the location of a Poisson process with intensity λ, and an observer on the master time axis sees a Poisson process with intensity $n\lambda$. Experimentation is carried out for a fixed period of time t_0, so the random variable is $N(t)$, the total number of failures from all sockets. We can write

$$P[N(t_0) = r] = \frac{e^{-n\lambda t_0}(n\lambda t_0)^r}{r!}, \quad r = 0, 1, 2, \ldots \quad (5.51)$$

For a single experiment the failure count r is the MLE of the Poisson parameter $n \lambda t_0$, so we set

$$n \hat{\lambda} t_0 = r$$

Then

$$\hat{\lambda} = \frac{r}{n t_0}$$

or

$$\hat{\theta} = \frac{n t_0}{r} \tag{5.52}$$

Recognizing that $n t_0 = T$ is the total life of all items on test, we find ourselves with the familiar formula

$$\hat{\theta} = \frac{T}{r}$$

Confidence limits for θ can also be constructed, but the method of doing so is different from before. In Type II censoring (index truncation), the confidence bounds were formed by applying the familiar pivot method to a probability statement based on the gamma distribution of the total life T. Here the *total life* is a *constant*; it is the failure count $N(t) = r$ which is the random variable, so the procedure is different. Based on the Poisson distribution of the failures, we have

$$P[N(t_0) \leq r] = \sum_{k=0}^{r} \frac{e^{-n \lambda t_0}(n \lambda t_0)^k}{k!} \tag{5.53}$$

$$= 1 - \sum_{k=r+1}^{\infty} \frac{e^{-n \lambda t_0}(n \lambda t_0)^k}{k!} \tag{5.54}$$

Now we invoke the relationship, which was proved in Section 2.6, between the tails of the Poisson and gamma distributions. Using $G(m, 2) = \chi^2(2m)$, we must choose m as the lower Poisson summation index $r + 1$, and choose the gamma integration limit so that x/β is the Poisson parameter. Indicating the χ^2 density by $h(\cdot)$, we require

$$P[N(t_0) \leq r] = 1 - \int_0^{2n \lambda t_0} h(\chi^2; 2r + 2) \, d\chi^2 \tag{5.55}$$

$$= \int_{2n \lambda t_0}^{\infty} h(\chi^2; 2r + 2) \, d\chi^2 \tag{5.56}$$

$$= P[\chi^2(2r + 2) \geq 2n \lambda t_0] \tag{5.57}$$

$$= P\left[\chi^2(2r + 2) \geq \frac{2n t_0}{\theta}\right] \tag{5.58}$$

Since (5.53) is continuous in the Poisson parameter $n\lambda t_0$—that is, in θ—we can find a value of θ, say θ_1, for which (5.58) is exactly equal to $\alpha/2$. For θ values smaller than θ_1 we have $2nt_0/\theta > 2nt_0/\theta_1$ and the right-hand tail area of (5.58) will be less than $\alpha/2$. Thus we can say that θ_1 is the smallest "reasonable" value of θ, because any smaller value is associated with $P[N(t_0) \le r] < \alpha/2$. But if θ_1 is the solution of

$$P\left[\chi^2(2r+2) \ge \frac{2nt_0}{\theta_1}\right] = \frac{\alpha}{2} \tag{5.59}$$

it must be true that

$$\frac{2nt_0}{\theta_1} = \chi^2_{\alpha/2}(2r+2)$$

and thus the smallest "reasonable" value of θ is

$$\theta_1 = \frac{2nt_0}{\chi^2_{\alpha/2}(2r+2)} \tag{5.60}$$

In a similar manner, we can write

$$P[N(t_0) \ge r] = \sum_{k=r}^{\infty} \frac{e^{-n\lambda t_0}(n\lambda t_0)^k}{k!} \tag{5.61}$$

$$= \int_0^{2n\lambda t_0} h[\chi^2(2r)]d\chi^2$$

$$= P[\chi^2(2r) \le 2n\lambda t_0] \tag{5.62}$$

$$= P\left[\chi^2(2r) \le \frac{2nt_0}{\theta}\right] \tag{5.63}$$

We can find a θ value, say θ_2, for which this will exactly equal $\alpha/2$. For any larger value of θ, it will be true that $2nt_0/\theta < 2nt_0/\theta_2$ and the left-hand tail area in (5.63) will be less than $\alpha/2$. Thus we may say that θ_2 is the largest "reasonable" value of θ, because any larger value is associated with $P[N(t_0) \ge r] < \alpha/2$. But θ_2 is the solution of

$$P\left[\chi^2(2r) \le \frac{2nt_0}{\theta_2}\right] = \frac{\alpha}{2}$$

so it must be true that

$$\frac{2nt_0}{\theta_2} = \chi^2_{1-\alpha/2}(2r)$$

or

$$\theta_2 = \frac{2nt_0}{\chi^2_{1-\alpha/2}(2r)} \tag{5.64}$$

We can summarize with the following statements.

From (5.64) if $N(t_0) \geq r$ there is a small probability ($< \alpha/2$) that θ exceeds θ_2. From (5.60) if $N(t_0) \leq r$ there is a small probability ($< \alpha/2$) that θ is less than θ_1. If $N(t_0) = r$ then both these statements must be true. Hence a $100(1 - \alpha)$ percent confidence interval for θ is of the form

$$\theta_L = \frac{2nt_0}{\chi^2_{\alpha/2}(2r + 2)} \leq \theta \leq \frac{2nt_0}{\chi^2_{1-\alpha/2}(2r)} = \theta_U \tag{5.65}$$

If only a lower bound is desired, it can be taken as

$$\theta_L^* = \frac{2nt_0}{\chi^2_{\alpha}(2r + 2)} \tag{5.66}$$

It is interesting to note that (5.66) can be used even though no failures occur during the experiment.

Example 5.6

Ten sockets are used in a life test with replacement, and the test continues for one week (168 hours). During this period there are (a) three failures, (b) no failures. Find point estimates and 50 percent confidence intervals in the two cases.

Solution:

(a) We have $\hat{\theta} = (10)(168)/3 = 560$ hours and

$$\theta_L = \frac{2(10)(168)}{\chi^2_{.25}(8)} = \frac{3360}{10.219} = 328.8 \text{ hours}$$

$$\theta_U = \frac{2(10)(168)}{\chi^2_{.75}(6)} = \frac{3360}{3.455} = 972.5 \text{ hours}$$

(b) No point estimate is possible—nor is an upper bound—when $r = 0$, but a 50 percent lower confidence bound is

$$\theta_L^* = \frac{3360}{\chi^2_{.5}(2)} = \frac{3360}{1.386} = 2424.2 \text{ hours}$$

Recall that in this section we have been dealing with time truncation *replacement* tests, in which it is not necessary to note failure times as long as failed items are instantly replaced. In Section 5.2 we discussed time truncation *without replacement* and with the failure times not noted. Each item tested constituted a Bernoulli trial, and all that mattered was survival or nonsurvival for the test period t_0. What if the failure times had been noted—would it have increased the validity of the estimates? The derivation of exact results is not possible, but it is suggested in an army course book (U.S. Army, 1970, page IV-55) that the formulae of this section can be used as approximations to the exact expressions given in Section 5.2. Let us demonstrate the results of both methods with an example.

Example 5.7

Twenty items were put on test for a period of 75 hours. Failed items were not replaced. Failures occurred at

   17, 25, 27, 30, 50, 66 hours

Construct a point estimate and a 90 percent confidence estimate of θ, assuming the components are CFR.

Solution:
Method A: Using the method of Section 5.2 (with $n = 20$, $r = 6$), we have

$$\hat{\theta} = \frac{-75}{\ln(14/20)} = 210.28$$

For the lower bound we find, from Equation 5.5,

$$R_L = (1 + 7F_2/14)^{-1} = .492$$

since $F_2 = F_{.05}(14, 28) = 2.07$. From Equation 5.10 we have the lower bound

$$\theta_L = \frac{-75}{\ln(.492)} = 105.74$$

Similarly, the upper bound is obtained using (5.7), whereby

$$R_U = 2.47(2.47 + 6/15)^{-1} = .861$$

since $F_1 = F_{.05}(30, 12) = 2.47$. Then from (5.10)

$$\theta_U = \frac{-75}{\ln(.861)} = 501.13$$

Method B: Using the method of this section, which will be an approximation, we get

$$\hat{\theta} = \frac{T}{r} = \frac{17 + 25 + 27 + 30 + 50 + 66 + 14(75)}{6} = \frac{1265}{6} = 210.83$$

Since $\chi^2_{.05}(14) = 23.685$ and $\chi^2_{.95}(12) = 5.226$, we have

$$\theta_L = \frac{2(1265)}{23.685} = 106.82$$

TABLE 5.6 Comparison of Two Estimation Methods

Statistic	A. Exact Method (Section 5.2)	B. Approximate Method (Section 5.6)
$\hat{\theta}$	210.28 hours	210.83 hours
θ_L	105.74	106.82
θ_U	501.13	484.12
	Some information ignored	All information used

and

$$\theta_U = \frac{2(1265)}{5.226} = 484.12$$

To compare results more easily, we summarize them in Table 5.6.

We see that here the two point estimates are almost identical. (If n is very small, will this be as likely?) The lower confidence bounds θ_L are also very close. The confidence interval of method B is narrower. Usually, "narrower is better" where confidence intervals are concerned, but here we do not know whether the improvement is real. It is unknown whether the apparent improvement is due to using more information, the failure times, or to a crudeness in the approximation. My personal vote, all things considered (including ease of calculation), is for method B. Limited experience with a few other examples has shown that the greatest discrepancy in the two methods seems to occur in the upper confidence bound.

EXERCISES 5.6

1. A batch of items is to be used in a replacement life test with $n = 20$ sockets and termination at 50 hours. The test results in the following failure times in hours: { 12.3, 17.1, 20.3, 20.5, 37.1, 47.8}. Assume CFR components.

   (a) Construct point and 90 percent confidence estimates for MTTF.

   (b) Do the same for $R(50)$, the reliability for a mission 50 hours long.

2. Repeat the previous problem assuming that the test was a nonreplacement test and using both methods indicated in this section. Compare the results.

3. A replacement test is carried out for 1000 hours, keeping five components constantly on test. During the test, failures occur at 85, 203, 502, 598, and 869 hours. Assume the exponential distribution for lifetimes.

   (a) Construct point and 80 percent confidence estimates for the mean life.

   (b) Do the same for $R(100)$.

4. Repeat the previous problem assuming that the test was a nonreplacement test and using both methods indicated in this section. Compare the results.

5. Repeat Exercises 1 and 2 using the data set for a 700-hour test: { 160, 295, 386.8, 412.3, 415.0, 473.5, 611.5}.

6. Repeat Exercises 1 and 2 using the data set from a 1000-hour test: { 129, 193, 219, 356, 515, 668, 806, 834, 919, 990}.

7. Repeat Exercises 1 and 2 using the data set from a 1000-hour test: {4, 55, 126, 138, 175, 618, 754, 880, 918, 938}.

5.7 HYPOTHESIS TESTING FOR CFR COMPONENTS AND TYPE II CENSORING

In the preceding sections, the emphasis was on forming estimates, point or interval, on the mean life θ of a CFR component. In this section the emphasis will be on a different aspect of the same problem, that of deciding whether the mean life of a product is good enough. The notation we will use is

$\theta_0 =$ an acceptably good mean life,

$\theta_1 =$ an unacceptably poor mean life ($< \theta_0$),

$\alpha =$ the (low) probability of rejecting product of mean life θ_0,

$=$ "producer's risk,"

$\beta =$ the (low) probability of accepting product with mean life θ_1,

$=$ "consumer's risk,"

$C =$ a suitably chosen constant.

The test procedure is to be as follows.

The number of items to be put on test is n. As soon as r of them have failed, the sample mean life $\hat{\theta}$ is computed. If $\hat{\theta} \geq C$, the lot is accepted. If $\hat{\theta} < C$, the lot is rejected.

Letting $P_a(\theta) = P(\text{accept lot}|\theta) = P(\hat{\theta} \geq C|\theta)$, we want to choose the details of the test so that

$$P_a(\theta_0) = 1 - \alpha \atop P_a(\theta_1) = \beta \Biggr\} \qquad (5.67)$$

The plot of $P_a(\theta)$ against θ is called the OC (operating characteristic) curve, and the requirements of (5.67) are equivalent to constructing the OC curve to pass through the points $(\theta_0, 1 - \alpha)$ and (θ_1, β). If it is not possible to force the OC curve exactly through those two points, because of the discreteness of some of the quantities, it is traditional to replace (5.67) with

$$\left.\begin{array}{l} P_a(\theta_0) = 1 - \alpha \\ P_a(\theta_1) \le \beta \end{array}\right\} \qquad (5.68)$$

Of course, the testing procedure can also be phrased in the usual statistical terminology of the simple-versus-simple hypothesis,

$H_0: \quad \theta = \theta_0$

$H_1: \quad \theta = \theta_1$

with $\alpha = P$ (Type I error) $= P$ (reject $H_0 \mid H_0$ true) and $\beta = P$ (Type II error) $= P$ (accept $H_0 \mid H_1$ true). The ideal OC curve is a step function of the form shown in Figure 5.7a, but such a curve obviously cannot be obtained through sampling; the best that can be realized is a curve of the form shown in Figure 5.7b. The questions to be asked are (a) how shall the criterion value C be chosen, (b) how many units will be observed (n = ?), and (c) how many units shall be destroyed (r = ?). Question a depends on the α level chosen, since

$$\begin{aligned} \alpha &= P(\text{reject } H_0 \mid H_0 \text{ true}) \\ &= P(\theta < C \mid H_0 \text{ true}) \\ &= P\left(\frac{T}{r} < C \mid H_0 \text{ true}\right) \\ &= P(T < rC \mid H_0 \text{ true}) \\ &= P\left(\frac{2T}{\theta} < \frac{2rC}{\theta} \mid H_0 \text{ true}\right) \\ &= P\left[\chi^2(2r) < \frac{2rC}{\theta} \mid \theta = \theta_0\right] \\ &= P\left[\chi^2(2r) < \frac{2rC}{\theta_0}\right] \end{aligned}$$

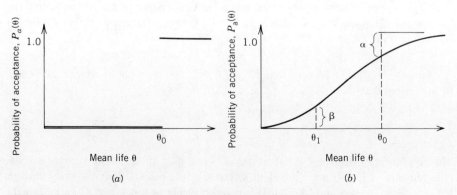

Figure 5.7 Operating characteristic curves: (a) Ideal, (b) Actual.

Then C must be chosen so that

$$\frac{2r\,C}{\theta_0} = \chi^2_{1-\alpha}(2r)$$

(5.69)

or

$$C = \theta_0 \frac{\chi^2_{1-\alpha}(2r)}{2r}$$

(5.70)

Notice that C depends only on θ_0, α, and the termination number r, and not at all on θ_1, β, or sample size n. The size of n, as we saw earlier, determines how soon a decision may be made but does not affect the power of that decision. The independence of (5.70) from θ_1 and β means, of course, that the same value of C can be used for the simple-versus-composite hypothesis that

$$H_0 : \theta = \theta_0$$
$$H_1 : \theta < \theta_0$$

To force the OC curve through the point (θ_1, β), we require that

$$\beta = P(\text{accept } H_0 \mid H_1 \text{ true})$$
$$= P(\hat{\theta} \geq C \mid H_1 \text{ true})$$
$$= P\left[\chi^2(2r) \geq \frac{2r\,C}{\theta}\,\middle|\,\theta = \theta_1\right]$$
$$= P\left[\chi^2(2r) \geq \frac{2r\,C}{\theta_1}\right]$$

(5.71)

Thus it is required that C be chosen so that

$$\frac{2r\,C}{\theta_1} = \chi^2_{\beta}(2r)$$

(5.72)

and this means that

$$C = \theta_1 \frac{\chi^2_{\beta}(2r)}{2r}$$

(5.73)

Equating this with Equation 5.70, we have the requirement that r must be chosen to satisfy the condition

$$\frac{\theta_0}{\theta_1} = \frac{\chi^2_{\beta}(2r)}{\chi^2_{1-\alpha}(2r)}$$

(5.74)

If the relation cannot be exactly satisfied, and it usually cannot, the equality should be replaced by the inequality

$$\frac{\theta_0}{\theta_1} > \frac{\chi^2_{\beta}(2r)}{\chi^2_{1-\alpha}(2r)}$$

(5.75)

To evaluate $P_a(\theta)$ for values other than θ_0 and θ_1, we can use the following trick: since $\tilde{T} \sim G(r, \theta)$, it is true that

$$P_a(\theta) = P(\hat{\theta} \geq C|\theta)$$
$$= P(\tilde{T} \geq rC|\theta)$$

$$= \int_{rC}^{\infty} dG(r, \theta) \qquad (5.76)$$

Then, invoking the familiar Poisson–gamma relationship of Section 2.6, we have

$$P_a(\theta) = 1 - \int_0^{rC} dG(r, \theta)$$

$$= 1 - \sum_{k=r}^{\infty} \text{poi}\left(k|\frac{rC}{\theta}\right) \qquad (5.77)$$

$$= \sum_{k=0}^{r-1} \text{poi}\left(k|\frac{rC}{\theta}\right) \qquad (5.78)$$

Example 5.8

Management wishes to accept with probability .95 any product with a mean life of 10,000 hours, and it has been decided that $r = 5$ items can be spared for destruction. How should the test be carried out? What is the probability of accepting product with a mean life of only 4000 hours?

Solution: We calculate $C = [10,000\chi^2_{.95}(10)]/10 = 1000(3.940) = 3490$. Any lot with a sample mean life at least as great as 3940 hours will be accepted. For $\theta_1 = 4000$ we have

$$P_a(4000) = P[T \leq 5(3940)] = P[T \leq 19,700]$$

$$= \sum_{k=0}^{4} \text{poi}\left(k; \frac{19,700}{4000}\right) = \sum_{k=0}^{4} \text{poi}(k; 4.925)$$

$$= e^{-4.925}\left[1 + 4.925 + \frac{(4.925)^2}{2} + \frac{(4.925)^3}{6} + \frac{(4.925)^4}{24}\right]$$

$$= .4538$$

The Poisson sum could be obtained more easily by interpolation from a cumulative Poisson table like Table T8.

In this example we calculated a single extra point on the OC curve, for $\theta = 4000$. Sometimes we are more interested in the general shape of the curve than in particular values. In such instances the following technique is helpful. Write $P_a(\theta)$ as

$$P_a(\theta) = P(\tilde{T} \geq rC|\theta)$$

$$= P\left(\frac{2\tilde{T}}{\theta} \geq \frac{2rC}{\theta}\bigg|\theta\right)$$

$$= P\left[\chi^2(2r) \geq \frac{2rC}{\theta}\right]$$

Now set $P_a(\theta) = \gamma$ for tabled probabilities $\{\gamma\}$ so that

$$\frac{2rC}{\theta} = \chi_\gamma^2(2r) \tag{5.79}$$

and solve for θ as

$$\theta_\gamma = \frac{2rC}{\chi_\gamma^2(2r)} \tag{5.80}$$

Example 5.9

Using the data from Example 5.8 where $\theta_0 = 10,000$, $r = 5$ and $C = 3940$, we set γ equal in turn to all the probability values in the χ^2 table (Table T3) for ten degrees of freedom. Thus $\gamma = \{.999, .995, .99, .975, .95, .90, .75, .50, .25, .10, .05, .025, .01, .005, .001\}$. The corresponding values of $\chi^2(10)$ are $\{1.497, 2.156, 2.558, 3.247, 3.940, 4.865, 6.737, 9.342, 12.549, 15.987, 18.307, 20.483, 23.209, 25.188, 29.588\}$, and the associated θ values are $\{26,319, 18,275, 15,403, 12,134, 10,000, 8099, 5848, 4218, 3140, 2465, 2152, 1924, 1698, 1564, 1332\}$. The graph is shown in Figure 5.8. This simple computer plot is quite adequate for understanding

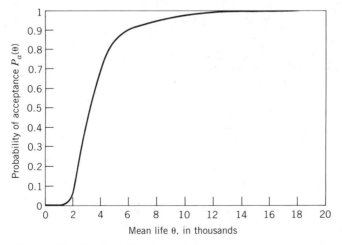

Figure 5.8 OC curve for Example 5.9.

how the sampling plan responds in a general way to various possible values of mean life.

Example 5.10

A manufacturer wishes to accept with probability .95 any lot with a mean life of 10,000 hours and to reject with probability .75 any lot with a mean life as poor as 4000 hours. Design the test.

Solution: Since $\alpha = .05$, $\beta = .25$, and $\theta_0 / \theta_1 = 2.5$, we consult a χ^2 table (Table T3) to construct Table 5.7 for determining r. The discrimination ratio $\theta_0 / \theta_1 = 2.5$ lies between successive entries, so r must be chosen as the larger value, $r = 8$. The acceptance criterion C is $10,000(7.96)/16 = 4975$ hours.

Comment. We see that both $\chi^2_{1-\alpha}$ and χ^2_β increase with increasing degrees of freedom, and that the distance between them widens because variance increases with degrees of freedom. But the ratio decreases monotonically so that any discrimination ratio will have a solution in r.

In the event that the discrimination ratio is small, with θ_1 not much less than θ_0, it will be necessary to have a large value of r in order to achieve the required power. Since the usual χ^2 tables seldom go beyond 30 degrees of freedom ($r = 15$) and even the best do not go beyond 100 degrees of freedom ($r = 50$) except in gross steps, some approximation will be necessary in (5.75). Two approximations for χ^2 are commonly used,

$$\sqrt{2\chi^2} - \sqrt{2\nu - 1} \sim N(0, 1) \tag{5.81}$$

and

$$\frac{\chi^2 - \nu}{\sqrt{2\nu}} \sim N(0, 1) \tag{5.82}$$

TABLE 5.7 Work Table for Example 5.10

r	$2r$	$X^2_{.25}(2r) = a$	$X^2_{.95}(2r) = b$	**Ratio** a/b
1	2	2.77	0.103	26.9
2	4	5.39	0.711	7.6
3	6	7.84	1.64	4.8
4	8	10.2	2.73	3.7
5	10	12.5	3.94	3.2
6	12	14.8	5.23	2.8
7	14	17.1	6.57	2.6 ←
8	16	19.4	7.96	2.4
9	18	21.6	9.39	2.3
10	20	23.8	10.9	2.2

Approximate critical points for $\chi^2(\nu)$ based on Equation 5.81 are

$$\chi^2_{1-\alpha} = (\sqrt{2\nu - 1} - z_\alpha)^2 / 2 \qquad (5.83a)$$

$$\chi^2_\beta = (\sqrt{2\nu - 1} + z_\beta)^2 / 2 \qquad (5.83b)$$

Approximate critical points based on Equation 5.82 are

$$\chi^2_{1-\alpha} = \nu - z_\alpha \sqrt{2\nu} \qquad (5.84a)$$

$$\chi^2_\beta = \nu + z_\beta \sqrt{2\nu} \qquad (5.84b)$$

The pair 5.83a–b seems to give better results and will be used in this work. The criterion 5.74 can be written as

$$\frac{\theta_0}{\theta_1} = \frac{(\sqrt{4r - 1} + z_\beta)^2}{(\sqrt{4r - 1} - z_\alpha)^2} \qquad (5.85)$$

Letting $\theta_0 / \theta_1 = K$ and solving for r, then squaring the result, gives

$$r \geq \left[\frac{z_\beta + z_\alpha \sqrt{K}}{2(-1 + \sqrt{K})} \right]^2 + \frac{1}{4} \qquad (5.86)$$

Equation set 5.84a–b gives

$$r \geq \left[\frac{z_\beta + K z_\alpha}{K - 1} \right]^2 \qquad (5.87)$$

For selected values of α, β, and K, the corresponding r values are given in Table 5.8.

TABLE 5.8 Termination Values from Two Approximations

α	β	K	r from (5.85)	r from (5.86)
.05	.05	2	24	25
		1.5	67	68
		1.25	219	220
.01	.05	2	36	40
		1.5	101	106
		1.25	324	332
.01	.10	2	31	34
		1.5	85	83
		1.25	271	267
.05	.25	2	14 *	16
		1.5	37	40
		1.25	114	120

*The exact value is 13.

EXERCISES 5.7

1. Design a test to use as the basis for deciding whether or not to accept a large lot, if only ten items may be spared for destruction. The purchaser wants to accept the lot with probability .95 if its mean life is 500 hours.

2. (a) Plot the OC curve for the test in Exercise 1. (*Hint.* Use the technique of Example 5.9.)
   (b) Find P_a for $\theta = 300$ hours, using the technique of Example 5.8. Does it match your graphed value?

3. Design a test for distinguishing H_0: $\theta = 1000$ versus H_1: $\theta = 300$ when $\alpha = .10$ and $\beta = .25$.

4. Repeat Exercise 3 for H_1: $\theta = 900$.

5. What is your decision in Exercise 1 if the sample values are as follows?
   (a) { 197, 16, 137, 166, 230, 19, 512, 516, 1737, 2853}.
   (b) { 750, 366, 1631, 133, 406, 485, 1334, 291, 121, 330}.

6. For the test you designed for Exercise 3, what would you conclude for the samples whose values are as follows.
   (a) { 426, 820, 237, 3486, 640, 397, 327, 399, 122, 1361}.
   (b) { 288, 125, 317, 257, 169, 212, 453, 1158, 30, 1053}.
   (c) { 173, 75, 190, 154, 101, 127, 272, 695, 18, 632}.
   Comment. The true means are 1000, 500, and 300 respectively.

7. Design a test for deciding whether to accept a large lot if 15 items may be spared for destruction and the product will be accepted with probability .90 if it has a mean life of 5000 hours.

8. Using the technique of Example 5.9, plot the OC curve for the test you designed in the previous exercise.

9. Create a table similar to Table 5.8 for determining the termination number (destruction index) for various values of K when $\alpha = .10$ and $\beta = .25$. Wherever possible (using Table T3) determine the exact value of r for comparison.

5.8 COST CONSIDERATIONS

So far, in our discussion of life testing, the question of cost has been ignored. We must now investigate how it affects the testing procedure by limiting the choice of n and r. It is reasonable to assume a cost model with the following factors:

$$TC = \text{total cost of testing (a random variable)},$$
$$E(TC) = \text{expected total cost of testing } (= V_n \text{, say}),$$
$$C_1 = \text{cost per hour of waiting time},$$

C_2 = cost per item used in the test (setup and monitoring),

C_3 = cost per item destroyed,

n = number of sockets used in test,

r = number of items destroyed.

Recall that $n + r - 1$ is the total number of items used in replacement testing, not n as in nonreplacement testing. For replacement testing we can write, for expected testing cost,

$$V_n = E\,(\mathrm{TC}) = C_1\,\mathrm{ETT}_{rn} + C_2(n + r - 1) + C_3r \qquad (5.88)$$

The value of r may already have been determined by the choice of OC curve, that is, by how much discrimination is needed between some acceptable θ_0 and some unacceptable θ_1. To obtain test results promptly, we should make n as large as possible. Relationship 5.88 determines what "possible" means from the economic point of view.

For *replacement* testing Equation 5.88 becomes

$$V_n = \frac{C_1\theta r}{n} + C_2(n + r - 1) + C_3r \qquad (5.89)$$

The minimum V_n is obtained by the usual differencing method of finding an n_0 for which

$$V_n < V_{n-1} \qquad (5.90)$$

and

$$V_n \le V_{n+1} \qquad (5.91)$$

Relationship 5.90 gives

$$C_2 < C_1\theta r\left(\frac{1}{n-1} - \frac{1}{n}\right) \qquad (5.92)$$

which leads to

$$n(n-1) < \frac{C_1\theta r}{C_2} = D \ (\text{say}) \qquad (5.93)$$

This equation can be solved by completing the square on the left-hand side,

$$n^2 - n + 1/4 < D + 1/4 \qquad (5.94)$$

so that

$$n - 1/2 < \sqrt{D + 1/4} \qquad (5.95)$$

Relationship 5.91 leads to

$$C_2 \ge C_1\theta r\left(\frac{1}{n} - \frac{1}{n+1}\right)$$

which in turn yields

$$n(n + 1) \geq D$$

and hence

$$n + 1/2 \geq \sqrt{D + 1/4} \qquad (5.96)$$

Relationships 5.95 and 5.96 can be combined in the following rule.

   Rule.   Compute $L_1 = (D + 1/4)^{1/2}$ and round it to the nearest integer. The result is n_0. (If L_1 is a semi-integer, round down.)

   In the *nonreplacement* case, the term ETT_{rn} of (5.88) is replaced by

$$\theta\left(\frac{1}{n} + \frac{1}{n-1} + \cdots + \frac{1}{n-r+1}\right) \qquad (5.97)$$

and n replaces $n + r - 1$ in the second term. Inserting these in (5.88) yields

$$V_n = C_1\theta\left(\frac{1}{n} + \cdots + \frac{1}{n-r+1}\right) + nC_2 + rC_3 \qquad (5.98)$$

Using $D = C_1\theta r/C_2$ as before, we must solve (5.90) and (5.91), which reduce to

$$(n + 1)(n + 1 - r) \geq D$$

and

$$n(n - r) < D$$

If we let

$$L_2 = r/2 + \sqrt{D + r^2/4}$$

the former yields $n < L_2$ and the latter yields $n \geq L_2 - 1$. Thus n_0 is the integer that lies in the half-open interval $[L_2 - 1, \ L_2)$.

   The value of θ needed to compute the expected testing cost from Equation 5.88 or 5.98 must be presumed or estimated. We frequently use the θ_0 value specified by H_0 in the same hypothesis test which determined r. If the product is much worse than θ_0, the testing procedure will be cheaper because the first term will be smaller than expected.

   If the testing costs are given in terms of usage costs for the testing equipment, rather than per item used, the term $n + r - 1$ in replacement testing should be replaced by n. The optimizing process gives the same result since it is equivalent to merely adding a constant, but the resulting total cost must be adjusted.

Example 5.11

Suppose that $r = 6$ has been decided for the test H_0: $\theta = 1200$ hours versus H_1: $\theta < 1200$. Testing costs per item are $C_2 = \$7$ per setup and $C_1 = \$10$/hour. Find the optimal n and the resulting total cost for (a) replacement testing and (b) nonreplacement testing.

Solution:

(a) $D = C_1 r\,\theta/C_2 = 10(6)1200/7 = 10,285.71$ and $L_1 = (D + 1/4)^{1/2} = 101.4$ so $n_0 = 101$ items. The expected total cost is

$$E(TC) = 10(1200)6/101 + 7(101 + 6 - 1) + 100(6) = 2054.87$$

(b) $D = 10,285.71$ as before, so $L_2 = 3 + \sqrt{10,294.71} = 104.5$ and $n_0 = 104$. The expected total cost is

$$E(TC) = 10(1200)(1/104 + 1/103 + \cdots + 1/99) + 7(104) + 100(6) = 2037.56$$

Example 5.12

Assume the same data as in Example 5.11 except that $C_2 = \$150$ per item. Find n_0 for (a) replacement testing and (b) nonreplacement testing.

Solution: (a) $L_1 = [10(6)1200/150 + 1/4]^{1/2} = 21.9$ so that $n_0 = 22$ items. (b) $L_2 = 3 + [480 + 9]^{1/2} = 25.1$ so that $n_0 = 25$ items.

In the replacement case, ignoring the term $+1/4$ in the expression for L_1, we see that n_0 is proportional to the ratio C_1/C_2. It might be convenient for the practitioner to construct tables for various r and C_1/C_2 values that would give the multiplier for $\sqrt{\theta}$.

EXERCISES 5.8

1. A large lot of electrical components has been presented for inspection. A mean life of 1500 hours is to be accepted with probability .90. Testing is to be continued until the tenth failure. Design a nonreplacement life test that minimizes total cost when the cost of setting up a component for testing is $12, the cost of waiting for results is $5 per hour, and the cost of a destroyed item is negligible. What is the expected total cost?

2. Rework Exercise 1 assuming that the cost of each destroyed component is $5.

3. Design a replacement life test for the situation in Exercise 1. Compare the resulting costs of the two tests.

4. Design a life test that will accept with probability .95 products with a mean life of 1000 hours and will reject with probability .75 products with a mean life of 500 hours. Determine the number of units to put on test if the setup and test equipment

charge for each test station is $100 and the cost of waiting is $10 per hour. Do this for

(a) a replacement test, and

(b) a nonreplacement test.

5. If the following data represent lifetimes for the previous exercise, will the lot be accepted or rejected? (Use only as many data values as needed.)

$$\{ 248, 650, 1101, 311, 975, 888, 400, 178, 462, 581,$$
$$484, 921, 393, 685, 1428, 806, 2335, 777, 520, 723\}.$$

5.9   MIXED STOPPING RULE LIFE TESTS

Epstein has suggested (1960c,d) that it may be convenient to combine the advantages of index truncation with time truncation, by proposing the following rule:

Rule.   A lot of items is put on test ($H_0: \theta = \theta_0$). Choose a stopping index r_0, and a time t_0 beyond which the test will not be allowed to run. If r_0 failures occur before t_0, stop testing and reject the lot; that is, reject H_0. If fewer than r_0 failures have occurred by time t_0, stop testing and accept the lot.

Stated briefly, the test will be terminated at $\min\{z_{r_0}, t_0\}$ hours. This mixed life test is mathematically more difficult than the previous life tests and will not be dealt with so exhaustively. The quantities of interest are

$E(r)$, the expected number of items failing during the test,

ETT, the expected time required for a decision,

$P_a(\theta)$, the probability of accepting the lot.

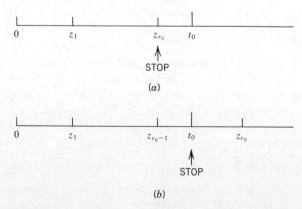

Figure 5.9   Depiction of two stopping rules. (a) Event E_1: Stopping time is z_{r_0}. (b) Event E_2: Stopping time is t_0.

The test has two mutually exclusively outcomes, which can be depicted graphically as shown in Figure 5.9. We can write $P(E_1) = P(z_{r_0} < t_0) = G_r(t_0)$ where $G_r(.)$ indicates the CDF of the order statistic z_r. But $P(E_1)$ is, for the nonreplacement test, simply the binomial probability of at least r_0 failures in a set of n Bernoulli trials of length t_0 where the probability of failure is

$$P_\theta = 1 - e^{-t_0/\theta} \tag{5.99}$$

In the following pages, the quantities $bi(x;n,p)$ and $Bi(x;n,p)$ for binomial mass function and CDF are abbreviated as $bi(x)$ and $Bi(x)$ where no confusion can result. Then

$$P(E_1) = P(\text{at least } r_0 \text{ failures in time } t_0)$$
$$= 1 - Bi(r_0 - 1) \tag{5.100}$$

$$= \sum_{k=r_0}^{n} bi(k) \tag{5.101}$$

On the other hand, the testing stops at time t_0 if the number of failures in n Bernoulli trials is strictly less than r_0, so that

$$P(E_2) = \sum_{k=0}^{r_0-1} bi(k) \tag{5.102}$$

which of course is $1 - P(E_1)$. The expected number of failures is

$$E(\tilde{r}) = \sum_{k=0}^{r_0-1} k \, bi(k) + r_0 \sum_{k=r_0}^{n} bi(k) \tag{5.103}$$

Students can verify that, in general, the "partial expectation" of the binomial in the first term can be simplified to a binomial CDF for the argument-minus-one; the technique is identical with that used in deriving the mean of the binomial distribution. Thus (5.103) simplifies to the form

$$E(\tilde{r}) = nP_\theta \, Bi(r_0 - 2) + r_0[1 - Bi(r_0 - 1)] \tag{5.104}$$

Example 5.13

Twenty items are put on nonreplacement test to decide whether mean life is at least 1500 hours. The test is to be terminated at the third failure ($r = 3$), unless it requires more than 200 hours, at which time it will be terminated. Find the average number of failures required for a decision.

Solution: The probability of a failure during the 200-hour test is $P_\theta = 1 - e^{-200/1500} = .1248$, if H_0 is true. The cumulative binomial probabilities are $Bi(1; 20, .125) = .2677$ and $Bi(2; 20, .125) = .5363$ so that $E(r) = 20(.1248)(.2677) + 3(1 - .5363) = 2.06$.

To find the expected test length, we follow Epstein's (1959) arguments. Let r be the number of failures that would occur in n Bernoulli trials with probability of success P_θ and with no truncation rule.

Let $\tilde{t}$ be the time required for a decision, and ETT $= E(\tilde{t})$. In the following, recall that z_r is shortened notation for z_{rn}; that is, when the second subscript does not appear, it is understood to be n. In some steps the number of trials is not n, and then the second subscript is displayed explicitly. From the definition of the expected test time, it can be written in the form

$$E(\tilde{t}) = t_0 \, \text{Bi}(r_0 - 1) + \sum_{k=r_0}^{n} E(z_r | r = k) \, \text{bi}(k) \tag{5.105}$$

In addition, the unconditional waiting time $E(z_{r_0})$ can be written as a mixture of conditional waiting times:

$$E(z_{r_0}) = \sum_{k=0}^{n} E(z_{r_0} | k) \, \text{bi}(k)$$

$$= \sum_{k=0}^{r_0-1} E(z_r | k) \, \text{bi}(k) + \sum_{k=r_0}^{n} E(z_r | k) \, \text{bi}(k) \tag{5.106}$$

Solving (5.105) and (5.106) for the second term on the right-hand side and equating the two expressions gives

$$E(\tilde{t}) = t_0 \, \text{Bi}(r_0 - 1) + E(z_r) - \sum_{k=0}^{r_0-1} E(z_r | k) \, \text{bi}(k) \tag{5.107}$$

$$= E(z_r) + \sum_{k=0}^{r_0-1} [t_0 - E(z_r | k)] \, \text{bi}(k) \tag{5.108}$$

Now consider the quantity $E(z_r | k)$ for $k = 0, 1, \ldots, r_0 - 1$. The situation can be depicted graphically, with k representing some intermediate number of failures before the r_0th failure, which we presume occurs after time t_0 as shown in Figure 5.10. The conditional waiting time until z_{r_0} is hereby decomposed into two pieces, one of constant length t_0 and the other representing the balance of time it would take to see $r_0 - k$ failures for the reduced sample size $n - k$. Thus Epstein claims that

$$E(z_{r_0,n} | k) = t_0 + E(z_{r_0-k, n-k}) \tag{5.109}$$

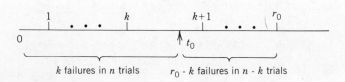

Figure 5.10   The situation for a mixed stopping rule.

Substituting this result into (5.108) gives

$$E(\tilde{t}) = E(\tilde{z}_r) = \sum_{k=0}^{r_0-1} E(z_{r_0-k,n-k}) \, bi(k) \tag{5.110}$$

But the format of

$$E(\tilde{z}_{r_0-1,n-k}) = \frac{1}{n-k} + \frac{1}{n-k-1} + \cdots + \frac{1}{n-r_0+1} \tag{5.111}$$

is such that adding and subtracting appropriate terms, namely

$$\frac{1}{n} + \frac{1}{n-1} + \cdots + \frac{1}{n-k+1} \tag{5.112}$$

leads to the identity

$$E(z_{r_0-k,n-k}) = E(z_{r_0,n}) - E(z_{k,n}) \tag{5.113}$$

and this is to be substituted under the summation sign in (5.110). The result is

$$E(\tilde{t}) = E(\tilde{z}_r) - \sum_{k=0}^{r_0-1} E(\tilde{z}_r) \, bi(k) + \sum_{k=0}^{r_0-1} E(\tilde{z}_k) \, bi(k)$$

$$= \sum_{k=0}^{r_0-1} E(\tilde{z}_k) \, bi(k) + E(\tilde{z}_{r_0})[1 - Bi(r_0 - 1)] \tag{5.114}$$

It is reasonable to define $E(\tilde{z}_0)$ as zero, since its form is $\sum_{i=0}^{0} 1/(n-i)$, so that (5.114) can be written as

$$E(\tilde{t}) = \sum_{k=1}^{r_0-1} E(\tilde{z}_k) \, bi(k) + E(\tilde{z}_r)[1 - Bi(r_0 - 1)] \tag{5.115}$$

Example 5.14

What is the expected test duration for the test of Example 5.1 where $n = 20$, $r = 3$, and $P_\theta = .1248$?

Solution: The needed values are $E(\tilde{z}_1/\theta) = 1/20 = .05$, $E(\tilde{z}_2)/\theta = 1/20 + 1/19 = .1026$, $E(\tilde{z}_3)/\theta = 1/20 + 1/19 + 1/18 = .1582$, $bi(1) = .1982$, $bi(2) = .2686$, $Bi(2) = .5363$. Thus $E(\tilde{t})/\theta = E(\tilde{z}_1)bi(1) + E(\tilde{z}_2)bi(2) + E(\tilde{z}_3)[1 - Bi(2)] = (.05)(.1982) + (.1025)(.2686) + (.1582)(1 - .5363) = .1108$. The resulting expected test time for $\theta = \theta_0 = 1500$ hours is $E(\tilde{t}) = .1108\theta = 166.2$.

The probability of acceptance is easily obtained as

$$P_a(\theta) = P(k < r_0) = \text{Bi}(r_0 - 1; n, P_\theta) \tag{5.116}$$

The problem of determining optimum values of n based on a cost model is beyond the scope of this chapter.

For the replacement test, a Poisson process occurs at each socket, as well as on the master time axis. (Recall that our notation is poi($\cdot$) for the Poisson mass function and POI($\cdot$) for the Poisson CDF.) The probability of an accept decision with exactly k failures ($k < r_0$) is

$$P(r = k) = \frac{e^{-n\lambda t_0}(n\lambda t_0)^k}{k!}, \qquad k = 0, 1, \ldots, r_0 - 1$$

$$= \text{poi}(k|\lambda'), \text{ say} \qquad \lambda' = n\lambda t_0$$

and the probability of a reject decision is

$$P(r \geq r_0) = \sum_{k=r_0}^{\infty} \text{poi}(k|\lambda')$$

$$= 1 - \sum_{k=0}^{r_0-1} \text{poi}(k) = 1 - \text{POI}(r_0 - 1) \tag{5.117}$$

The expected number of failures for reaching a decision is then

$$E(\tilde{r}) = r_0[1 - \text{POI}(r_0 - 1)] + \sum_{k=0}^{r_0-1} k \, \text{poi}(k|\lambda') \tag{5.118}$$

The partial expectation of the second term on the right-hand side can be simplified by using the same technique one uses for finding the Poisson mean, so that

$$E(\tilde{r}) = r_0[1 - \text{POI}(r_0 - 1)] + \lambda'\text{POI}(r_0 - 2) \tag{5.119}$$

The expected time for reaching a decision is merely

$$E(\tilde{T}) = \theta \frac{E(\tilde{r})}{n} \tag{5.120}$$

the proof offered by Epstein (1954, 1959) being analogous to that given earlier for the nonreplacement test. The probability of acceptance is

$$P_a(\theta) = P(\tilde{r} \leq r_0 - 1) = \text{POI}(r_0 - 1) \tag{5.121}$$

If the decision rule is different from the one mentioned, then $P_a(\theta)$ requires a formula different from (5.121) (or from Equation 5.116 in the nonreplacement test), but the expressions for $E(\tilde{r})$ and $E(\tilde{t})$ are not affected. To illustrate, suppose that r_0 has been chosen as seven and t_0 as 100 hours. Suppose the decision rule is as follows.

Rule. Keep testing until seven failures or 100 hours, whichever comes first. If $r \leq 3$ during the 100 hours, the lot will be accepted, otherwise rejected.

In short, the truncation rule need not necessarily coincide with the decision rule, although ordinarily it will do so.

The optimal test has its r_0 value determined by the familiar relationship from Section 5.7,

$$\frac{\chi_\beta^2(2r)}{\chi_{1-\alpha}^2(2r)} \leq \frac{\theta_0}{\theta_1}$$

The value of n is based on the OC curve requirement of acceptance whenever total life exceeds rC. But total life under acceptance is nt_0 in the replacement test. Hence n must satisfy

$$nt_0 \geq rC$$
$$\geq r\,\theta_0 \frac{\chi_{1-\alpha}^2(2r)}{2r} \tag{5.122}$$

from which

$$n \geq \theta_0 \frac{\chi_{1-\alpha}^2(2r)}{2t_0} \tag{5.123}$$

and n is inversely proportional to test time.

For the nonreplacement test a different kind of reasoning may be applied, as follows. If $\hat{\theta} < C$, it is unlikely (probability $< \alpha$) that $\theta \geq \theta_0$ and thus rejection is the correct decision. The test is designed so that if the number of failures in time t_0 exceeds or equals r_0, there will be rejection. Thus the events $\{r < r_0\}$ and $\{\theta < C\}$ can be considered equivalent, and the sample size n should be chosen so that $E(\tilde{r}) = nP_\theta$ is consistent with both. Since P_θ is not known, it must be estimated, and rather than use the point estimate θ, it is sensible to use the value C, which is a one-sided lower confidence bound (LCB), so that we obtain the "safest" value of n consonant with $P(\text{accept } H_0 | H_0) = 1 - \alpha$. The result is to set

$$nP_\theta \geq r_0$$

so that

$$n \geq r_0(1 - e^{-t_0/C})^{-1} \tag{5.124}$$

For small values of t_0/C the series expansion of $e^{-t_0/C}$ yields $n \geq r_0C/t_0$, which is equivalent to (5.123).

EXERCISES 5.9

1. Ten items are put on test, without replacement, to decide whether mean life is at least 500 hours. The test is to be terminated at the third failure or 100 hours,

whichever comes first. What is the average number of failures required for a decision?

2. Repeat Exercise 1 for the replacement test.

5.10   MISCELLANEOUS CONSIDERATIONS

Some of the formulae presented in this chapter can be of service in applications which are only peripherally related to life testing.

Suppose that a number of identical CFR units are placed in service at time $t = 0$. An observer may suspect that there is a change in the mean life θ as time goes on. How can such changes be detected? It has been proved in earlier sections of this chapter that the partial lives $\tilde{w}_i$ (in the nonreplacement test) and $\tilde{w}_i^*$ (in the replacement test) are distributed $G(1, \theta)$, so that the $2\tilde{w}/\theta$ (or $2\tilde{w}^*/\theta$) is distributed $\chi^2(2)$. This fact can be used in a variety of ways, as suggested by Epstein in (1960c,d). Let $T(z_i)$ stand for the partial life generated by components until the ith failure. In other words

$$\left.\begin{aligned}
T(z_1) &= w_1 = nz_1 \\
T(z_2) &= w_1 + w_2 = nz_1 + (n-1)(z_2 - z_1) \\
&= z_1 + (n-1)z_2 \\
&\;\;\vdots \\
T(z_r) &= w_1 + \cdots + w_r \\
&= z_1 + z_2 + \cdots + z_r + (n-r)z_r
\end{aligned}\right\} \tag{5.125}$$

Additionally, let

$$T(z_m - z_i) = T(z_m) - T(z_i) = w_{i+1} + \cdots + w_m \tag{5.126}$$

be the contribution to total life between the ith and the mth failure. Because the w_i are statistically independent, the quantity

$$\frac{2T(z_m - z_i)}{\theta} \tag{5.127}$$

is distributed as $\chi^2(2m - 2i)$. Recall that Snedecor's F is the ratio of two independent χ^2 variables with their degrees of freedom divided out, to wit

$$F = F(\nu_1, \nu_2) = \frac{\chi_1^2(\nu_1)/\nu_1}{\chi_2^2(\nu_2)/\nu_2} \tag{5.128}$$

and this is the basis of some convenient tests.

If there is contamination by duds, leading to an unexpectedly small value for z_1, the quantity $T(z_1)/2$ will be small compared with $T(z_r - z_1)/(2r - 2)$, and the ratio of the latter to the former will be "too large," say in excess of $F_\alpha^* = F_\alpha(2r - 2, 2)$, which is the upper $(1 - \alpha)$ critical point of the F distribution. Thus

$$T(z_1) < \frac{T(z_r - z_1)}{(r - 1)F_\alpha^*} \tag{5.129}$$

is an indication of an abnormally early first failure. On the other hand, if, after the components are put in service, a long period elapses before the first failure, then $T(z_1)$ is large compared with $T(z_r - z_1)/(r - 1)$. Here

$$\frac{(r - 1)T(z_1)}{T(z_r - z_1)} > F_\alpha(2, \ 2r - 2) \tag{5.130}$$

may be taken as an indication that the first failure is unusually delayed, and that the correct model for the life distribution of the components may in fact be the shifted exponential density $f(t) = \lambda \exp[-\lambda(t - A)]$. The second as well as the first failure may be included in the test just given by computing the statistic

$$\frac{(r - 2) T(z_2)}{2T(z_r - z_2)} \tag{5.131}$$

and comparing it with $F_\alpha(4, \ 2r - 4)$ or $F_{1-\alpha}(4, \ 2r - 4)$, as appropriate.

If it is suspected that the mean life changes significantly at some time during the test, say between the mth and the $(m + 1)$st failure, we examine the ratio

$$\frac{(r - m) T(z_m)}{mT(z_r - z_m)} \tag{5.132}$$

and compare it with $F_\alpha(2m, \ 2r - 2m)$ or $F_{1-\alpha}(2m, \ 2r - 2m)$, as appropriate.

The technique described here is clever in theory, but so insensitive that it is of little practical worth, as shown in Example 5.15.

Epstein suggests other tests based on the quantities $T(z_m - z_i)$ and applied to Bartlett's test for homogeneity of variance, Hartley's maximum F test, and so on.

Example 5.15

Twenty items are placed on test and the test is discontinued after eight failures. The failure times are

{ 123, 173, 222, 258, 363, 374, 464, 598 }

To test whether the first failure is unusually delayed, we construct $(r - 1)T(z_1)/ T(z_8 - z_1) = 7(2460)/7291 = 2.36$ with $(2,14)$ degrees of freedom. The result is far from significant. This is a disappointment since the data are in fact the result of a simulation based on generating an exponential sample of size 20, and adding 100 to all observations. Thus the model was

$$f(t) = .001 \exp[-.001(t - 100)]$$

Based on $F_{.05}(2, 14) = 3.74$, it would be necessary for $T(z_1)$ to exceed $(7291)(3.74)/7 = 3895$, so that $z_1(= w_1/20)$ would need to exceed 195 for the test to be significant. A statistical test is hardly necessary to detect a shift of that size when the subsequent $z_i - z_{i-1}$ values are { 50, 49, 36, 105, 11, 90, 134 }.

Order Statistics

We will develop the basic concepts of order statistics using a specific example to fix ideas; then we will generalize. Let us consider the population which consists of the 10 integers

$$\mathfrak{X} = \{0, 1, \ldots, 9\} \tag{5A.1}$$

and examine some results associated with drawing samples of size $n = 4$ without replacement. We use the notation

$$x_1, x_2, x_3, x_4 \tag{5A.2}$$

to indicate the observations in the order they are drawn, x_1 being the first observation, and so on. Thus the list $(5, 2, 1, 8)$ means that $x_1 = 5$ was drawn first, $x_2 = 2$ was drawn second, and so on. If the observations are arranged in ascending order, we create an *order statistic*, indicated with the notation

$$x_{(1)}, x_{(2)}, x_{(3)}, x_{(4)} \tag{5A.3}$$

Thus the sample point $(5, 2, 1, 8)$ corresponds to the order statistic

$$(1, 2, 5, 8)$$

where $x_{(1)} = 1$ is the smallest element, $x_{(2)} = 2$ is the next smallest, and so on. Obviously, for nonreplacement sampling from a finite population, it is true that

$$x_{(1)} < x_{(2)} < x_{(3)} < \ldots < x_{(n)} \tag{5A.4}$$

and in general, for sampling from an arbitrary population,

$$x_{(1)} \leq x_{(2)} \leq \ldots \leq x_{(n)} \tag{5A.5}$$

Since the smaller of two observations can be expected to have a probability distribution different from that of the smallest of five, say, precise notation would necessitate two subscripts. Thus $x_{(i,n)}$ would indicate the ith ordered observation out of a sample of size n. This double subscript notation is sometimes used in the literature, but where no confusion of sample size is likely, it is convenient to suppress the second subscript.

Returning to our ten-element population, we see that for sample size four, there are $(10)(9)(8)(7) = 5040$ possible distinct outcomes, or 5040 equally likely points in the sample space. Let us calculate the probability of a few selected outcomes.

Example 5A.1

For the ten-element population just given, calculate $P(X_4 = 3)$.

Solution: We can immediately write the answer as the marginal probability (which ignores the x_1, x_2, x_3 values), $P(x_4 = 3) = 1/10$. For readers who are uncomfortable with this approach, we recognize that an alternative more detailed formulation is

$$P(X_4 = 3) = P(X_1 \neq 3, \; X_2 \neq 3, \; X_3 \neq 3, \; X_4 = 3)$$
$$= (9/10)(8/9)(7/8)(1/7) = 1/10$$

as before. Readers should practice until they become convinced that $P(X_1 = 3) = P(X_2 = 3) = P(X_3 = 3) = P(X_4 = 3)$.

Example 5A.2

For the ten-element population given, calculate $P(X_1 = 4, \; X_3 = 6)$.

Solution: By the marginal method, $P(x_1 = 4, \; x_3 = 6) = (1/10)(1/9) = 1/90$. By the more detailed method, $P = P(X_1 = 4, \; X_2 \neq 4 \text{ and } X_2 \neq 6, \; X_3 = 6, \; X_4 \neq 4 \text{ and } X_4 \neq 6) = (1/10)(8/9)(1/8)(7/7) = 1/90$, as before.

Example 5A.3

For the ten-element population, calculate $P(X_{(1)} = 3)$.

Solution: Here we are dealing for the first time with an ordered value, and it is not correct to use the marginal approach. Specifically, $X_{(1)} = 3$ means that the value 3 is the smallest of the four observations and that it might have been any one of the four. There are thus four cases to consider, so the total probability is

$$\begin{aligned} P(X_{(1)} = 3) = \; & P(X_1 = 3, X_2 > 3, X_3 > 3, X_4 > 3) \\ & + P(X_1 > 3, X_2 = 3, X_3 > 3, X_4 > 3) \\ & + P(X_1 > 3, X_2 > 3, X_3 = 3, X_4 > 3) \\ & + P(X_1 > 3, X_2 > 3, X_3 > 3, X_4 = 3) \end{aligned} \qquad (5A.6)$$

Since the four terms are all equal, we can write

$$\begin{aligned} P(X_{(1)} = 3) &= 4\,P(X_1 = 3, \; X_2 > 3, \; X_3 > 3, \; X_4 > 3) \\ &= 4(1/10)(6/9)(5/8)(4/7) = 480/5040 \end{aligned} \qquad (5A.7)$$

The rule of formulation could be stated as follows.

> **Rule.** Find the simplest outcome (sample point) that satisfies the given condition (in this case, that the first element drawn equals 3 and all others are greater) and compute its probability. Then multiply this probability by the number of equivalently satisfactory sample points.

Note that $P(X_{(1)} = 3) \neq P(X_{(2)} = 3) \neq \cdots \neq P(X_{(4)} = 3)$, in contrast with the problem of Example 5A.1.

The result could have been obtained by enumeration of the permutations in the sample space that have 3 as the smallest element. Letting L_1, L_2, and L_3 stand for the three observations that are larger than 3 (there are six such numbers), there are $\frac{4!}{1!3!}(1)(6)(5)(4) = 480$ allowable sample points. The factor $\frac{4!}{1!3!}$ is the permutation factor that results when counting the number of permutations of four objects where one is distinct (the 3) and the other three are "alike" in being larger than 3. Readers should recall that the number of permutations of n objects when r_1 are alike, r_2 are alike, $\ldots$, r_k are alike is

$$P(n,n;r_1,\ldots,r_k) = \frac{n!}{r_1!r_2!\ldots r_k!} \tag{5A.8}$$

Example 5A.4

For the ten-element population, calculate $P(X_{(1)} = 3, X_{(2)} = 5)$.

Solution: Once more, to fix ideas, we go through the enumeration process. Letting T stand for three, F for five, and L for any of $\{6, 7, 8, 9\}$, we have the desired probability

$$P(TFLL) + P(FTLL) + \cdots + P(LLFT) \tag{5A.9}$$

consisting of twelve equal terms which sum to

$$12P(TFLL) = 12(1/10)(1/9)(4/8)(3/7) = 144/5040 \tag{5A.10}$$

More directly we could write

$$\frac{4!}{1!1!2!}P(X_1 = 3, X_2 = 5, X_3 > 5, X_4 > 5) = 144/5040 \tag{5A.11}$$

The key to the general method is to write down the probability for any one realization, preferably the easiest, and then multiply it by the permutation coefficient corresponding to the number of outcomes specified and the number that are "alike" in being "greater than," "less than," or "between" the specified values.

Example 5A.5

For the ten-element population, verify the following results.

(a) $P(X_{(2)} = 4) = \frac{4!}{1!1!2!}(4/10)(1/9)(5/8)(4/7)$

(b) $P(X_{(1)} = 3, X_{(4)} = 8) = \frac{4!}{1!2!1!}(1/10)(4/9)(3/8)(1/7)$

(c) $P(X_{(4)} = 8) = \frac{4!}{3!1!}(8/10)(7/9)(6/8)(1/7)$

Going now to a general random variable X with probability mass function $f(x)$, the probability that the first, or smallest, r observations of a size n sample will have the given numerical values $(x_{(1)}, \ldots, x_{(r)})$, when the upper $n - r$ values are unspecified, is

$$P(X_{(1)} = x_{(1)}, \ldots, X_{(r)} = x_{(r)})$$
$$= \frac{n!}{(n-r)!} f(x_{(1)})f(x_{(2)}) \ldots f(x_{(r)})[1 - F(x_{(r)})]^{n-r} \tag{5A.12}$$

The last factor is the probability that all of $n - r$ observations are larger than $x_{(r)}$. Students should compare (5A.12) with (5A.10). For a continuous random variable with density $f(x)$, we write the probability element (p.e.) that the smallest observation out of n will lie in the $dx_{(1)}$ interval around the specified $x_{(1)}$, the next smallest "in $dx_{(2)}$ at $x_{(2)}$," up to $x_{(r)}$, when the largest $n - r$ values are unspecified. The probability element, then, is

$$P(x_{(1)} \le X_{(1)} \le x_{(1)} + dx_{(1)}, \ldots, x_{(r)} \le X_{(r)} \le x_{(r)} + dx_{(r)})$$
$$= \frac{n!}{(n-r)!} f(x_{(1)}) \, dx_{(1)} f(x_{(2)}) \, dx_{(2)} \ldots$$
$$f(x_{(r)}) \, dx_{(r)}[1 - F(x_r)]^{n-r} \tag{5A.13}$$

Refer to Figure 5.11 for a graphical interpretation of the various factors. It is important to notice that in formulae 5A.12 and 5A.13 we are considering only the "partial order statistic" for the r smallest values out of n, because that is the only one needed in

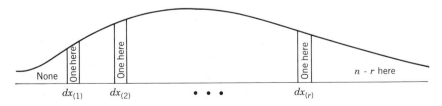

Figure 5.11   A density function labeled for the partial order statistic $(x_{(1)} \ldots x_{(r)})$.

the life testing work of Chapter 5. Many others could have been considered as well. Wilks (1962) is a good reference for students who want to learn more about order statistics.

Another way to achieve the result of equation 5A.13 is to start with the joint density for the random sample $(x_1, \ldots, x_n)$, which for independent observations is

$$f(x_1, \ldots, x_n) = f(x_1)f(x_2)\ldots f(x_n) \tag{5A.14}$$

Then, using the argument just given regarding equivalent samples that lead to the same order statistic, we can write

$$g(x_{(1)}, \ldots, x_{(n)}) = n! f(x_{(1)})\ldots f(x_{(n)})$$

or, in the easier notation that appears in Chapter 5,

$$g(z_1, \ldots, z_n) = n! f(z_1)\ldots f(z_n) \tag{5A.15}$$

with $z_1 \le z_2 \le \ldots \le z_{n-1} \le z_n < \infty$. To arrive at the joint density of the partial order statistic, we perform successive integrations on the variables in descending order. To fix ideas, suppose that $n = 7$. Then

$$g(z_1, \ldots, z_6) = \int_{z_6}^{\infty} g(z_1, \ldots, z_7)\, dz_7$$

$$= 7! f(z_1)\ldots f(z_6) \int_{z_6}^{\infty} f(z_7)\, dz_7 \tag{5A.16}$$

$$= 7! f(z_1)\ldots f(z_6) F(z_7) \Big|_{z_6}^{\infty} \tag{5A.17}$$

$$= 7! f(z_1)\ldots f(z_6)[1 - F(z_6)] \tag{5A.18}$$

To find $g(z_1, \ldots, z_5)$ another integration is performed.

$$g(z_1, \ldots, z_5) = \int_{z_5}^{\infty} g(z_1, \ldots, z_6)\, dz_6$$

$$= 7! f(z_1)\ldots f(z_5) \int_{z_5}^{\infty} f(z_6)[1 - F(z_6)]\, dz_6$$

$$= -7! f(z_1)\ldots f(z_5)[1 - F(z_6)]^2 \Big|_{z_5}^{\infty} (1/2)$$

$$= (7!/2) f(z_1)\ldots f(z_5)[1 - F(z_5)]^2 \tag{5A.19}$$

After a few more iterations it is obvious that for general n and r, the density for the partial order statistic is

$$g(z_1, \ldots, z_r) = \frac{n! f(z_1)\ldots f(z_r)[1 - F(z_r)]^{n-r}}{(n-r)!} \tag{5A.20}$$

Maximum Likelihood Estimators

Students are already familiar with the concept that samples are taken from a population primarily for the purpose of estimating the parameters of that population. Thus, if we consider a random variable X whose density function is given by $f(x;\theta)$, and a sample of size n is observed, yielding observations $(x_1, \ldots, x_n)$, the question becomes the following: How do we combine the n observations into a statistic whose role is to estimate θ? Formally speaking, we say

$$\hat{\theta} = T(x_1, x_2, \ldots, x_n) \tag{5B.1}$$

In elementary work we have used the method of moments as our criterion for estimating parameters. Thus, for example, in the normal distribution with which students are familiar, it is traditional to estimate the parameters μ and σ^2 by means of

$$\hat{\mu} = \bar{x}$$

$$\hat{\sigma}^2 = \mathscr{S}^2 = \sum_{i=1}^{n} \frac{(x_i - \bar{x})^2}{n} \tag{5B.2}$$

or

$$\hat{\sigma}^2 = S^2 = \sum_{i=1}^{n} \frac{(x_1 - \bar{x})^2}{n-1}$$

In other words, the sample mean is used as an estimate of a population mean and the sample variance is used as an estimate of the population variance. These are very reasonable things to do—intuitively appealing and in many ways very satisfactory. In populations in which the parameters are not the simple moments, as for example in the gamma distribution where the parameters are α and β, the relationship between the parameters and the mean and standard deviation is more complex, as given by

$$\mu = \alpha\beta$$
$$\sigma^2 = \alpha\beta^2 \tag{5B.3}$$

227

By setting $\hat{\mu} = \bar{x}$, $\sigma^2 = S^2$ (or $\mathcal{S}^2$) in Equation 5B.3 and solving for the resulting α and β values, we get

$$\hat{\alpha} = \bar{x}^2/S^2 \quad (\text{or } \bar{x}^2/\mathcal{S}^2) \tag{5B.4}$$

$$\hat{\beta} = S^2/\bar{x} \quad (\text{or } \mathcal{S}^2/\bar{x}) \tag{5B.5}$$

Now the following question comes up: Is there another method of forming estimates that may be based on a somewhat more mathematical principle than this? The answer is yes; the method is called the maximum likelihood technique. The process is as follows.

We write down the joint density function or joint probability mass function for the random sample that has in fact been observed. This is

$$L = L(x_1, \ldots, x_n; \theta) = f(x_1; \theta)f(x_2; \theta)\ldots f(x_n; \theta) \tag{5B.6}$$

Because the observations have been made, this function is a function of θ, the unknown quantity. The x's themselves are simply constants that have been observed.

We treat this as a calculus problem. Choose a θ estimator that maximizes the likelihood of obtaining the sample that was in fact obtained. Thus we may often simply solve

$$\frac{dL}{d\theta} = 0 \tag{5B.7}$$

Example 5B.1   The Negative Exponential Distribution

We illustrate the procedure by means of a simple example involving a density function which is already familiar—that of the negative exponential distribution. In this case $f(x; \lambda)$ is

$$f(x; \lambda) = \lambda e^{-\lambda x} \quad \text{for } x \geq 0 \tag{5B.8}$$

The likelihood function for a sample of size n is

$$L = L(\lambda) = (\lambda e^{-\lambda x_1})(\lambda e^{-\lambda x_2})\ldots(\lambda e^{-\lambda x_n}) \tag{5B.9}$$

which simplifies to

$$L = \lambda^n e^{-\lambda B} \tag{5B.10}$$

where B is the sum of observations,

$$B = x_1 + \cdots + x_n \tag{5B.11}$$

Because of the product form of (5B.10), before differentiating and equating to zero, it is convenient to realize that the logarithm is a monotone increasing function of its argument. Thus we replace Equation 5B.10 by its logarithmic form

$$\ln L = n \ln \lambda - \lambda B \tag{5B.12}$$

which is much more convenient mathematically. Differentiating with respect to the unknown quantity λ gives

$$\frac{d \ln L}{d\lambda} = \frac{n}{\lambda} - B \qquad (5B.13)$$

which can be solved for λ, yielding the result

$$\hat{\lambda} = \frac{n}{B} = \frac{n}{\sum x_i} = \frac{1}{\bar{x}} \qquad (5B.14)$$

The maximum likelihood estimate of failure rate, or intensity, from the negative exponential distribution is thus the reciprocal of the sample mean. This result seems strange at first, until we realize that for the negative exponential distribution the mean is λ^{-1}; the result of Equation 5B.14 is perfectly in keeping with this fact.

Example 5B.2   The Poisson Distribution

The probability mass function for the Poisson random variable with intensity λ and experiment length t is given by

$$f(x;\lambda,t) = \frac{e^{-\lambda t}(\lambda t)^x}{x!} \qquad \text{for } x = 0, 1, 2, \ldots \qquad (5B.15)$$

The resulting likelihood function for a sample of size n is given by

$$L(\lambda) = \frac{e^{-\lambda t_1}(\lambda t_1)^{x_1}}{x_1!} \frac{e^{-\lambda t_2}(\lambda t_2)^{x_2}}{x_2!} \cdots \frac{e^{-\lambda t_n}(\lambda t_n)^{x_n}}{x_n!} \qquad (5B.16)$$

which simplifies to

$$L(\lambda) = \frac{e^{-\lambda(t_1 + \cdots + t_n)} \lambda^{x_1 + \cdots + x_n} t_1^{x_1} t_2^{x_2} \cdots t_n^{x_n}}{x_1! x_2! \ldots x_n!} \qquad (5B.17)$$

For simplification, it is convenient to let TF (the time factor), FP (the factorial product), B (the sum of observations), and ST (the sum of times) be written as

$$\text{TF} = t_1^{x_1} t_2^{x_2} \ldots t_n^{x_n}$$
$$\text{FP} = x_1! x_2! \ldots x_n!$$
$$B = x_1 + \cdots + x_n$$
$$\text{ST} = t_1 + \cdots + t_n$$

Then the log-likelihood function assumes the simple form

$$\ln L = -\lambda(\text{ST}) + B \ln \lambda + \ln(\text{TF}) - \ln(\text{FP}) \qquad (5B.18)$$

Differentiating with respect to λ, and equating to zero, we arrive at the maximum likelihood estimator of λ, given by

$$\hat{\lambda} = \frac{B}{ST} = \frac{x_1 + \cdots + x_n}{t_1 + \cdots + t_n} \qquad (5B.19)$$

Notice that this result is dimensionally correct. If only one experiment were to be performed, (5B.19) would reduce simply to x/t—the number of failures per time— which is in accord with our understanding of the intensity of a Poisson process.

Example 5B.3   The Normal Distribution

Here we have a two-variable problem with a density given by

$$f(x;\mu,\sigma^2) = (\sqrt{2\pi}\sigma)^{-1} \exp\left[\frac{-(x-\mu)^2}{2\sigma^2}\right] \qquad \text{for } -\infty < x < \infty \qquad (5B.20)$$

The likelihood function for the sample size n is given by

$$L(\mu,\sigma^2) = \prod_{i=1}^{n} (\sqrt{2\pi}\sigma)^{-n} \exp\left[\frac{-(x_i-\mu)^2}{2\sigma^2}\right] \qquad (5B.21)$$

Simplifying to the logarithmic form yields

$$\ln L = -\frac{n}{2}\ln 2\pi - n\ln\sigma - \frac{1}{2\sigma^2}\sum_{i=1}^{n}(x_i-\mu)^2 \qquad (5B.22)$$

There are two unknown quantities, μ and σ^2; differentiation of (5B.22) with respect to μ leads to

$$\frac{\partial \ln L}{\partial \mu} = \frac{1}{\sigma^2}\sum_{i=1}^{n}(x_i-\mu) \qquad (5B.23)$$

and with respect to σ leads to

$$\frac{\partial \ln L}{\partial \sigma} = -\frac{n}{\sigma} + \frac{1}{\sigma^3}\sum_{i=1}^{n}(x_i-\mu)^2 \qquad (5B.24)$$

These two equations must be solved simultaneously. Setting the right-hand side of (5B.23) equal to zero yields the estimator

$$\hat{\mu} = \bar{x} \qquad (5B.25)$$

Substituting this into Equation 5B.24 yields the biased variance estimator,

$$\hat{\sigma}^2 = \frac{\sum_{i=1}^{n}(x_i - \bar{x})^2}{n} = \mathscr{S}^2 \qquad (5B.26)$$

This is a biased estimate of population variance, showing that a maximum likelihood estimator need not be unbiased.

Example 5B.4   The Continuous Uniform Distribution

We present this example not because it is of interest in the current work, but as a warning to students that the maximum likelihood estimator cannot always be found by the process of simple differentiation. The density function for the random variable which is uniform over the interval from zero to an upper end point θ is given by

$$f(x; \theta) = \frac{1}{\theta} \qquad \text{for } 0 \le x \le \theta \qquad (5B.27)$$

The corresponding likelihood function is

$$L(\theta) = \left(\frac{1}{\theta}\right)\left(\frac{1}{\theta}\right)\cdots\left(\frac{1}{\theta}\right) = \frac{1}{\theta^n} \qquad (5B.28)$$

which upon differentiation with respect to θ yields

$$\frac{dL}{d\theta} = \frac{-n}{\theta^{n+1}} \qquad (5B.29)$$

It is apparent that setting this latter equation equal to zero does not yield any fruitful estimate of θ. This is a situation in which the differentiation method does not work, so we go back to first principles. Looking at Equation 5B.28 and recognizing that we want to choose some function of the observations which will maximize the likelihood function, or minimize the denominator, we simply ask the question, "How can we minimize the expression θ^n?" Obviously θ^n can be minimized by taking

$$\hat{\theta} = 0 \qquad (5B.30)$$

Recall, however, that the estimator must be consistent with the data values that have in fact been observed; if even a single observation is greater than zero, an estimator of the form 5B.30 is foolish. Therefore it is intuitively obvious that θ, the upper bound of the distribution, must exceed or be at least equal to all the observations. The maximum observed x is thus the smallest value that can reasonably be inserted into (5B.28). The resulting maximum likelihood estimator is

$$\hat{\theta} = x_{\max} = x_{(n)} \qquad (5B.31)$$

The reasoning in this example must be used any time a parameter defines either end of the support region of the distribution.

Example 5B.5   The Gamma Distribution

The density function for the gamma variable X with shape parameter α and scale parameter β is given by

$$f(x:\alpha, \beta) = [\Gamma(\alpha)\beta^{\alpha}]^{-1}x^{\alpha-1}e^{-x/\beta} \qquad \text{for } x > 0 \qquad (5B.32)$$

After carrying out the mathematical manipulation for writing a likelihood function for the sample, and then revising it to the logarithmic form, we arrive at

$$\ln L = (\alpha - 1) \ln(x_1 \ldots x_n) - (x_1 + \cdots + x_n)/\beta - n \ln \Gamma(\alpha) - n\alpha \ln \beta \quad (5B.33)$$

Here, as in the normal case, there are two unknowns to solve for, α and β. Students may verify that

$$\frac{\partial \ln L}{\partial \alpha} = \ln \prod_{i=1}^{n} x_i - \frac{n\Gamma'(\alpha)}{\Gamma(\alpha)} - n \ln \beta \qquad (5B.34)$$

and

$$\frac{\partial \ln L}{\partial \beta} = \frac{x_1 + \cdots + x_n}{\beta^2} - \frac{n\alpha}{\beta} \qquad (5B.35)$$

Equation 5B.35 is easily solvable for β in terms of α, but when the result is put back into Equation 5B.34, the equation does not yield a ready analytic solution because of the presence of the factors $\Gamma(\alpha)$ and $\Gamma'(\alpha)$. There is no analytic way of finding the joint maximum likelihood estimators for the gamma distribution. Estimates can be found using numerical techniques, but no closed formulae can be developed.

If the parameter α is known, only (5B.35) need be solved, and this is easily done, yielding

$$\hat{\beta} = \frac{x_1 + \cdots + x_n}{n\alpha} = \frac{\bar{x}}{\alpha} \qquad (5B.36)$$

which is the same as the method-of-moments estimate.

Transformation of Variables

Consider the situation in which some basic random variable X with density $f(x)$ has been found to be of less interest than a derived variable Y which is functionally and deterministically related to it. A physicist may decide, for example, that kinetic energy is of more interest to him than velocity (with K.E. $= mv^2/2$), or a physiologist may be less concerned with body length than with body volume (usually proportional to length cubed). In such cases, a transformation of variable is indicated. The problem is posed formally this way.

If X is a random variable with a known density $f(x)$, and $Y = h(X)$ is a known transformation, how may the density function $g(y)$ of the new random variable Y be obtained?

We consider three cases and then illustrate with some examples. We let $f(x)$ and $g(y)$ stand for the density functions, and let $F(x)$ and $G(y)$ stand for the CDFs of the old and new variables, respectively.

Case 1: Y Is a Monotone Nondecreasing Function of X The relationship $Y = h(X)$ for a monotone nondecreasing function can be depicted graphically as in Figure 5.12. The important thing to notice is that the derivative $Y' = dY/dX$ is always nonnegative. As a result we have

$$G(y) = P(Y \leq y) = P(X \leq x) = F(x) \qquad (5C.1)$$

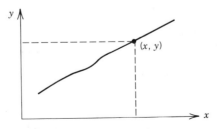

Figure 5.12 Transformation of variable, case 1.

Differentiation with respect to y, and appropriate use of the chain rule where it is called for, gives

$$g(y) = \frac{dG(y)}{dy} = \frac{dF(x)}{dy} = \frac{dF(x)}{dx}\frac{dx}{dy} = f(x)\frac{dx}{dy} \qquad (5C.2)$$

Case 2: Y Is a Monotone Nonincreasing Function of X   The situation is depicted graphically in Figure 5.13. Here $Y' = dY/dX$ is everywhere negative, and the technique used in Case 1 is applied *mutatis mutandis*. We have

$$G(y) = P(Y \le y) = P(X \ge x) = 1 - F(x) \qquad (5C.3)$$

so that

$$g(y) = \frac{dG(y)}{dy} = \frac{-dF(x)}{dy} = \frac{-dF(x)}{dx}\frac{dx}{dy} = \frac{dF(x)}{dx}\frac{-dx}{dy}$$

$$= f(x)\left|\frac{dx}{dy}\right| \qquad (5C.4)$$

Since (5C.4) contains (5C.2) as a special case, we use (5C.4) as the general formula of transformation for single-valued, or monotone, transformations. Of course, it is understood that the right-hand side of (5C.4) must be expressed in terms of y, using the transformation equation

$$Y = h(X) \qquad (5C.5)$$

and its inverse

$$X = h^{-1}(Y) \qquad (5C.6)$$

so that the new density could be more correctly and formally expressed as

$$g(y) = f[h^{-1}(Y)]\left|\frac{d}{dy}[h^{-1}(Y)]\right| \qquad (5C.7)$$

The examples to be given will make this clear. Equation 5C.4 is easily remembered by means of the mnemonic that equates probability elements (p.e.'s) for the old

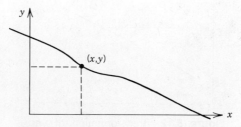

Figure 5.13 Transformation of variable, case 2.

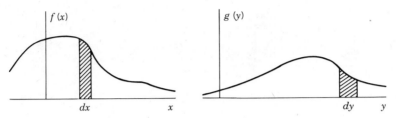

Figure 5.14 Density curves for old and new variables.

and new random variables, as suggested in Figure 5.14. If dy is the interval around $Y = y$ that corresponds to the dx interval around $X = x$, the two probability elements $f(x)\,dx$ and $g(y)\,dy$ are equal in absolute value and the relationship 5C.4 follows.

Case 3: Y Is Not a Monotone Function of X We consider only a very simple case, illustrated in Figure 5.15, but the technique can be generalized as needed. Notice that to the left of x_0, the curve is monotone decreasing so that $dY/dX < 0$, and to the right of x_0 it is monotone increasing, so that $dY/dX > 0$. Thus

$$G(y) = P(Y \le y) = P(x_1 \le X \le x_2)$$
$$= F(x_2) - F(x_1) \tag{5C.8}$$

Differentiating and applying the chain rule where needed gives

$$g(y) = \frac{dG(y)}{dy} = \frac{dF(x_2)}{dy} - \frac{dF(x_1)}{dy} \tag{5C.9}$$
$$= \frac{dF(x_2)}{dx_2}\frac{dx_2}{dy} - \frac{dF(x_1)}{dx_1}\frac{dx_1}{dy}$$
$$= f(x_2)\frac{dx_2}{dy} + f(x_1)\frac{-dx_1}{dy}$$
$$= f(x_1)\left|\frac{dx_1}{dy}\right| + f(x_2)\left|\frac{dx_2}{dy}\right| \tag{5C.10}$$

Figure 5.16 illustrates the relationship between the various probability elements.

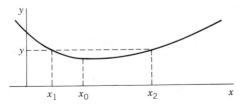

Figure 5.15 A nonmonotone transformation.

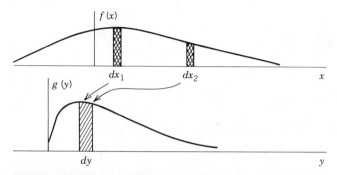

Figure 5.16   Addition of probability elements for a two-valued transformation.

Example 5C.1

Let $X \sim G(\alpha, \beta)$ so that

$$f(x) = [\Gamma(\alpha)\beta^{\alpha}]^{-1}x^{\alpha-1}e^{-x/\beta} \qquad (5C.11)$$

Now let $Y = CX$, so that $dx/dy = 1/C$. Then

$$
\begin{aligned}
g(y) &= [\Gamma(\alpha)\beta^{\alpha}]^{-1}(y/C)^{\alpha-1}e^{-(y/C)/\beta}(1/C) \\
&= [\Gamma(\alpha)(C\beta)^{\alpha}]^{-1}y^{\alpha-1}e^{-y/C\beta}
\end{aligned}
\qquad (5C.12)
$$

This result is so important and useful that we state it as a theorem.

THEOREM 5C.1

If $X \sim G(\alpha, \beta)$, then $Y = CX \sim G(\alpha, C\beta)$. As an immediate corollary we have the important case where C is chosen to have the value $C = 2/\beta$.

Corollary 5C.1   If $X \sim G(\alpha, \beta)$ then $Y = 2X/\beta \sim G(\alpha, 2) = \chi^2(2\alpha)$.

Example 5C.2

Let Z be the standard normal random variable with density

$$f(z) = (\sqrt{2\pi})^{-1}e^{-z^2/2} \qquad \text{for } -\infty < z < \infty \qquad (5C.13)$$

and let the new variable W be defined by

$$W = Z^2 \qquad (5C.14)$$

Then Z is a two-valued function of W with

$$Z_1 = -\sqrt{W}$$

and

$$Z_2 = +\sqrt{W}$$

so that

$$\frac{dZ_1}{dW} = \frac{-1}{2\sqrt{W}}$$

and

$$\frac{dZ_2}{dW} = \frac{1}{2\sqrt{W}}$$

Substituting these expressions into (5C.10) gives

$$g(w) = \frac{e^{-(-\sqrt{w})^2/2}}{\sqrt{2\pi}2\sqrt{w}} + \frac{e^{-(+\sqrt{w})^2/2}}{\sqrt{2\pi}2\sqrt{w}}$$

$$= \frac{e^{-w/2}}{\sqrt{2\pi}\sqrt{w}}$$

$$= \frac{w^{(1/2)-1}e^{-w/2}}{\Gamma(1/2)2^{1/2}} \tag{5C.15}$$

Equation 5C.15 can be recognized as the density function of a gamma random variable with $\alpha = 1/2$ and $\beta = 2$, which is the chi-square type with one degree of freedom. We summarize this as a theorem.

THEOREM 5C.2

If $Z \sim N(0,1)$ then $W = Z^2 \sim \chi^2(1)$. In words, the square of a standard normal variable is a chi-square variable with one degree of freedom.

In all the foregoing, we have assumed the univariate case, where a single old random variable was being transformed into a single new random variable. This transformation is too restricted for our work in this book, since it frequently happens that a collection of old variables $X_1, \ldots, X_r$ with joint density $f(x_1, \ldots, x_r)$ must be transformed into a collection of new variables $Y_1, \ldots, Y_r$ with density $g(y_1, \ldots, y_r)$. The form for $g(y_1, \ldots, y_r)$ is given by

$$g(y_1, \ldots, y_r) = f(x_1, \ldots, x_r) |J| \tag{5C.16}$$

where J is the Jacobian determinant of partial derivatives, defined as

$$J = \begin{vmatrix} \dfrac{\partial x_1}{\partial y_1} & \cdots & \dfrac{\partial x_r}{\partial y_1} \\ \vdots & & \vdots \\ \dfrac{\partial x_1}{\partial y_r} & \cdots & \dfrac{\partial x_r}{\partial y_r} \end{vmatrix} \tag{5C.17}$$

No proof will be given for this result; interested readers may consult an advanced calculus book. Relationship 5C.16 is seen to be a generalization of (5C.4). The Jacobian is familiar to calculus students in the conversion from Cartesian to polar or spherical coordinates; the reasoning is based on the geometry of the appropriate differential elements. In two dimensions the transformation

$$(x, y) \rightarrow (r, \theta)$$

is accompanied by replacing the differential element $dx\,dy$ with $r\,dr\,d\theta$. Here the multiplier r is the Jacobian. In three dimensions, the transformation

$$(x, y, z) \rightarrow (r, \theta, \phi)$$

is accompanied by replacing $dx\,dy\,dz$ with $r^2 \sin\phi\,dr\,d\theta\,d\phi$, where the multiplier $r^2 \sin\phi$ is the Jacobian.

Confidence Limits for the Binomial Parameter

The following notation will be used in this appendix:

$\mathrm{bi}(x; n, p) = $ binomial probability mass function

$$= \binom{n}{x} p^x (1 - p)^{n-x} \qquad \text{for } x = 0, 1, \ldots, n \qquad (5D.1)$$

$\mathrm{Bi}(x; n, p) = $ cumulative binomial probability

$$= \sum_{k=0}^{x} \mathrm{bi}(k; n, p) \qquad (5D.2)$$

$\mathrm{be}(z; a, b) = $ beta density function

$$= \frac{z^{a-1}(1 - z)^{b-1}}{B(a, b)} \qquad \text{for } 0 < z < 1 \qquad (5D.3)$$

$B(a, b) = $ beta function (a constant)

$$= \frac{\Gamma(a)\Gamma(b)}{\Gamma(a + b)} \qquad (5D.4)$$

Starting from the definition

$$\mathrm{Bi}(x; n, p) = \sum_{k=0}^{x} \mathrm{bi}(k; n, p) = \sum_{k=0}^{x} \frac{n!}{k!(n - k)!} p^k (1 - p)^{n-k} \qquad (5D.5)$$

and recognizing that $\mathrm{Bi}(x; n, p)$ is continuous in the argument p, (5D.5) can be differentiated with respect to p, yielding

$$\frac{d\,\mathrm{Bi}(x; n, p)}{dp} = \sum_{k=0}^{x} \frac{n!}{k!(n - k)!} [p^k (n - k)(1 - p)^{n-k-1}(-1) + kp^{k-1}(1 - p)^{n-k}]$$

$$= -\sum_{k=0}^{x} \frac{n!}{k!(n - k - 1)!} p^k (1 - p)^{n-k-1}$$

$$+ \sum_{k=1}^{x} \frac{n!}{(k - 1)!(n - k)!} p^{k-1}(1 - p)^{n-k} \qquad (5D.6)$$

If we let $k - 1 = r$ in the second summation we get

$$\frac{d\,\text{Bi}(x;n,p)}{dp} = \text{same first term} + n \sum_{r=0}^{x-1} \frac{(n-1)!}{r!(n-1-r)!}p^r(1-p)^{n-1-r} \quad (5D.7)$$

Notice that if we write $n! = n(n-1)!$, the structure of the two summations in (5D.7) is the same except for the symbol used as the dummy variable of summation, and the second summation contains one term less than the first. Canceling the common terms yields

$$\frac{d\text{Bi}(x;n,p)}{dp} = -n\frac{(n-1)!}{x!(n-1-x)!}p^x(1-p)^{n-1-x}$$

$$= \frac{-\Gamma(n+1)}{\Gamma(x+1)\Gamma(n-x)}p^x(1-p)^{n-1-x}$$

$$= \frac{-p^x(1-p)^{n-x-1}}{B(x+1,n-x)} \quad (5D.8)$$

Now integrate both sides of the equation from 0 to p, with z as dummy variable of integration, to arrive at

$$\int_0^p \frac{d\text{Bi}(x;n,z)}{dz}\,dz = -\int_0^p \frac{z^x(1-z)^{n-x-1}}{B(x+1,n-x)}\,dz \quad (5D.9)$$

whereupon

$$\text{Bi}(x;n,p) - \text{Bi}(x;n,0) = -\int_0^p \frac{z^x(1-z)^{n-x-1}}{B(x+1,n-x)}\,dz \quad (5D.10)$$

Since $\text{Bi}(x;n,0) = 1$ we have

$$\text{Bi}(x;n,p) = 1 - \int_0^p \frac{z^x(1-z)^{n-x-1}}{B(x+1,n-x)}\,dz = \int_p^1 \text{be}(z;x+1,n-x)\,dz \quad (5D.11)$$

In words, the left-hand tail of the binomial distribution has the same probability as the right-hand tail of a beta distribution when the integration limit is suitably chosen.

Now consider a binomial experiment that has yielded x successes, so that x and n are known quantities, and consider the equation

$$\text{Bi}(x;n,p) = \alpha/2 \quad (5D.12)$$

Since $\text{Bi}(x;n,p)$ is continuous in p, for any assigned significance level α an exact solution is possible for some value of p, say p_2. This is depicted in Figure 5.17a, where the left-hand tail probability is exactly $\alpha/2$. For any value of p larger than p_2, the distribution shifts to the right, pulling some of the probability with it so that $\text{Bi}(x;n,p)$ is less than $\alpha/2$. We may state the following.

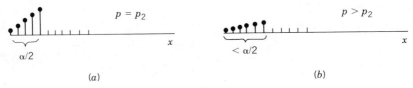

Figure 5.17 Binomial left-hand tail probability: (a) For $p = p_2$. (b) For $p > p_2$.

Observing no more than x successes leads us to regard p_2 as the largest "reasonable" value of p, since for our observed x those p's larger than p_2 are associated with probabilities less than $\alpha/2$.

The manner of finding p_2 (which will be the upper confidence limit) is to use the relationship 5D.11 in the form

$$\int_{p_2}^{1} \frac{z^x(1-z)^{n-x-1}}{B(x+1, n-x)}\, dz = \frac{\alpha}{2} \tag{5D.13}$$

If we had good tables of the incomplete beta function, we could use this directly, just as we could use (5D.12) directly if we had sufficiently good tables of the binomial distribution. Because we do not, we make a transformation that converts (5D.13) into a beta integral of the second kind. Letting

$$z = \frac{w}{1+w}$$

$$1 - z = \frac{1}{1+w}$$

$$dz = \frac{dw}{(1+w)^2} \tag{5D.14}$$

the limits $(p_2, 1)$ transform to the new limits (w_2, ∞). The new integral is

$$\int_{w_2}^{\infty} \frac{w^x\, dw}{B(x+1, n-x)(1+w)^{n+1}} = \frac{\alpha}{2} \tag{5D.15}$$

which suggests another transformation

$$w = \frac{\nu_1 F}{\nu_2} \tag{5D.16}$$

to convert the integral into Snedecor's F distribution with $\nu_1 = 2(x+1)$ and $\nu_2 = 2(n-x)$ degrees of freedom. The corresponding lower integration limit is

$$F_2 = F_{\alpha/2}(2x+2, 2n-2x)$$

Substituting this into (5D.16) to get w_2, and w_2 into $p_2 = w_2/(1 + w_2)$, we arrive at the upper confidence limit

$$p_2 = \frac{\nu_1 F_2/\nu_2}{1 + \nu_1 F_2/\nu_2} = \frac{F_2}{\nu_2/\nu_1 + F_2} \tag{5D.17}$$

$$= \frac{F_{\alpha/2}(2x + 2, 2n - 2x)}{\dfrac{n - x}{x + 1} + F_{\alpha/2}(2x + 2, 2n - 2x)} \tag{5D.18}$$

Now we consider the right-hand tail of the binomial distribution in a similar manner, as shown in Figure 5.18. There is some p, call it p_1, for which the upper tail probability is exactly $\alpha/2$, that is, for which

$$\mathrm{Bi}(x - 1; n, p_1) = 1 - \alpha/2 \tag{5D.19}$$

and any value of p smaller than p_1 will pull the distribution to the left so that less than $\alpha/2$ will be in the upper tail. We may state the following.

> Observing at least x successes leads us to regard p_1 as the smallest "reasonable" value of p, since for our observed x those p's smaller than p_1 are associated with probabilities less than $\alpha/2$.

Making the same sequence of transformations as before on (5D.19), and keeping in mind that $x - 1$ and $1 - \alpha/2$ have replaced x and $\alpha/2$, we arrive at

$$p_1 = \frac{w_1}{1 + w_1} = \frac{\nu_1 F_1/\nu_2}{1 + \nu_1 F_1/\nu_2} \tag{5D.20}$$

but this time the degrees of freedom are $\nu_1 = 2x$ and $\nu_2 = 2(n - x + 1)$ and

$$F_1 = F_{1-\alpha/2}(2x, 2n - 2x + 2) = 1/F_3$$

where

$$F_3 = F_{\alpha/2}(2n - 2x + 2, 2x)$$

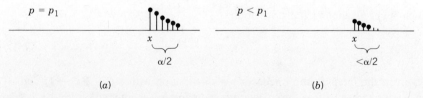

$p = p_1$ $p < p_1$

x x

$\alpha/2$ $<\alpha/2$

(a)   (b)

Figure 5.18   Binomial right-hand tail probability. (a) For $p = p_1$. (b) For $p < p_1$.

Multiplying numerator and denominator of the right-hand side of (5D.20) by v_2/v_1F_1, we get

$$p_1 = \frac{1}{1 + \dfrac{n - x + 1}{x}F_3} \tag{5D.21}$$

The results of (5D.18) and (5D.21) are combined as follows. If

$$P(p \geq p_2 \mid X = x) = \alpha/2 \tag{5D.22}$$

and

$$P(p \leq p_1 \mid X = x) = \alpha/2 \tag{5D.23}$$

then

$$P[(p \geq p_2) \text{ or } (p \leq p_1)|X = x] = \alpha \tag{5D.24}$$

and thus we have the important exact confidence statement

$$P(p_1 \leq p \leq p_2) = 1 - \alpha \tag{5D.25}$$

We recognize that p is usually taken as the probability of a failed item. In this work we deal more often with component reliability

$$R = 1 - p$$

so we write $1 - p_1 = R_U$ and $1 - p_2 = R_L$

$$P(1 - p_1 \geq 1 - p \geq 1 - p_2) = P(R_U \geq R \geq R_L) \tag{5D.26}$$

$$R_U = 1 - \frac{1}{1 + \dfrac{n - x + 1}{x}\dfrac{1}{F_3}}$$

$$= \frac{F_{\alpha/2}(2n - 2x + 2, 2x)}{F_{\alpha/2}(2n - 2x + 2, 2x) + x/(n - x + 1)} \tag{5D.27}$$

Similarly, the lower bound on R is

$$R_L = \frac{1}{1 + \dfrac{x + 1}{n - x}F_{\alpha/2}(2x + 2, 2n - 2x)} \tag{5D.28}$$

When the binomial sample size is so large that the exact method using the F distribution is inconvenient, and the x is so small that the normal approximation is poor because x/n is near zero, we use a technique based on the Poisson approximation to the binomial. Under these conditions only an upper bound is wanted, because an obvious lower bound is zero. As before, we set

$$\text{Bi}(x; n, p_2) = \alpha \tag{5D.29}$$

and solve for p_2. For $p > p_2$ the distribution shifts to the right and $\mathrm{Bi}(x; n, p) < \alpha$. Thus, if no more than x successes are observed, then p_2 is the largest "reasonable" value of p. Now

$$\mathrm{Bi}(x; n, p) = \sum_{k=0}^{x} \binom{n}{k} p^k q^{n-k}$$

$$\sim \sum_{k=0}^{x} \frac{e^{-np}(np)^k}{k!} \tag{5D.30}$$

$$= 1 - \sum_{k=x+1}^{\infty} \frac{e^{-np}(np)^k}{k!} \tag{5D.31}$$

Using the Poisson-gamma relationship of Theorem 2.6, we can convert this Poisson sum into the integral under a gamma density curve, and for convenience we choose the form $G(\nu/2, 2) = \chi^2(\nu)$, where the value of $\nu/2$ must be the lower summation limit of $x + 1$ in (5A.30). Thus

$$\mathrm{Bi}(x; n, p_1) \sim 1 - \int_{0}^{2np_2} dG(w; x + 1, 2)$$

$$= \int_{2np_2}^{\infty} d\chi^2(2x + 2) \tag{5D.32}$$

Since this is to equal α, it must be true that

$$2np_2 = \chi_{\alpha}^2(2x + 2) \tag{5D.33}$$

and thus the upper $100(1 - \alpha)$ percent confidence limit on p is

$$p_2 = \frac{\chi_{\alpha}^2(2x + 2)}{2n} \tag{5D.34}$$

Occurrence Times of a Poisson Process

The Poisson process is often called the random process, or the model of complete randomness, because it represents the distribution at random of an infinite number of points over the infinite interval $(0, \infty)$. More precisely we have the following theorems paraphrased from Parzen (1962, pages 139 and 143).

THEOREM 5E.1

Let $\{N(t), t > 0\}$ be a Poisson process with intensity λ. Under the condition that $N(T) = n$, the n times $\{\tau_1, \ldots, \tau_n\}$ in the interval 0 to T at which events occur are random variables distributed as though they were the order statistics corresponding to n independent random variables $\{U_1, \ldots, U_n\}$ uniformly distributed on the interval $(0, T)$.

Proof   We first note that if $\{U_1, \ldots, U_n\}$ are independently uniformly distributed on the interval $(0, T)$, they have joint density function

$$f(u_1, \ldots, u_n) = \frac{1}{T^n} \qquad \text{for } 0 \le u_i \le T \tag{5E.1}$$

$$= 0 \qquad \text{otherwise}$$

The order statistics $(\tau_1, \ldots, \tau_n)$ corresponding to $\{U_1, \ldots, U_n\}$ then have joint density function

$$f(\tau_1, \ldots, \tau_n) = \frac{n!}{T^n} \qquad \text{for } 0 \le \tau_1 \le \ldots \le \tau_n \le T \tag{5E.2}$$

Getting back to the Poisson process, the conditional probability element, *given that* n events have occurred in $(0, T)$, and that in each of n nonoverlapping subintervals

$(\tau_1, \tau_1 + h_1], \ldots, (\tau_n, \tau_n + h_n]$ of the interval $(0, T]$ exactly one event occurs and elsewhere no events occur, is

$$P[\tau_1 < t_1 \le \tau_1 + h_1, \ldots, \tau_n < t_n \le \tau_n + h_n \mid N(T) = n]$$

$$\frac{P(\tau_1 < t_1 \le \tau_1 + h_1) \ldots P(\tau_n < t_n \le \tau_n + h_n) P(\text{no events in } T - h_1 - \cdots - h_n)}{\dfrac{e^{-\lambda T}(\lambda T)^n}{n!}}$$

$$= \frac{(\lambda h_1 e^{-\lambda h_1})(\lambda h_2 e^{-\lambda h_2}) \ldots (\lambda h_n e^{-\lambda h_n}) e^{-\lambda(T - h_1 - \cdots - h_n)})}{\dfrac{e^{-\lambda T}(\lambda T)^n}{n!}} \tag{5E.3}$$

$$= \frac{n!}{T^n} h_1 h_2 \ldots h_n \tag{5E.4}$$

This is divided by $h_1 h_2 \ldots h_n$ to convert the probability element to a density. Comparing (5E.4) with (5E.2), we see that the assertion in the theorem is justified.

THEOREM 5E.2

If we observe a Poisson process until exactly r events occur, where r is a preassigned integer, and if the events occur at times $\tau_1, \ldots, \tau_r$, the $r - 1$ random variables $\{\tau_k\}$, with $k = 1, \ldots, r - 1$, can be considered to be the order statistic of $r - 1$ independent observations on a random variable that is uniformly distributed over $(0, \tau_r)$.

Note. The events occurring "at" τ_r and "in" interval h are brief ways to say that they occur in a short interval $\tau < t \le \tau + h$ where $h = \Delta\tau$.

Proof The conditional probability element for the occurrence of the failure times "in" the given intervals $\{h_1, \ldots, h_r\}$, *given* that the rth failure occurred "at" time τ_r, is of the form $P(\text{one event in } h_1, \ldots, \text{ one event in } h_r, \text{ no events elsewhere} \mid r\text{th event at } \tau_r)$, and so can be written as

$$\frac{P(\text{one event in } h_1, \ldots, \text{ one event in } h_r, \text{ no events elsewhere})}{P(r - 1 \text{ events before } \tau_r \text{ and one more in } h_r)}$$

$$= \frac{(\lambda h_1 e^{-\lambda h_1})(\lambda h_2 e^{-\lambda h_2}) \ldots (\lambda h_{r-1} e^{-\lambda h_{r-1}})(\lambda h_r e^{-\lambda h_r}) e^{-\lambda(\tau_r - h_1 - \cdots - h_{r-1})}}{\dfrac{e^{-\lambda \tau_r}(\lambda \tau_r)^{r-1}}{(r - 1)!} \lambda h_r e^{-\lambda h_r}}$$

$$= \frac{(r - 1)!}{\tau_r^{r-1}} h_1 h_2 \ldots h_{r-1} \tag{5E.5}$$

Converting this probability element to a density, and comparing this with (5E.2), we see that it corresponds to the order statistics for $r - 1$ observations from the continuous uniform distribution over $(0, \tau_r)$.

CHAPTER 6

The Sequential Probability Ratio Test

6.1 THE FORMULATION OF THE TEST CRITERIA

The sequential probability ratio test (SPRT) was developed primarily by Wald (1947). The test is based on the very appealing concept that sometimes the future outcome of a testing procedure becomes obvious before the planned test is completed. Suppose, for example, that it has been decided to accept a lot if a sample of size 20 shows no more than two defective items, and to reject it otherwise. Let G indicate a good item and D a defective item. If the sequence G G D G D G G D . . . occurs, it is apparent that a reject decision can be made on the eighth trial. (If testing is continued, it will be solely for the purpose of constructing a data base or history of the process.) If the sequence G G G G G G G G G G . . . occurs, there is no absolute assurance that the lot will be accepted, but the inspector may logically feel that acceptance is in the offing, since no defectives in the first 10 items seem to indicate good quality and the last 10 items are not likely to contain three or more defectives. Even if the inspector dares not stop testing and accept the lot at this point, there will be the possibility that he can do so with complete certainty on trial 17 or 18 or 19, depending on what has happened by that time.

To illustrate the basic ideas of the probability ratio test, consider the following simple case, involving an urn containing 10 balls, some black, some white. On the basis of a single observation a decision must be made between the two following possibilities:

H_0: 8 black balls

2 white balls

H_1: 1 black ball

9 white balls

The sample space is the set $\mathfrak{X} = \{B, W\}$. Obviously, if the ball drawn is black, then H_0 should be chosen because $X = B$ is much less likely under H_1 than under H_0. This can be stated formally as $P(B \mid H_0) = .8$ and $P(B \mid H_1) = .1$, so the probability ratio for the outcome $X = B$ is

$$\frac{P(B \mid H_1)}{P(B \mid H_0)} = \frac{.1}{.8} = 0.125$$

For $X = W$ we would have

$$\frac{P(W \mid H_1)}{P(W \mid H_0)} = \frac{.9}{.2} = 4.5$$

and H_1 would be preferred. Suppose the two possible urns specified in the hypotheses are not so disparate as in the first case. Let the new possibilities be indicated as

H_0^*: 6 black balls

4 white balls

H_1^*: 5 black balls

5 white balls

Then the probability ratios for one observation are

$$\frac{P(B \mid H_1)}{P(B \mid H_0)} = \frac{.5}{.6} = 0.833$$

and

$$\frac{P(W \mid H_1)}{P(W \mid H_0)} = \frac{.5}{.4} = 1.25$$

Here the outcome is not so clear, and we can see that the following rule could be promulgated.

Rule.   If $P(x \mid H_1)/P(x \mid H_0)$ is large, choose H_1; if it is small, choose H_0. If it is in the neighborhood of unity, get more data.

This is the SPRT in a nutshell, and the rest of the presentation is refinement for deciding how large is "large," how small is "small," and what "the neighborhood of unity" means. We define the needed terms as follows for the single parameter case:

$$\theta = \text{the parameter under scrutiny}$$
$$\theta_0 = \text{the parameter value under } H_0$$
$$\theta_1 = \text{the parameter value under } H_1$$
$$f(x \mid \theta) = \text{pdf or pmf of the random variable } X \text{ with parameter } \theta$$

$$p_{im} = \prod_{j=1}^{m} f(x_j \mid \theta_i), \qquad i = 0, 1$$

$$A = \text{the largeness criterion}$$
$$B = \text{the smallness criterion}$$
$$\alpha = \text{probability of Type I error (producer's risk)}$$
$$\beta = \text{probability of Type II error (consumer's risk)}$$

The testing procedure may now be stated.

Make observations $x_1, x_2, \ldots, x_m, \ldots$, and compute p_{1m}/p_{0m} at each step. If $p_{1m}/p_{0m} \geq A$, stop testing and accept H_1. If $p_{1m}/p_{0m} \leq B$, stop testing and accept H_0. If $B < p_{1m}/p_{0m} < A$, continue testing.

As will be seen later, the numerical values of A and B depend on the operating characteristic (OC) curve specified, that is, on the α and β values chosen for θ_0 and θ_1.

To illustrate the process, consider the Bernoulli trials with $H_0: p = p_0$ versus $H_1: p = p_1(> p_0)$. Each trial has outcome from the set $\mathcal{X} = \{0, 1\}$ defects, say. Usually we write

$$P(\text{def}) = P(X = 1) = p$$
$$P(\text{good}) = P(X = 0) = 1 - p$$

but for the SPRT these must be expressed in a single expression as

$$f(x; p) = p^x (1 - p)^{1-x}, \qquad x = 0, 1 \tag{6.1}$$

Then the probability ratio has the form

$$\frac{p_{1m}}{p_{0m}} = \frac{p_1^{x_1}(1 - p_1)^{1-x_1} p_1^{x_2}(1 - p_1)^{1-x_2} \ldots p_1^{x_m}(1 - p_1)^{1-x_m}}{p_0^{x_1}(1 - p_0)^{1-x_1} p_0^{x_2}(1 - p_0)^{1-x_2} \ldots p_0^{x_m}(1 - p_0)^{1-x_m}} \tag{6.2}$$

$$= \frac{p_1^{x_1 + \cdots + x_m}(1 - p_1)^{m - (x_1 + \cdots + x_m)}}{p_0^{x_1 + \cdots + x_m}(1 - p_0)^{m - (x_1 + \cdots + x_m)}} \tag{6.3}$$

$$= \frac{p_1^d (1 - p_1)^{m-d}}{p_0^d (1 - p_0)^{m-d}} \tag{6.4}$$

where $d = x_1 + \cdots + x_m$ is the total defect count for the m trials. Testing is continued as long as (6.4) lies strictly between B and A, that is, as long as

$$B < \left(\frac{p_1}{p_0}\right)^d \left(\frac{1 - p_1}{1 - p_0}\right)^{m-d} < A \tag{6.5}$$

or, taking logarithms,

$$\ln B < d \ln \frac{p_1}{p_0} + (m - d) \ln \frac{1 - p_1}{1 - p_0} < \ln A$$

or

$$\ln B < d \left(\ln \frac{p_1}{p_0} - \ln \frac{1 - p_1}{1 - p_0} \right) + m \ln \frac{1 - p_1}{1 - p_0} < \ln A \tag{6.6}$$

For compactness, let the positive constants C_1 and C_2 be defined as

$$C_1 = \ln \frac{p_1}{p_0} - \ln \frac{1 - p_1}{1 - p_0} = \ln \frac{p_1(1 - p_0)}{p_0(1 - p_1)} \tag{6.7}$$

$$C_2 = -\ln \frac{1 - p_1}{1 - p_0} = \ln \frac{1 - p_0}{1 - p_1} \tag{6.8}$$

Continue testing, then, as long as

$$\ln B + C_2 m < C_1 d < \ln A + m C_2$$

or as long as

$$\frac{\ln B + m C_2}{C_1} < d < \frac{\ln A + C_2 m}{C_1} \tag{6.9}$$

If $A > 1$ then $\ln A > 0$ and if $B < 1$ then $\ln B < 0$, so the rule states that testing is continued as long as d, the total defect count, when plotted against the number of trials m, lies between two parallel lines. It is traditional to write the slope and intercepts as

$$s = \frac{C_2}{C_1} \tag{6.10}$$

$$h_2 = \frac{\ln A}{C_1} \tag{6.11}$$

$$h_1 = \frac{-(\ln B)}{C_1} \tag{6.12}$$

so that the rule may be stated as follows.

Rule. Continue testing as long as

$$-h_1 + sm < d < h_2 + sm \tag{6.13}$$

As soon as $d \geq h_2 + sm$, stop testing and reject H_0 (accept H_1), or as soon as $d \leq -h_1 + sm$, stop testing and accept H_0.

This is depicted graphically in Figure 6.1.

For determining the value of A and B, the infinite sample space Ω of all possible outcomes is partitioned into the following subsets:

Ω_0 : Outcomes leading to the acceptance of H_0.

Ω_1 : Outcomes leading to the rejection of H_1.

Ω_C : Outcomes leading to infinite continuation.

Wald had proved that Ω_C is empty, that is, that sampling must eventually terminate in a clear decision. Thus we write

$$\Omega = \Omega_0 \cup \Omega_1$$

In the accept region Ω_0, it will be true that $p_{1m}/p_{0m} \leq B$ so that $p_{1m} \leq Bp_{0m}$ on Ω_0 can be written more fully as

$$f(x_1 \mid \theta_1)\ldots f(x_m \mid \theta_1) \leq B f(x_1 \mid \theta_0)\ldots f(x_m \mid \theta_0) \qquad (6.14)$$

Integrating over the subspace Ω_0 gives

$$\int \ldots \int_{\Omega_0} f(x_1 \mid \theta_1)\ldots f(x_m \mid \theta_1)\, dx_1 \ldots dx_m \leq$$

$$B \int \ldots \int_{\Omega_0} f(x_1 \mid \theta_0)\ldots f(x_m \mid \theta_0)\, dx_1 \ldots dx_m \qquad (6.15)$$

Recall that the integral of a density function over some region is simply the probability that the associated random variable or outcome will lie in that region. Thus (6.15) can be expressed as

$$P(\Omega_0 \mid \theta_1) \leq B P(\Omega_0 \mid \theta_0) \qquad (6.16)$$

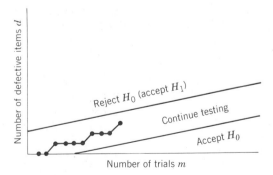

Figure 6.1 General form of the SPRT plot for Bernoulli trials.

But $P(\Omega_0 \mid \theta_1) = P(\text{accept } H_0 \mid H_1) = \beta$ and $P(\Omega_0 \mid \theta_0) = P(\text{accept } H_0 \mid H_0) = 1 - \alpha$, so (6.16) is

$$\beta \le B(1 - \alpha) \tag{6.17}$$

In an exactly analogous way, in the subspace Ω_1 it will be true that $p_{1m}/p_{0m} \ge A$ and the resulting equation

$$f(x_1 \mid \theta_1)\ldots f(x_m \mid \theta_1) \ge A f(x_1 \mid \theta_0)\ldots f(x_m \mid \theta_0) \tag{6.18}$$

when integrated over the subspace Ω_1, yields

$$P(\Omega_1 \mid \theta_1) \ge A\, P(\Omega_1 \mid \theta_0) \tag{6.19}$$

or

$$1 - \beta \ge A\alpha \tag{6.20}$$

The relationships 6.17 and 6.20 give the desired solution

$$A \le \frac{1 - \beta}{\alpha} \tag{6.21}$$

$$B \ge \frac{\beta}{1 - \alpha} \tag{6.22}$$

For practical purposes the inequalities are replaced by equalities, the result of which is to change the effective (or resultant) α and β to something slightly different from the nominal α and β. Wald (1947, page 42) discusses the matter in some detail.

Example 6.1

Let p be the unknown defective fraction of a manufacturing process. Sampling is to be carried out until a decision can be made about the alternatives,

H_0: $p = .10$
H_1: $p = .30$

with $\alpha = .05$ and $\beta = .25$. Describe how to carry out the test.

Solution: From (6.21) and (6.22) we have $A = .75/.05 = 15$, $B = .25/.95 = 5/19$. From (6.7) and (6.8) we have $C_1 = \ln[(.30)(.90)/(.10)(.70)] = 1.3499$ and $C_2 = \ln[(.90)/(.70)] = .2513$. Then from (6.10) through (6.12), the constants needed are $s = (.2513)/(1.3499) = .1862$, $h_2 = (\ln 15)/(1.2499) = 2.0061$ and $h_1 = -(\ln 5/19)/(1.3499) = .9889$. Thus testing continues as long as

$$-.99 + .186m < d < 2.01 + .186m$$

The accept line and reject lines can be plotted as in Figure 6.2 and the inspection process plotted point by point. Or a table can be constructed like Table 6.1, where

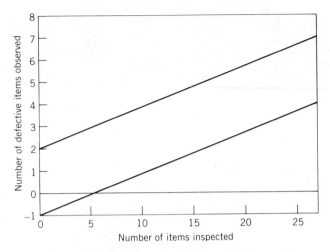

Figure 6.2   The SPRT lines for the binomial case of Example 6.1.

the accept and reject numbers a and r are computed from the equations

$$a = -0.99 + .186m$$
$$r = 2.01 + .186m$$

Since only integer values are possible for the defect count, the values of a and r are listed only to the nearest tenth.

The columns labeled a' and r' are the effective (integerized) values of a and r. The value a must be rounded downward and r must be rounded upward, in order to violate the continue inequalities of (6.23). Furthermore, as inspection of the table shows, no *accept* decision is possible until at least six trials have been made, and no *reject* decision until the third trial—and then only if all the first three items tested are bad.

Consider next the SPRT for a normally distributed variable with unknown mean and known variance. The probability ratio for H_0: $\mu = \mu_0$ versus H_1: $\mu = \mu_1$ (after canceling all the $\sigma\sqrt{2\pi}$ factors) is

$$\frac{p_{im}}{p_{0m}} = \frac{e^{-(x_1-\mu_1)^2/2\sigma^2} \ldots e^{-(x_m-\mu_1)^2/2\sigma^2}}{e^{-(x_1-\mu_0)^2/2\sigma^2} \ldots e^{-(x_m-\mu_0)^2/2\sigma^2}} \qquad (6.23)$$

$$= e^{-Q/2\sigma^2}$$

TABLE 6.1   Accept and Reject Numbers for Example 6.1

m	a	r	a'	r'
1	−0.8	2.2	*	*
2	−0.6	2.4	*	*
3	−0.4	2.6	*	3
4	−0.2	2.8	*	3
5	−0.1	2.9	*	3
6	0.1	3.1	0	4
7	0.3	3.3	0	4
8	0.5	3.5	0	4
9	0.7	3.7	0	4
10	0.9	3.9	0	4
11	1.1	4.1	1	5
12	1.2	4.2	1	5
13	1.4	4.4	1	5
14	1.6	4.6	1	5
15	1.8	4.8	1	5
16	2.0	5.0	2	5
17	2.2	5.2	2	6
18	2.4	5.4	2	6
19	2.5	5.5	2	6
20	2.7	5.7	2	6
21	2.9	5.9	2	6
22	3.1	6.1	3	7
23	3.3	6.3	3	7
24	3.5	6.5	3	7
25	3.7	6.7	3	7
26	3.8	6.8	3	7
27	4.0	7.0	4	7

*Means no decision is possible at this stage.

where

$$Q = \sum_{i=1}^{m} (x_i - \mu_1)^2 - \sum_{i=1}^{m} (x_i - \mu_0)^2 \tag{6.24}$$

$$= \sum_{i=1}^{m} x_i^2 - 2\mu_1 \sum_{i=1}^{m} x_i + m\mu_1^2 - \sum_{i=1}^{m} x_i^2 + 2\mu_0 \sum_{i=1}^{m} x_i - m\mu_0^2$$

$$= 2(\mu_0 - \mu_1) \sum_{i=1}^{m} x_i - m(\mu_0^2 - \mu_1^2) \tag{6.25}$$

Testing is to continue as long as

$$\ln B < \frac{1}{2\sigma^2}[2(\mu_0 - \mu_1)\sum x_i - m(\mu_0^2 - \mu_1^2)] < \ln A \qquad (6.26)$$

Suppose that $\mu_0 > \mu_1$. Then (6.26) when solved for $\sum x_i$ assumes the form: continue testing as long as $-h_1 + sm < \sum x_i < h_2 + sm$ where

$$s = \frac{\mu_0 + \mu_1}{2} \qquad (6.27a)$$

$$h_1 = \frac{-\sigma^2 \ln A}{\mu_0 - \mu_1} \qquad (6.27b)$$

$$h_2 = \frac{-\sigma^2 \ln B}{\mu_0 - \mu_1} \qquad (6.27c)$$

As soon as the left-hand inequality is violated, stop testing and reject H_0, and so on. Notice that here it is the lower line that leads to a reject decision, in contrast with the binomial situation. This is so because rejecting H_0 is always associated with $p_{im}/p_{0m} > A$, and algebraic manipulations involving the negative multipliers within the inequalities led to the A criterion appearing on the left member of the inequality. The student is urged to repeat the derivation for $\mu_0 < \mu_1$.

We come now to the SPRT for CFR components. To start with, assume the simplest case in which there is only one test socket, so that items must be tested sequentially. The probability ratio for $H_0: \theta = \theta_0$ versus $H_1: \theta = \theta_1(< \theta_0)$ is

$$\frac{p_{im}}{p_{0m}} = \frac{\theta_1^{-1}e^{-x_1/\theta_1}\ldots\theta_1^{-1}e^{-x_m/\theta_1}}{\theta_0^{-1}e^{-x_1/\theta_0}\ldots\theta_0^{-1}e^{-x_m/\theta_0}} \qquad (6.28)$$

$$= \left(\frac{\theta_0}{\theta_1}\right)^m e^{-\sum x_i(1/\theta_1 - 1/\theta_0)} \qquad (6.29)$$

Thus testing is continued as long as

$$\ln B < m \ln\frac{\theta_0}{\theta_1} - \left(\frac{1}{\theta_1} - \frac{1}{\theta_0}\right)\sum x_i < \ln A$$

that is, as long as

$$\frac{-\ln A}{1/\theta_1 - 1/\theta_0} + m\frac{\ln(\theta_0/\theta_1)}{1/\theta_1 - 1/\theta_0} < \sum_{i=1}^{m} x_i < \frac{-\ln B}{1/\theta_1 - 1/\theta_0} + m\frac{\ln(\theta_0/\theta_1)}{1/\theta_1 - 1/\theta_0} \qquad (6.30)$$

This can be written as

$$-h_1 + sm < \sum x_i < h_2 + sm \qquad (6.31)$$

where

$$h_1 = \frac{\ln A}{1/\theta_1 - 1/\theta_0} \tag{6.32}$$

$$h_2 = \frac{-\ln B}{1/\theta_1 - 1/\theta_0} \tag{6.33}$$

$$s = \frac{\ln(\theta_0/\theta_1)}{1/\theta_1 - 1/\theta_0} \tag{6.34}$$

Consider next the situation in which there is no replacement and n items are placed on test at time $t = 0$. The shortest-lived (poorest) components will fail first, so the probability ratio will be the quotient of two order-statistic densities of the familiar form (from Section 5.2)

$$\frac{n!}{(n-m)!} f(z_1 \mid \theta) \ldots f(z_m \mid \theta)[1 - F(z_m \mid \theta)]^{n-m}$$

After the factorials are canceled, the probability ratio is

$$\frac{p_{im}}{p_{0m}} = \frac{\theta_1^{-m} e^{-T/\theta_1}}{\theta_0^{-m} e^{-T/\theta_0}} = \left(\frac{\theta_0}{\theta_1}\right)^m e^{-T(1/\theta_1 - 1/\theta_0)} \tag{6.35}$$

where $T = z_1 + \cdots + z_m + (n-m)z_m$ is the total time lived by all components by the time of the mth failure. It is obvious that (6.35) is identical in structure with (6.29) and so must lead to the same SPRT.

Consider next the replacement test with n sockets kept full at all times. On the master time axis there is a Poisson process with intensity $n\lambda = n/\theta$. The probability ratio could be expressed in terms of the Poisson probability function for clock time t. After canceling the common factor $m!$, we have

$$\frac{p_{im}}{p_{0m}} = \frac{e^{-n\lambda_1 t}(n\lambda_1 t)^m}{e^{-n\lambda_0 t}(n\lambda_0 t)^m} \tag{6.36}$$

$$= \left(\frac{\lambda_1}{\lambda_2}\right)^m e^{-nt(\lambda_1 - \lambda_0)}$$

$$= \left(\frac{\theta_0}{\theta_1}\right)^m e^{-nt(1/\theta_1 - 1/\theta_0)} \tag{6.37}$$

If we let $T = nt$, this is identical to (6.29) and (6.35). Thus it appears that all our familiar life tests have the same SPRT format; the only difference between them is the structure of the total life $V(t) = T$, that is, whether it comes from a single socket or many, and with or without replacement.

Example 6.2

Find a SPRT that will accept a lot with a mean life of 1000 hours 90 percent of the time, and that will accept a lot with a mean life of 500 hours only five percent of the time. The constant number of items to be kept on test at all times is $n = 20$.

Solution: We have $\theta_0 = 1000$, $\theta_1 = 500$, $\alpha = .10$, $\beta = .05$. The continue criterion, in terms of total life T, is

$$-2251 + 693m < T < 2890 + 693m \tag{6.38}$$

In terms of clock time, $t = T/20$, the continue criterion is

$$-112.6 + 34.7m < t < 144.5 + 34.7m \tag{6.39}$$

Figure 6.3 shows the lines as well as the data of the next example.

Example 6.3

Suppose that the failure times in hours for the test of Example 6.2 are { 22, 32, 39, 146, 167, 171, 201, 403, 500, 607, 821, 878, 914, 999, 1016, 1066, 1129, 1282, 1345, 1381}. Plot the conduct of the test, and make a table that could be used in place of a plot, if desired.

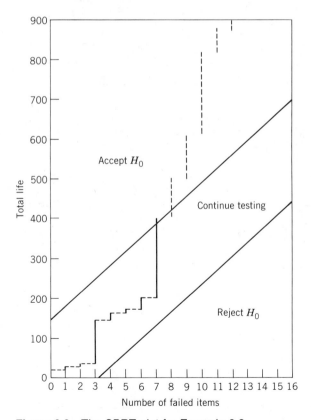

Figure 6.3   The SPRT plot for Example 6.2.

Solution: Figure 6.3 depicts the plot and Table 6.2 gives the corresponding tabular version. Inspection of the plot shows that the time of the mth failure must be compared with the accept number a_{m-1} rather than a_m. This is easiest to understand in terms of the initial period. The traditional procedure is to start at the origin, running the time line up the ordinate, or time, axis. We are plotting the time *before* the first failure; that is, time accrued *toward* the first ($m = 1$) lifetime is plotted against $m = 0$.

In plotting the outcome of the test procedure, is it ever possible to reach a decision on the coordinate axes? Yes, as reference to Figure 6.3 shows. If the first failure had not occurred until after 145 hours, the accept decision could have been made before the first failure was observed.

Example 6.2 represents an easy case because the total life T is merely twenty times the clock time t and the problem could have been handled with equal facility in terms of either clock time or total life. For *nonreplacement* testing, the graph or table must be formulated in terms of total life, so an extra set of calculations is necessary. To illustrate, suppose the failure times of Example 6.3 had referred to a *non*replacement test with $n = 20$. Then the continue criterion is that given in (6.38).

TABLE 6.2   Accept and Reject Criteria for Examples 6.2 and 6.3

Failure Times T_m	Failure Count m	Reject Line $r_m = -112 + 34.7m$	Accept Line $a_{m-1} = 110 + 34.7m$
22	1	-78	145
32	2	-43	179
39	3	-8	214
146	4	26	249
167	5	61	284
171	6	96	318
201	7	130	353
403 STOP	8	165	388
500	9	200	422
607	10	234	457
821	11	269	492
878	12	304	526
914	13	339	561
999	14	373	596
1016	15	408	631
1066	16	443	665
1129	17	477	700
1282	18	512	735
1345	19	547	769
1381	20	581	804

TABLE 6.3 Calculations for Example 6.3 Viewed as a Non-replacement Test

z_m	$n - m + 1$	$z_i - z_{i-1}$	$T(z_m)$ (Total Life)
22	20	22	$20(22) = 440$
32	19	10	$440 + 19(10) = 630$
39	18	7	$630 + 18(7) = 756$
146	17	107	$756 + 17(107) = 2575$
167	16	21	$2575 + 16(21) = 2911$
171	15	4	$2911 + 15(4) = 2971$
201	14	30	$2971 + 14(30) = 3391$
etc.			

Pretending for convenience that the failure times given in Example 6.3 were those for a nonreplacement test, we must manipulate the data as shown in Table 6.3 in order to construct total lives from failure times.

EXERCISES 6.1

1. Compute and plot the SPRT lines for Bernoulli sampling if it is desired to distinguish between $H_0: p = .01$ and $H_1: p = .05$ with $\alpha = .10$ and $\beta = .30$. Construct an action table like Table 6.1. Use it to make a decision about the following three samples which may contain more data than you need.

   (a) All items were good except item numbers { 12, 18, 20, 35, 37, 48}.

   Comment. The true probability of a defective for the simulation was $p = .2$.

   (b) All items were good except item numbers 26 and 101.

   Comment. The true probability of a defective for the simulation was $p = .04$.

   (c) All items were good except item numbers { 16, 21, 37, 65}.

   Comment. The true probability of a defective for the simulation was again $p = .04$.

   (d) No defective items were found among the first eighty items inspected.

   Comment. The true probability of a defective for the simulation was $p = .01$.

2. Create additional simulated random samples for the binomial case of Exercise 1 by using a random number table or the RAND function of an electronic spreadsheet. Let PVAL stand for the true fraction defective of the simulation, and RAND the random number. Use the transformation

   If RAND $\leq$ PVAL then item is defective.

   If RAND $>$ PVAL then item is good.

   Assign several different values to PVAL and test each of your simulated samples to see how the SPRT treats it.

3. Repeat Exercise 1 using $p_0 = .01$ and $p_1 = .10$ with $\alpha = .05$ and $\beta = .25$. Use the same samples.

4. Compute the SPRT lines for the exponential model where $n = 20$ items are kept on test at all times, and it is desired to distinguish between $H_0: \theta = 1000$ versus $H_1: \theta = 500$ with $\alpha = \beta = .10$. Use clock time on the ordinate axis. Construct the SPRT lines and the corresponding table. Use them to test the sample consisting of the failure times $\{51, 103, 105, 226, 272, 301, 314, 335, 463, 498, 516, 569, 616, 630, 675, 715, 852, 875, 1080, 1326\}$. The given number of observed failures may be more than are needed to make a decision.

5. Repeat Exercise 4 for the case where $\theta_0 = 1000$ and $\theta_1 = 400$, with $\alpha = .05$ and $\beta = .20$. Assume that the testing is nonreplacement with $n = 10$ sockets, and the failure times are as follows. (Use as many values as you need.)

   (a) $\{90, 329, 669, 677, 945, 1069, 1618, 3372, 3475, 3879\}$

   (b) $\{85, 123, 153, 162, 166, 584, 610, 739, 1827, 2247\}$

   (c) $\{11.0, 26.5, 27.3, 29.7, 83.0, 113.8, 151.4, 164.7, 254.6, 275.0\}$

   (d) $\{79.1, 81.2, 249.9, 260.5, 283.2, 340.8, 640.8, 723.0, 741.3, 768.7\}$

   Comment. Samples a and b were simulated from $\theta = 1050$ populations, and samples c and d were from $\theta = 500$ populations.

6. Set up an electronic spreadsheet to perform the calculations needed to analyze data like that in the previous exercise. You will want named cells for the values of α, β, θ_0, θ_1 from which you can create other named cells, calling them "slope," "h-one," and "h-two" (say). You will also want the following columns: index (the m-value), accept number, reject number, multiplier (the values n, $n - 1$, $n - 2$, etc.), and total life. An added convenience is a "decision" column containing an IF-statement with output "continue" or "quit." When the data are entered into an appropriate column and "recalculate" is invoked, the correct decision will appear.

7. Design the SPRT for testing the normal population mean $H_0: \mu = 100$ versus $H_1: \mu = 80$ with $\alpha = .10$ and $\beta = .20$. Assume that $\sigma = 30$.

6.2 THE OC CURVE

To obtain the OC curve for the sequential test, we need a formula for $P_a(\theta)$ for points other than θ_0 and θ_1. We will use Wald's notation of $L(\theta)$ for the accept probability $P_a(\theta)$. His derivation for $L(\theta)$, which is creative, clever, and a little bit weird, is somewhat lengthy for inclusion here, so it is presented in Appendix 6A.

As might be expected, the formula for $L(\theta)$ could be, in general, very complicated, so Wald's technique is to write both θ and $L(\theta)$ in parametric form in terms of a parameter h. He found the general formula for $L(\theta)$ to be

$$L(\theta) = \frac{A^h - 1}{A^h - B^h} \qquad \text{for} \ -\infty < h < \infty \qquad (6.40)$$

The formula for θ in terms of h must be found individually for each distribution, based on the fact that $h = h(\theta)$ is defined as a solution to the equation

$$\int_{-\infty}^{\infty} \left(\frac{f(x \mid \theta_1)}{f(x \mid \theta_0)}\right)^{h(\theta)} f(x \mid \theta) \, dx = 1 \tag{6.41}$$

for continuous distributions, or of

$$\sum^{\text{all } x} \left(\frac{f(x \mid \theta_1)}{f(x \mid \theta_0)}\right)^{h(\theta)} f(x \mid \theta) = 1 \tag{6.42}$$

for discrete distributions. The rationale for these requirements is supplied in the derivation in the appendix. To illustrate, we examine the three cases studied earlier: the binomial, the normal (mean unknown, variance known), and the negative exponential.

For a Bernoulli trial, (6.42) assumes the form

$$\left(\frac{f(0 \mid p_1)}{f(0 \mid p_0)}\right)^{h} f(0 \mid p) + \left(\frac{f(1 \mid p_1)}{f(1 \mid p_0)}\right)^{h} f(1 \mid p) = 1$$

which becomes

$$\left(\frac{1 - p_1}{1 - p_0}\right)^{h} (1 - p) + \left(\frac{p_1}{p_0}\right)^{h} p = 1 \tag{6.43}$$

and this is easily solved for p to give

$$p = \frac{1 - \left(\dfrac{1 - p_1}{1 - p_0}\right)^{h}}{\left(\dfrac{p_1}{p_0}\right)^{h} - \left(\dfrac{1 - p_1}{1 - p_0}\right)^{h}} \tag{6.44}$$

Combining this with (6.40) permits the construction of the OC curve.

Example 6.4

Compute the values of $L(p)$ for $h = -2, -1, 0, 1,$ and 2 for the data of Example 6.1.

Solution: As computed previously, $A = 15$, $B = 5/19$, $p_1/p_0 = 3$ and $(1 - p_1)/(1 - p_0) = 7/9$, so the parametric equations to be used are

$$p = \frac{1 - (7/9)^{h}}{3^{h} - (7/9)^{h}} \tag{6.45}$$

$$L(p) = \frac{15^{h} - 1}{15^{h} - (5/19)^{h}} \tag{6.46}$$

These give the results summarized in Table 6.4 and shown in Figure 6.4.

TABLE 6.4   Selected Values for the OC
Curve of Examples 6.1 and 6.4

h	p	$L(p)$
-2	.42	.06
-1	.30($= p_1$)	.25 ($= \beta$)
0	indeterminate, but see comment following (6.59)	
1	.10 ($= p_0$)	.95 ($= 1 - \alpha$)
2	.047	.996

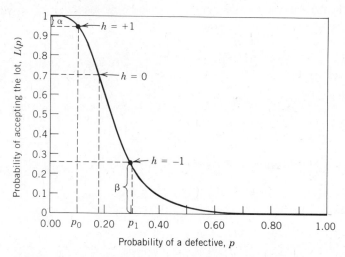

Figure 6.4   OC curve for the binomial case in Example 6.1.

For the normal distribution we must obtain the solution of

$$\frac{1}{\sigma\sqrt{2\pi}} \int_{-\infty}^{\infty} \exp\left[-\frac{1}{2\sigma^2}(x - \mu)^2\right]\left(\frac{\exp\left[-\frac{1}{2\sigma^2}(x - \mu_1)^2\right]}{\exp\left[-\frac{1}{2\sigma^2}(x - \mu_0)^2\right]}\right)^h dx = 1 \qquad (6.47)$$

This can be written as

$$\frac{1}{\sigma\sqrt{2\pi}} \int_{-\infty}^{\infty} e^{-Q/2\sigma^2} dx = 1 \qquad (6.48)$$

where

$$Q = (x - \mu)^2 + h[(x - \mu_1)^2 - (x - \mu_0)^2] \qquad (6.49)$$

After several algebraic steps of squaring, canceling, and regrouping, this can be written as

$$Q = (x - C)^2 + hD \tag{6.50}$$

where

$$C = \mu + h(\mu_1 - \mu_0) \tag{6.51}$$

$$D = \mu_1^2 - \mu_0^2 - 2\mu\mu_1 + 2\mu\mu_0 - h(\mu_1 - \mu_0)^2$$
$$= (\mu_1 - \mu_0)[\mu_1 + \mu_0 - 2\mu - h(\mu_1 - \mu_0)] \tag{6.52}$$

Now since the factor hD in (6.50) does not contain any terms involving x, Equation 6.48 can be written as

$$e^{-hD/2\sigma^2} \frac{1}{\sigma\sqrt{2\pi}} \int_{-\infty}^{\infty} e^{-(x-C)^2/2\sigma^2} \, dx = 1$$

In order for this to be valid, the quantity D must be zero. Thus from (6.52) we get

$$h = \frac{\mu_0 + \mu_1 - 2\mu}{\mu_1 - \mu_0} \tag{6.53}$$

or, to compute μ as a function of h, we can write

$$\mu = \frac{\mu_0 + \mu_1 - h(\mu_1 - \mu_0)}{2} \tag{6.54}$$

Finally, for the negative exponential distribution the solution h must be obtained from the requirement that

$$\int_0^{\infty} \left(\frac{\theta_0 e^{-x/\theta_1}}{\theta_1 e^{-x/\theta_0}} \right)^h \frac{e^{-x/\theta}}{\theta} \, dx = 1 \tag{6.55}$$

This can be written in the form

$$\frac{1}{\theta M} \left(\frac{\theta_0}{\theta_1} \right)^h \int_0^{\infty} e^{-Mx} M \, dx = 1 \tag{6.56}$$

where

$$M = \frac{1}{\theta} + h \left(\frac{1}{\theta_1} - \frac{1}{\theta_0} \right) \tag{6.57}$$

Since the integral equals unity, it must be true that the coefficient

$$\frac{1}{\theta M} \left(\frac{\theta_0}{\theta_1} \right)^h = 1 \tag{6.58}$$

and this yields

$$\left(\frac{\theta_0}{\theta_1}\right)^h = \theta M = \theta\left[\frac{1}{\theta} + h\left(\frac{1}{\theta_1} - \frac{1}{\theta_0}\right)\right]$$

$$= 1 + \theta h\left(\frac{1}{\theta_1} - \frac{1}{\theta_0}\right)$$

from which it follows that the needed parametric equation is

$$\theta = \frac{\left(\dfrac{\theta_0}{\theta_1}\right)^h - 1}{h\left(\dfrac{1}{\theta_1} - \dfrac{1}{\theta_0}\right)} \tag{6.59}$$

The OC curve for the exponential case of Example 6.2 is shown in Figure 6.5.

For plotting the OC curve from the parametric equations, values of h from $-\infty$ to ∞ may be used. Note that the values $h = \pm 1$ yield the two determining points of the curve. The value $h = 0$ gives an indeterminate result; it is thus necessary to use l'Hôpital's rule. The limiting formula for $L(\theta)$ for all distributions is

$$\lim_{h\to 0} L(\theta) = \lim_{h\to 0} \frac{A^h - 1}{A^h - B^h}$$

$$= \lim_{h\to 0} \frac{A^h \ln A}{A^h \ln A - B^h \ln B} \tag{6.60}$$

$$= \frac{\ln A}{\ln A - \ln B} = \frac{\ln \dfrac{1-\beta}{\alpha}}{\ln\left(\dfrac{1-\beta}{\alpha}\dfrac{1-\alpha}{\beta}\right)} \tag{6.61}$$

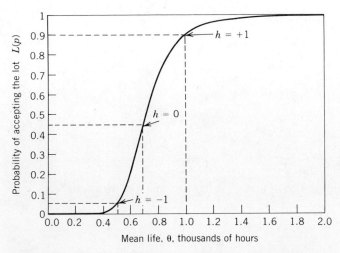

Figure 6.5   OC curve for the exponential case in Example 6.2.

For the parameter θ (or p or μ) we offer without proof a useful and interesting result, as follows.

THEOREM 6.1

For the parameter itself, the value corresponding to $h = 0$ is equal to the slope of the SPRT lines.

EXERCISES 6.2

1. Verify and expand Table 6.4 for additional h-values and plot the resulting OC curve.

2. Verify the correctness of the result in Equation 6.61 starting from (6.60).

3. Verify Theorem 6.1 for the three distributions treated in this chapter: the binomial, the normal, and the exponential.

4. Compute a table and plot the OC curve for the hypothesis test $H_0: p = .001$ versus $H_1: p = .005$ for consumer's risk .20 and producer's risk .10.

5. Repeat Exercise 4 using $H_1: p = .01$.

6. Repeat Exercise 5 cutting both the consumer's and the producer's risk in half. Superimpose the revised plot on the original and compare.

7. Construct the OC curve for CFR components when $H_0: \theta = 10,000$ versus $H_1: \theta = 5000$ is tested using $\alpha = .05$, $\beta = .30$.

8. Compute and plot the OC curve for Exercise 6.1.7.

Derivation of the OC Curve for the SPRT

The value of A and B have been chosen so that the OC curve will pass through the points $(\theta_0, 1 - \alpha)$ and (θ_1, β). How may the probability of acceptance (of H_0), here designated as $L(\theta)$, be determined for other arbitrary values of θ? Wald (1947, page 48) developed the following technique.

Consider the expression

$$\left(\frac{f(x \mid \theta_1)}{f(x \mid \theta_0)}\right)^{h(\theta)} \tag{6A.1}$$

For each value of θ, a nonzero value of $h(\theta)$ is chosen so that the expression in (6A.1) has an expected value equal to unity. Is this always possible? In general, yes, because the nonnegative ratio function

$$g(x) = \frac{f(x \mid \theta_1)}{f(x \mid \theta_0)} \tag{6A.2}$$

can be raised to some power, positive or negative, that will have expectation unity, when the expectation is taken with respect to the density $f(x \mid \theta)$. In short, it should be possible to find a power $h = h(\theta)$ for which

$$\int_{-\infty}^{\infty} \left[\frac{f(x \mid \theta_1)}{f(x \mid \theta_0)}\right]^{h(\theta)} f(x \mid \theta) \, dx = 1 \tag{6A.3}$$

But this means that

$$\left[\frac{f(x \mid \theta_1)}{f(x \mid \theta_0)}\right]^{h(\theta)} f(x \mid \theta) = f^*(x \mid \theta) \text{ (say)} \tag{6A.4}$$

is a density function.

Since we don't know whether h is positive or negative, let us consider a case in which it is positive. We invent a new hypothesis test

H: The true density of X is $f(x \mid \theta)$.

H^*: The true density of X is $f^*(x \mid \theta)$.

and construct its SPRT. The continue criterion will be

$$B^* < \frac{f^*(x_1 \mid \theta) \ldots f^*(x_m \mid \theta)}{f(x_1 \mid \theta) \ldots f(x_m \mid \theta)} < A^* \tag{6A.5}$$

where A^* and B^* may be chosen as desired. If we let $B^* = B^{h(\theta)}$ and $A^* = A^{h(\theta)}$, we notice that the new test is exactly equivalent to the original test. Thus H versus H^* has the same accept and reject regions as H_0 versus H_1, though not the same α and β. Thus we can say

$$\begin{aligned} L(\theta) &= P(\text{accept } H_0 \mid \theta) \\ &= P(\text{accept } H \mid H \text{ true}) \\ &= 1 - \alpha^* \end{aligned} \tag{6A.6}$$

where α^* and β^* satisfy the relationships

$$\begin{aligned} A^* &= \frac{1 - \beta^*}{\alpha^*} \\ B^* &= \frac{\beta^*}{1 - \alpha^*} \end{aligned} \tag{6A.7}$$

which are equivalent to

$$\begin{aligned} \alpha^* A^* + \beta^* &= 1 \\ \alpha^* B^* + \beta^* &= B^* \end{aligned} \tag{6A.8}$$

and whose solution is

$$\alpha^* = \frac{1 - B^*}{A^* - B^*} \tag{6A.9}$$

This leads to the desired result

$$\begin{aligned} L(\theta) &= 1 - \alpha^* \\ &= \frac{A^* - 1}{A^* - B^*} \\ &= \frac{A^h - 1}{A^h - B^h} \end{aligned} \tag{6A.10}$$

The relationship between h and θ is obtained from (6A.3) and must be carried out separately for each distribution, as demonstrated in Chapter 6.

CHAPTER 7

Availability and Maintainability

7.1 SYSTEMS WITH A SINGLE REPAIRABLE ELEMENT

The previous chapters dealt with nonrepairable elements. Now we consider a variety of situations using repairable units. We begin with the simplest possible system, with a single unit, illustrating completely several approaches to the problem.

For repairable systems an important concept and performance criterion, or figure of merit, is the *availability* of the system $A(t)$, defined as the probability that the system functions at time t irrespective of its past history of breakdown and repair. Associated ideas frequently referred to are

$$\overline{A}(T) = \text{average availability during a mission of length } T$$

$$= \frac{1}{T}\left[\int_0^T A(t)\, dt \right] \tag{7.1}$$

and the long-term or steady-state availability

$$A(\infty) = \lim_{t \to \infty} A(t) \tag{7.2}$$

Because an exact formula for $A(t)$ is often difficult to find, this second expression is commonly used.

Let the unit be described as being in one of two possible states:

S_0: The unit is functioning.

S_1: The unit has failed in service and is under repair.

For mathematical simplicity, we usually assume that failure detection is immediate and that repair begins at once. When such an assumption is untenable, the nice derivations in the following pages must be replaced by "brute force" solutions using various numerical methods.

As usual, the failure law is assumed to be exponential with failure rate λ, and the time to repair (given that a repair is in progress) is also exponential with rate μ. Just as $1/\lambda$ is often called MTTF (mean time to failure), so is $1/\mu$ called MTTR (mean time to repair). The system can be depicted as shown in Table 7.1, where the possible state of the system is shown at an arbitrary time t and at a short ("infinitesimal") time later, $t + h$. The length h of the incremental time period can be regarded as equivalent to the differential Δt or dt.

Because the failures and repairs are the Poisson-type events considered in Section 2.3, we can write the transition probabilities, sometimes referred to as the forward Markov equations, in the familiar way as

$$P_0(t + h) = P(E_1 \text{ or } E_2)$$
$$= P_0(t)(1 - \lambda h) + P_1(t)\mu h$$

and

$$P_1(t + h) = P(E_3 \text{ or } E_4)$$
$$= P_0(t)\lambda h + P_1(t)(1 - \mu h)$$

Expanding each expression, transposing one special term from the right to the left-hand side, dividing by h, and then taking the limit as h approaches 0 yields the set of simultaneous differential equations

$$P_0'(t) = -\lambda P_0(t) + \mu P_1(t) \tag{7.3a}$$
$$P_1'(t) = \lambda P_0(t) - \mu P_1(t) \tag{7.3b}$$

These can be solved as an exercise in elementary differential equations, but the easiest way to carry out the solution is to use Laplace transforms. We assume that at time $t = 0$ the unit is functioning, so that $P_0(0) = 1$ and $P_1(0) = 0$. Then, designating the transform as

$$\mathcal{L}[P_i(t)] = Z_i(s) = Z_i, \qquad i = 0, 1$$

TABLE 7.1 System States for a Repairable Unit

Event	At Time t	During Increment h	At Time $t + h$
$E\,1$:	S_0	No failure	S_0
$E\,2$:	S_1	Repair	S_0
$E\,3$:	S_0	Failure	S_1
$E\,4$:	S_1	No repair	S_1

we get the transformed pair

$$sZ_0 - 1 = -\lambda Z_0 + \mu Z_1 \tag{7.4a}$$

$$sZ_1 = \lambda Z_0 - \mu Z_1 \tag{7.4b}$$

which can be written as

$$(s + \lambda)Z_0 - \mu Z_1 = 1 \tag{7.5a}$$

$$-\lambda Z_0 + (s + \mu)Z_1 = 0 \tag{7.5b}$$

Solution by Cramer's determinant rule gives

$$Z_0 = \frac{\begin{vmatrix} 1 & -\mu \\ 0 & s+\mu \end{vmatrix}}{\begin{vmatrix} s+\lambda & -\mu \\ -\lambda & s+\mu \end{vmatrix}} = \frac{s+\mu}{(s+\lambda)(s+\mu) - \lambda\mu}$$

$$= \frac{s+\mu}{s(s+\lambda+\mu)} \tag{7.6}$$

$$= \frac{A}{s} + \frac{B}{s+\lambda+\mu} \text{ (say)} \tag{7.7}$$

Solution of (7.7) by partial fraction decomposition shows that $A + B = 1$ and $A = \mu/(\lambda + \mu)$ whereby $B = \lambda/(\lambda + \mu)$. Thus

$$Z_0 = \frac{\mu}{\lambda+\mu}\frac{1}{s} + \frac{\lambda}{\lambda+\mu}\frac{1}{s+\lambda+\mu} \tag{7.8}$$

Taking inverse Laplace transforms gives the well-known result

$$P_0(t) = \frac{\mu}{\lambda+\mu} + \frac{\lambda}{\lambda+\mu}e^{-(\lambda+\mu)t} = A(t) \tag{7.9}$$

It is clear that for the simple one-component system, the availability $A(t)$ is the same as $P_0(t)$. Furthermore

$$A(\infty) = \frac{\mu}{\lambda+\mu} = \frac{\text{MTTF}}{\text{MTTF} + \text{MTTR}} \tag{7.10}$$

and

$$\overline{A}(T) = \frac{\mu}{\lambda+\mu} + \frac{\lambda}{(\lambda+\mu)^2 T}\left[1 - e^{-(\lambda+\mu)T}\right] \tag{7.11}$$

Remember that these results are reasonably general; although they pertain to a single unit, this unit may in fact be a very complex configuration of assemblies and subassemblies. The exponential failure and repair laws are obviously open to doubt in such configurations. Remember that nonseries configurations of CFR components are not CFR.

Clearly the equation for $P_1(T)$ can be obtained in the same way as (7.9) by starting from Cramer's rule with

$$Z_1 = \frac{\begin{vmatrix} s + \lambda & 1 \\ -\lambda & 0 \end{vmatrix}}{\begin{vmatrix} s + \lambda & -\mu \\ -\lambda & s + \mu \end{vmatrix}}$$

It is easier, however, to obtain $P_1(t)$ from the fact that

$$P_1(t) = 1 - P_0(t)$$

In order to generalize the results just given to more complex systems, it has been found convenient to adopt a slightly different formulation, that of the Markov transition matrix. To illustrate the technique, we can redo the simple problem as follows. A transition matrix **T** is constructed whose (i, j)th element represents the probability of a change from state i to state j in an infinitesimal period of length h. For the current case the matrix **T** is

Final State

$$\begin{array}{c} \\ \text{Initial State} \end{array} \begin{array}{cc} & \begin{array}{cc} 0 & \quad 1 \end{array} \\ \begin{array}{c} 0 \\ 1 \end{array} & \begin{bmatrix} 1 - \lambda h & \lambda h \\ \mu h & 1 - \mu h \end{bmatrix} \end{array} \qquad (7.12)$$

Notice that the elements of each row sum to unity, as with any stochastic (probability) matrix. The situation can also be depicted graphically as shown in Figure 7.1. The loops at each event are necessary so that the outgoing probabilities from each node sum to unity. Each outgoing arrow from a node corresponds to an entry in the corresponding row of the transition matrix. Students are warned not to fall into the trap of incorrectly labeling the figure of the transition matrix with the values $1 - \lambda$, λ, μ, $1 - \mu$ as some writers have done.

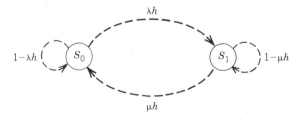

Figure 7.1 The transitions for a single repairable component.

The matrix $\mathbf{T}$ of (7.12) is used as a coefficient matrix for the state row vector (where the superscript t stands for transpose)

$$\mathbf{P}^t(t) = [P_0(t), P_1(t)]$$

in the matrix equation

$$\mathbf{P}^t(t)\, \mathbf{T} = \mathbf{P}^t(t + h)$$

Since it is more traditional to work with column vectors than with row vectors, this is written as

$$\mathbf{T}^t\, \mathbf{P}(t) = \mathbf{P}(t + h)$$

Subtracting $\mathbf{P}(t)$ from both sides gives

$$\mathbf{P}(t + h) - \mathbf{P}(t) = (\mathbf{T}^t - \mathbf{I})\, \mathbf{P}(t)$$

so that

$$\lim_{h \to 0} \frac{\mathbf{P}(t + h) - \mathbf{P}(t)}{h} = \left(\frac{\mathbf{T}^t - \mathbf{I}}{h} \right) \mathbf{P}(t)$$

Simplifying by letting $\mathbf{C} = (\mathbf{T}^t - I)/h$ leads to the matrix differential equation

$$\mathbf{P}'(t) = \mathbf{C}\, \mathbf{P}(t) \tag{7.13}$$

where the form of $\mathbf{C}$ for this problem is

$$\mathbf{C} = \begin{bmatrix} -\lambda & \mu \\ \lambda & -\mu \end{bmatrix} \tag{7.14}$$

The columns of $\mathbf{C}$ must add to zero because the rows of $\mathbf{T}$ added to one. Matrix equation 7.13 is seen to be exactly equivalent to equation set 7.3. The Laplace transform of (7.13) is

$$s\, \mathbf{Z}(s) - \mathbf{P}(0) = \mathbf{C}\mathbf{Z}(s) \tag{7.15}$$

which is equivalent to equation set 7.4. With a bit of practice it is possible to write down the transition matrix $\mathbf{T}$ from first principles, go immediately to matrix $\mathbf{C}$, and then to the transformed equation set 7.15. The latter is solved for $\mathbf{Z} = \mathbf{Z}(s)$ to give

$$s\mathbf{Z} - \mathbf{C}\mathbf{Z} = \mathbf{P}(0)$$

or

$$(s\mathbf{I} - \mathbf{C})\mathbf{Z} = \mathbf{P}(0)$$

Our procedure is to construct a matrix $\mathbf{C}^*$ by the rule

$$\mathbf{C}^* = s\mathbf{I} - \mathbf{C}$$

$$= s\mathbf{I} - \frac{\mathbf{T}^t - \mathbf{I}}{h} \tag{7.16}$$

and then find a vector $\mathbf{Z}$ which satisfies

$$\mathbf{C}*\mathbf{Z} = \mathbf{P}(0) \tag{7.17}$$

This is where the hard work of the problem will come for more complicated systems. The technique is illustrated in the next section. Many of the problems posed cannot be solved in a clear analytic way, so we often settle for the steady-state solution of (7.18), rather than the general, or *transient*, solution of (7.17).

In the steady state the *rate of change* of transitions from one state to another has settled down to zero. Do not be deceived: we are not claiming that there are no longer transitions from one state to another. We are saying that after a sufficiently long transient introductory period, the *rate* of these transitions is stable. This condition is described by setting $\mathbf{P}'(t) = \mathbf{0}$ whereby (7.13) reduces to

$$\mathbf{C}\,\mathbf{P}(\infty) = \mathbf{0}$$

which is usually written as

$$\mathbf{CP} = \mathbf{0} \tag{7.18}$$

with $\mathbf{P}^t = [P_0(\infty), P_1(\infty)] = [P_0, P_1]$. The time dependence is omitted for convenience since at steady state the elapsed time is no longer relevant. The homogeneous set of equations 7.18 is underdetermined, since the rows of $\mathbf{C}$ are linearly dependent (recall that each column sums to zero), and the determinant $|\mathbf{C}|$ vanishes. Thus there are an infinite number of solutions. The only one of interest is the one for which

$$P_0 + P_1 = 1 \tag{7.19}$$

so the last row, or any other convenient row, of (7.18) is replaced by (7.19). The resulting equations are

$$\begin{bmatrix} -\lambda & \mu \\ 1 & 1 \end{bmatrix} \begin{bmatrix} P_0 \\ P_1 \end{bmatrix} = \begin{bmatrix} 0 \\ 1 \end{bmatrix} \tag{7.20}$$

or

$$\begin{aligned} -\lambda P_0 + \mu P_1 &= 0 \\ P_0 + P_1 &= 1 \end{aligned} \tag{7.21}$$

whereby

$$P_0 = \frac{\mu}{\lambda + \mu}$$

the same result we obtained in (7.10).

Figure 7.2 shows the availability curves for three selected conditions for which the time units are scaled so that $\lambda = 1$ and $\mu = 10, 1, 0.1$, representing various relative speeds of repair. The value $\mu = 10$ can be considered "normal" because it describes a unit that can be repaired in about one-tenth as long as an average lifetime. The value $\mu = 1$ would have to be considered poor service, and $\mu = 0.1$ is execrably bad.

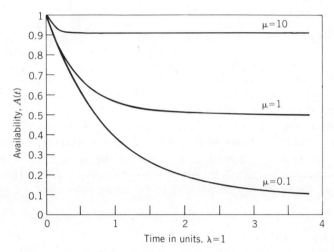

Figure 7.2 $A(t)$ for one component with $\mu = 10, 1, 0.1$.

For all but the last condition, one average lifetime could be considered the transient period.

EXERCISES 7.1

1. Find the average availability of a nonrepairable CFR component for a 100-hour mission if its failure rate is $\lambda = 0.01$. What is the reliability of the component at the end of the mission?

2. Repeat Exercise 1 for a ten-hour mission for the same component.

3. Master the details of proof that lead to Equations 7.3a and b.

4. Relying on your knowledge of elementary differential equations, solve Equations 7.3a and b without using Laplace transforms. Which method do you prefer?

5. Verify the partial fraction decomposition that yields Equation 7.8 from (7.6).

6. Perform the integration that carries Equation 7.9 to (7.11).

7. Show that Equation 7.15 is equivalent to (7.4).

8. Generate curves like those in Figure 7.2 using standard time units ($\lambda = 1$), with repair rates $\mu = 5$ and $\mu = 2$. How long do you consider the transient period to be?

   Note. With an electronic spreadsheet this problem can be solved in just a few minutes. I suggest using named ranges LAM and MU (say) with absolute addresses, so that curves for any λ and μ may be generated instantly.

7.2 TWO PARALLEL UNITS WITH A SINGLE REPAIRMAN

The second system to be considered is that of two identical or equivalent units in parallel, with a single repairman available to service the units. The possible states of the system are

S_0: Both units functioning.

S_1: One unit functioning, one failed and under repair.

S_2: Both units failed, one under repair.

The state space is depicted in Figure 7.3. It is assumed, as before, that failed units are noticed and that repair begins immediately if the repairman is free. It is also assumed that the Poisson-type postulates hold; that is, two events (failures, repairs, or both) cannot occur in the same infinitesimal interval of length h. Notwithstanding this assumption, we write the transition matrix as though double events were possible and then make a correction. We do this because it has been found safer from a pedagogical point of view. Students do not have to decide which events are "impossible"; they can write down all the events and then expunge those terms involving h^2 because they represent differentials of higher order. With this understanding, the exact transition matrix is seen to be (7.22)

$$\text{Final State}$$

$$\begin{array}{c} \text{Initial State} \end{array} \begin{array}{c} 0 \\ 1 \\ 2 \end{array} \begin{bmatrix} (1-\lambda h)^2 & 2\lambda h(1-\lambda h) & (\lambda h)^2 \\ \mu h(1-\lambda h) & (1-\lambda h)(1-\mu h)+(\lambda h)(\mu h) & (1-\mu h)\lambda h \\ \mu h(0) & \mu h & 1-\mu h \end{bmatrix} \quad (7.22)$$

Let us examine a few of the elements in detail. The northwest corner element is the joint probability that neither unit fails. The next element to the right is the probability that one unit fails (probability $= \lambda h$) and the other one doesn't (probability $= 1-\lambda h$), times the binomial coefficient $\binom{2}{1}$, which is the number of ways of selecting the failing and nonfailing components. Notice that when the term is expanded and the h^2 term expunged, it reduces to $2\lambda h$. The structure of the term is usually explained by saying that λh is the probability of a single item failing and, since there are two items, it should be doubled. Alert readers may feel nervous about such logic, as indeed they should—particularly since such casual reasoning is treacherous and unnecessary.

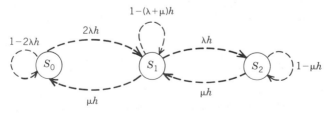

Figure 7.3 The state space diagram for two units, one repairman.

The center element of the matrix is the probability that (a) neither unit changes condition—the one functioning continues to function and the one under repair stays under repair; *or* (b) both units change condition—the functioning unit fails and the unit under repair is returned to service. When the differentials of higher order are removed, the entry reduces to $1 - (\lambda + \mu)h$. The element in the lower left corner is not an "approximate zero" like h^2, but an exact zero, since it is impossible for a single repairman to return the system to state S_0 from S_2 in an infinitesimal interval of length h. Notice that all the rows add exactly to one, as they should.

When all the h^2 terms are expunged, the transition matrix assumes the more traditional form

$$
\mathbf{T} = \begin{array}{c} \\ 0 \\ 1 \\ 2 \end{array}
\begin{array}{ccc} 0 & 1 & 2 \end{array} \atop
\begin{bmatrix} 1 - 2\lambda h & 2\lambda h & 0 \\ \mu h & 1 - (\lambda + \mu)h & \lambda h \\ 0 & \mu h & 1 - \mu h \end{bmatrix} \tag{7.23}
$$

The corresponding matrix $\mathbf{C} = (\mathbf{T}^t - \mathbf{I})/h$ is

$$
\mathbf{C} = \begin{bmatrix} -2\lambda & \mu & 0 \\ 2\lambda & -(\lambda + \mu) & \mu \\ 0 & \lambda & -\mu \end{bmatrix} \tag{7.24}
$$

Again, it will be assumed that the system starts in state S_0 at time $t = 0$, so the equation set has the form

$$
\begin{bmatrix} P_0'(t) \\ P_1'(t) \\ P_2'(t) \end{bmatrix} = \mathbf{C} \begin{bmatrix} P_0(t) \\ P_1(t) \\ P_2(t) \end{bmatrix}
$$

Taking Laplace transforms gives

$$
\begin{bmatrix} sZ_0 - 1 \\ sZ_1 \\ sZ_2 \end{bmatrix} = \mathbf{C} \begin{bmatrix} Z_0 \\ Z_1 \\ Z_2 \end{bmatrix} \tag{7.25}
$$

or

$$
(s\mathbf{I} - \mathbf{C})\mathbf{Z} = [\, 1, 0, 0 \,]^t
$$

Henceforth we refer to $\mathbf{C}^* = s\mathbf{I} - \mathbf{C}$ as the Laplace solution coefficient matrix. A more familiar form for the system is

$$
\begin{bmatrix} s + 2\lambda & -\mu & 0 \\ -2\lambda & s + \lambda + \mu & -\mu \\ 0 & -\lambda & s + \mu \end{bmatrix} \begin{bmatrix} Z_0 \\ Z_1 \\ Z_2 \end{bmatrix} = \begin{bmatrix} 1 \\ 0 \\ 0 \end{bmatrix} \tag{7.26}
$$

As an arithmetic check, notice that each column sums to the value s. The system can be solved, but not easily, using Cramer's rule. Writing $NUM(i)(i = 0, 1, 2)$ for the numerators and $DNOM$ for the denominator gives

$$Z_i = \frac{\text{NUM}(i)}{\text{DNOM}} \qquad i = 0, 1, 2 \qquad (7.27)$$

After some algebra the value for DNOM is found to be

$$\text{DNOM} = s^3 + s^2(3\lambda + 2\mu) + s(\mu^2 + 2\lambda\mu + 2\lambda^2) \qquad (7.28)$$
$$= s(s - r_1)(s - r_2) \qquad (7.29)$$

where r_1 and r_2 are the roots of the quadratic factor of DNOM. In particular,

$$r_{1,2} = \frac{-(3\lambda + 2\mu) \pm \sqrt{\lambda^2 + 4\lambda\mu}}{2} \qquad (7.30)$$

The value of NUM(0) is easily found to be

$$\text{NUM}(0) = s^2 + s(\lambda + 2\mu) + \mu^2 \qquad (7.31)$$
$$= (s - a_1)(s - a_2) \text{ (say)}$$

where

$$a_{1,2} = \frac{-(\lambda + 2\mu) \pm \sqrt{\lambda^2 + 4\lambda\mu}}{2} \qquad (7.32)$$

These roots for the numerator differ by λ from the roots of the denominator. We write

$$Z_0 = \frac{(s - a_1)(s - a_2)}{s(s - r_1)(s - r_2)} \qquad (7.33)$$
$$= \frac{(s - \lambda - r_1)(s - \lambda - r_2)}{s(s - r_1)(s - r_2)}$$
$$= \frac{c_0}{s} + \frac{c_1}{s - r_1} + \frac{c_2}{s - r_2} \qquad (7.34)$$

With much tedious algebra it can be proved that

$$c_0 = \frac{a_1 a_2}{r_1 r_2} = \frac{\mu^2}{\mu^2 + 2\mu\lambda + 2\lambda^2} \qquad (7.35)$$

$$c_1 = \frac{\lambda(\lambda + r_2 - r_1)}{r_1(r_1 - r_2)} \qquad (7.36)$$

$$c_2 = \frac{\lambda(\lambda + r_1 - r_2)}{r_2(r_2 - r_1)} \qquad (7.37)$$

The full general solution, then, for the probability of the system to be in state 0 is

$$P_0(t) = \frac{\mu^2}{\mu^2 + 2\mu\lambda + 2\lambda^2} + c_1 e^{r_1 t} + c_2 e^{r_2 t} \qquad (7.38)$$

Notice that r_1 and r_2 are negative, so that $P_0(t)$ converges for large t to the steady-state value

$$P_0 = \frac{\mu^2}{\mu^2 + 2\lambda\mu + 2\lambda^2} \qquad (7.39)$$

To derive the expression for $P_1(t)$, we use

$$Z_1 = \frac{\text{NUM}(1)}{\text{DNOM}}$$

$$= \frac{1}{\text{DNOM}} \begin{vmatrix} s + 2\lambda & 1 & 0 \\ -2\lambda & 0 & -\mu \\ 0 & 0 & s + \mu \end{vmatrix} \tag{7.40}$$

$$= \frac{2\lambda(s + \mu)}{\text{DNOM}} = \frac{2\lambda(s + \mu)}{s(s - r_1)(s - r_2)}$$

$$= \frac{d_0}{s} + \frac{d_1}{s - r_1} + \frac{d_2}{s - r_2} \tag{7.41}$$

where the coefficients d_i are found to be

$$d_0 = \frac{2\lambda\mu}{r_1 r_2} = \frac{2\lambda\mu}{\mu^2 + 2\lambda\mu + 2\lambda^2} \tag{7.42}$$

$$d_1 = \frac{2\lambda(r_1 + \mu)}{r_1(r_1 - r_2)} \tag{7.43}$$

$$d_2 = \frac{2\lambda(r_2 + \mu)}{r_2(r_2 - r_1)} \tag{7.44}$$

The formula for $P_1(t)$ is therefore

$$P_1(t) = \frac{2\lambda\mu}{\mu^2 + 2\lambda\mu + 2\lambda^2} + d_1 e^{r_1 t} + d_2 e^{r_2 t} \tag{7.45}$$

The availability of the system is given by

$$
\begin{aligned}
A(t) &= P_0(t) + P_1(t) \\
&= (c_0 + d_0) + (c_1 + d_1)e^{r_1 t} + (c_2 + d_2)e^{r_2 t} \\
&= \frac{\mu^2 + 2\lambda\mu}{\mu^2 + 2\lambda\mu + 2\lambda^2} + \frac{\lambda^2 + \lambda(r_1 + r_2) + 2\lambda\mu}{r_1 - r_2}\left(\frac{e^{r_1 t}}{r_1} - \frac{e^{r_1 t}}{r_2}\right)
\end{aligned} \tag{7.46}
$$

This simplifies nicely because $r_1 + r_2 = -(3\lambda + 2\mu)$ so that

$$\lambda^2 + \lambda(r_1 + r_2) + 2\lambda\mu = -2\lambda^2$$

and hence

$$A(t) = \frac{\mu^2 + 2\lambda\mu}{\mu^2 + 2\lambda\mu + 2\lambda^2} - \frac{2\lambda^2}{r_1 - r_2}\left(\frac{e^{r_1 t}}{r_1} - \frac{e^{r_2 t}}{r_2}\right) \tag{7.47}$$

It may be noted in the foregoing derivations that $c_0 + c_1 + c_2 = 1$ because $P_0(0) = 1$, and $d_0 + d_1 + d_2 = 0$ because $P_1(0) = 0$.

If $P_0(t)$ and $P_1(t)$ are of no interest and only $A(t)$ is wanted, it may be obtained directly from $A(t) = 1 - P_2(t)$. This will be easier than the preceding calculations because the form for the numerator of Z_2 is

$$\text{NUM}(2) = \begin{vmatrix} s + 2\lambda & -\mu & 1 \\ -2\lambda & s + \lambda + \mu & 0 \\ 0 & -\lambda & 0 \end{vmatrix} = 2\lambda^2 \tag{7.48}$$

which leads to an easy partial fractions decomposition, namely

$$Z_2 = \frac{e_0}{s} + \frac{e_1}{s - r_1} + \frac{e_2}{s - r_2}$$

with

$$e_0 = \frac{2\lambda^2}{r_1 r_2}$$

$$e_1 = \frac{2\lambda^2}{r_1(r_1 - r_2)}$$

$$e_2 = \frac{2\lambda^2}{r_2(r_2 - r_1)}$$

The form of $P_2(t)$ follows immediately as

$$P_2(t) = \frac{2\lambda^2}{\mu^2 + 2\lambda\mu + 2\lambda^2} + \frac{2\lambda^2}{r_1 - r_2}\left(\frac{e^{r_1 t}}{r_1} - \frac{e^{r_2 t}}{r_2}\right) \tag{7.49}$$

Example 7.1

Figures 7.4 through 7.8 display the graphs for the same three conditions as before, that is, where t is expressed in "standard" time units so that $\lambda = 1$ and $\mu = 10, 1, 0.1$.

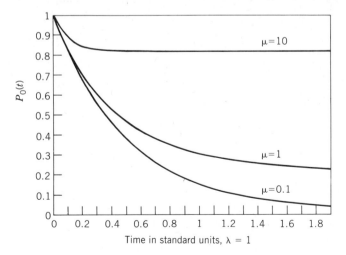

Figure 7.4 $P_0(t)$ for two units and a single repairman, $\mu = 10, 1, 0.1$.

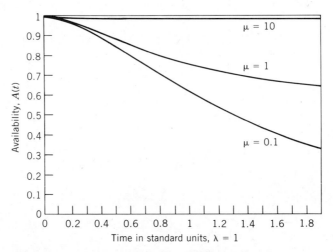

Figure 7.5 $A(t)$ for two units and a single repairman, $\mu = 10, 1, 0.1$.

The pertinent coefficients are shown in Table 7.2. Figure 7.4 displays the three plots for $P_0(t)$, and Figure 7.5 those for $A(t)$. Figures 7.6 through 7.8 display the same information in a different way by showing $P_0(t)$ and $A(t)$ for the separate μ conditions, 10, 1, and 0.1.

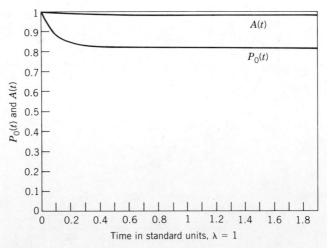

Figure 7.6 $A(t)$ and $P_0(t)$ for two units, one repairman, $\mu = 10$.

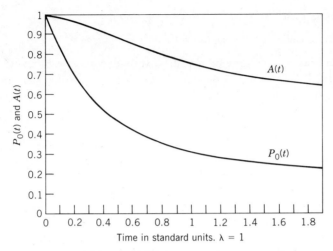

Figure 7.7 $A(t)$ and $P_0(t)$ for two units, one repairman, $\mu = 1$.

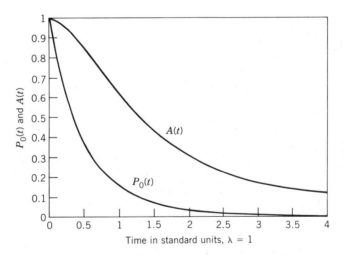

Figure 7.8 $A(t)$ and $P_0(t)$ for two units, one repairman, $\mu = 0.1$.

TABLE 7.2 Coefficients for Example 7.1 ($\lambda = 1$ in All Cases)

μ	r_1	r_2	c_0	c_1	c_2	d_0	d_1	d_2
10	-8.30	-14.7	.820	.102	.079	.164	-0.064	-0.100
1	-1.38	-3.62	.2	.4	.4	.4	0.247	-0.647
0.1	-1.01	-2.19	.00452	.154	.842	.0905	1.523	-1.613

EXERCISES 7.2

1. Verify the formula for $P_2(t)$ in (7.49).

2. Find the steady-state solution for the two-unit parallel system with a single repairman. Use the matrix method of the previous section, and show that it agrees with $P_i(\infty)$ ($i = 0, 1, 2$) from (7.38), (7.45), and (7.49).

3. Repeat the developments of this section by assuming that there are two equally competent repairmen available.

Comment. This situation is much easier than when there is only a single repairman.

4. Using standard time units (set $\lambda = 1$), re-create the graphs of this section for $\mu = 5$ and $\mu = 0.5$.

5. Verify the derivations of the following formulae in this chapter.
   a. (7.23) from (7.22).
   b. (7.25) from the previous (unnumbered) formula.
   c. (7.35) through (7.37) from (7.34).
   d. (7.42) through (7.44) from (7.41).
   e. (7.48) from (7.26).
   f. (7.49) from (7.48).

6. Verify the numerical values in Table 7.2 from the appropriate formulae.

7.3   TWO PARALLEL UNITS WITH TWO REPAIRMEN

For the sake of completeness, we present here the situation posed as Exercise 7.2.3, that in which two equally competent repairmen are available. The transition matrix (7.22) is valid, except for the last row which must be replaced by

$$[(\mu h)^2 \quad 2\mu h(1 - \mu h) \quad (1 - \mu h)^2] \tag{7.50}$$

and which simplifies for very small h to

$$[0 \quad 2\mu h \quad 1 - 2\mu h] \tag{7.51}$$

It is easily verified that the Laplace solution coefficient matrix $\mathbf{C}^* = s\mathbf{I} - \mathbf{C}$ has the form

$$\mathbf{C}^* = \begin{bmatrix} s + 2\lambda & -\mu & 0 \\ -2\lambda & s + \lambda + \mu & -2\mu \\ 0 & -\lambda & s + 2\mu \end{bmatrix} \tag{7.52}$$

and its determinant reduces to

$$\text{DNOM} = s(s + \lambda + \mu)[s + 2(\lambda + \mu)] \tag{7.53}$$

Application of Cramer's rule leads to

$$Z_0 = \frac{s^2 + s(\lambda + 3\mu) + 2\mu^2}{\text{DNOM}}$$

$$= \frac{c_0}{s} + \frac{c_1}{s + \lambda + \mu} + \frac{c_2}{s + 2(\lambda + \mu)} \tag{7.54}$$

where

$$c_0 = \frac{\mu^2}{(\lambda + \mu)^2} \tag{7.55a}$$

$$c_1 = \frac{2\lambda\mu}{(\lambda + \mu)^2} \tag{7.55b}$$

$$c_2 = \frac{\lambda^2}{(\lambda + \mu)^2} \tag{7.55c}$$

These sum to unity because of the usual assumption that $P_0(0) = 1$ because the system starts with both components good. The general (transient) solution for $P_0(t)$ is the Laplace inverse of (7.54) using (7.55a–c), to wit

$$P_0(t) = \frac{1}{(\lambda + \mu)^2} [\mu^2 + 2\lambda\mu e^{-(\lambda + \mu)t} + \lambda^2 e^{-2(\lambda + \mu)t}] \tag{7.56}$$

Using the same techniques, we find that

$$P_1(t) = \frac{1}{(\lambda + \mu)^2} [\mu^2 + 2\lambda\mu + 2\lambda(\lambda - \mu)e^{-(\lambda + \mu)t} - 2\lambda^2 e^{-2(\lambda + \mu)t}] \tag{7.57}$$

and adding these two gives

$$A(t) = \frac{1}{(\lambda + \mu)^2} [(2\lambda\mu + \mu^2) + 2\lambda^2 e^{-(\lambda + \mu)t} - \lambda^2 e^{-2(\lambda + \mu)t}] \tag{7.58}$$

If $P_0(t)$ and $P_1(t)$ are of no interest in themselves and only $A(t)$ is wanted, it is more easily obtained from the relationship $A(t) = 1 - P_2(t)$. This is so because the form of Z_2 is

$$Z_2 = \frac{2\lambda^2}{\text{DNOM}}$$

with a much easier partial fraction decomposition. Specifically,

$$Z_2 = \frac{\lambda^2}{(\lambda + \mu)^2} \left[\frac{1}{s} - \frac{2}{s + \lambda + \mu} + \frac{1}{s + 2(\lambda + \mu)} \right] \tag{7.59}$$

so that

$$P_2(t) = \frac{\lambda^2}{(\lambda + \mu)^2} [1 - 2e^{-(\lambda + \mu)t} + e^{-2(\lambda + \mu)t}] \tag{7.60}$$

   It is interesting to compare the availabilities in the one-repairman situation and in the two-repairman situation. This is done graphically in Figure 7.9 for our usual conditions $\lambda = 1$, $\mu = 10$, 1, 0.1. Notice that for $\mu = 10$ the repair rate is so good compared with the failure rate that it hardly matters whether there is one repairman or two. For $\mu = 0.1$ there is a large percentage difference in the two availabilities, but both are so low that again the number of repairmen hardly matters.

   An amplification of the situation just presented is the one in which one repairman can assist the other if only one unit has failed. Suppose that, with two repairmen working together on a single unit, the repair rate is $m\mu$ instead of μ, with $1 < m \leq 2$. The exact transition matrix is

$$
\begin{array}{c}
 \\
0 \\
1 \\
2
\end{array}
\begin{array}{ccc}
\quad 0 & \quad 1 & \quad 2 \\
\left[\begin{array}{ccc}
(1 - \lambda h)^2 & 2\lambda h(1 - \lambda h) & (\lambda h)^2 \\
m\mu h(1 - \lambda h) & (1 - m\mu h)(1 - \lambda h) + (m\mu h)(\lambda h) & \lambda h(1 - m\mu h) \\
(\mu h)^2 & 2\mu h(1 - \mu h) & (1 - \mu h)^2
\end{array}\right]
\end{array}
\tag{7.61}
$$

which reduces to the traditional form

$$
\mathbf{T} = \begin{bmatrix}
1 - 2\lambda h & 2\lambda h & 0 \\
m\mu h & 1 - (\lambda + m\mu)h & \lambda h \\
0 & 2\mu h & 1 - 2\mu h
\end{bmatrix}
\tag{7.62}
$$

leading, after the usual steps, to the transform coefficient matrix

$$
\mathbf{C}^* = \begin{bmatrix}
s + 2\lambda & -m\mu & 0 \\
-2\lambda & s + \lambda + m\mu & -2\mu \\
0 & -\lambda & s + 2\mu
\end{bmatrix}
\tag{7.63}
$$

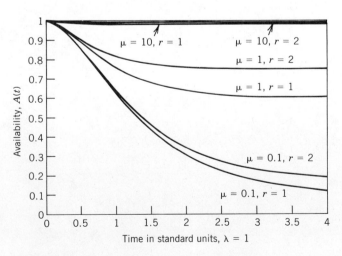

Figure 7.9 $A(t)$ for one versus two repairmen, $\mu = 10, 1, 0.1$.

The denominator determinant for the Cramer's rule solution is

$$\text{DNOM} = s^3 + s^2[3\lambda + \mu(m + 2)] + 2s(\lambda^2 + 2\lambda\mu + m\mu^2) \tag{7.64}$$

whose roots are zero and r_1, r_2 with

$$r_{1,2} = \frac{-[3\lambda + (m + 2)\mu] \pm \sqrt{\lambda^2 + 2\lambda\mu(3m - 2) + \mu^2(2 - m)^2}}{2} \tag{7.65}$$

The numerators for Z_0, Z_1, and Z_2 are

$$\text{NUM}(0) = s^2 + s(\lambda + 2\mu + m\mu) + 2m\mu^2 \tag{7.66}$$
$$\text{NUM}(1) = 2\lambda(s + 2\mu) \tag{7.67}$$
$$\text{NUM}(2) = 2\lambda^2 \tag{7.68}$$

Clearly $P_2(t)$ will be the easiest to calculate, and the availability can be obtained as

$$A(t) = 1 - P_2(t) \tag{7.69}$$

Following the procedures of Section 7.2, the solution is tedious, but it can be summarized in the form

$$P_2(t) = \frac{2\lambda^2}{r_1 r_2} + \frac{2\lambda^2}{r_1(r_1 - r_2)}e^{r_1 t} + \frac{2\lambda^2}{r_2(r_2 - r_1)}e^{r_2 t} \tag{7.70}$$

and plotted easily using an electronic spreadsheet. The behavior is similar to that shown in Figure 7.9, which corresponds to $m = 1$.

An interesting situation is that in which $m = 2$, when two repairmen working together can make the repair twice as fast as one working alone. The nonzero roots of DNOM simplify to

$$r_{1,2} = \frac{-(3\lambda + 4\mu) \pm \sqrt{\lambda^2 + 8\lambda\mu}}{2} \tag{7.71}$$

with similar simplification of (7.66) and 7.67).

EXERCISES 7.3

1. Starting from Equations 7.64 through 7.67, derive the solutions for $P_0(t)$ and $P_1(t)$ in a form similar to (7.70).

2. Verify that $P_2(t)$ from Equation 7.70, and $P_0(t)$ and $P_1(t)$ from Exercise 1 do indeed sum to unity.

3. For $\lambda = 1$ and $\mu = 10$ plot $A(t)$ for $m = \{1, 1.5, 2\}$. Do this using an electronic spreadsheet with named ranges so that the formulae for $A(t)$ need be entered only one time each.

4. Repeat Exercise 3 for $\mu = 1$.

5. Repeat Exercise 3 for $\mu = .1$.

6. Repeat Exercise 3, plotting $P_0(t)$ rather than $A(t)$
 (a) For $\mu = 10$,
 (b) For $\mu = 1$,
 (c) For $\mu = .1$.

7.4   TWO UNEQUAL UNITS, TWO REPAIRMEN

In this section we consider a system similar to that of the previous section but with an important difference: the two units, called A and B, have different failure rates. We assume that the two repairmen are equally competent but that the units are not equally easy to service. Thus there are failure rates λ_1 and λ_2 and repair rates μ_1 and μ_2. The possible states are

S_0: Both units functioning.
S_1: A functioning, B under repair.
S_2: B functioning, A under repair.
S_3: Both units under repair.

The system described is assumed to be parallel, but it could also refer to a series system provided that a unit can fail even when the system is down. The main difference between the two is that for the series system

$$A(t) = P_0(t)$$

whereas for the parallel system

$$A(t) = P_0(t) + P_1(t) + P_2(t)$$

The system diagram is given in Figure 7.10. Notice that if the repairmen were

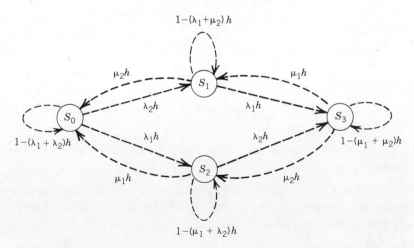

Figure 7.10 State diagram for two unlike components and two equally competent repairmen.

not equally competent, there would be four repair rates to consider with μ_{ij} being the service time for repairman i on unit j. This model is considered by Dhillon and Singh (1981).

Readers are encouraged to write out in full (see Exercise 1) the transition matrix for this situation. Setting $h^2 = 0$ we arrive at the reduced transition matrix

$$
\mathbf{T} = \begin{bmatrix}
1 - (\lambda_1 + \lambda_2)h & \lambda_2 h & \lambda_1 h & 0 \\
\mu_2 h & 1 - (\lambda_1 + \mu_2)h & 0 & \lambda_1 h \\
\mu_1 h & 0 & 1 - (\mu_1 + \lambda_2)h & \lambda_2 h \\
0 & \mu_1 h & \mu_2 h & 1 - (\mu_1 + \mu_2)h
\end{bmatrix}
\tag{7.72}
$$

Following the usual substitutions $\mathbf{C} = (\mathbf{T}^t - \mathbf{I})/h$ and $\mathbf{C}^* = s\mathbf{I} - \mathbf{C}$ yields the Laplace Markov matrix

$$
\mathbf{C}^* = \begin{bmatrix}
s + \lambda_1 + \lambda_2 & -\mu_2 & -\mu_1 & 0 \\
-\lambda_2 & s + \lambda_1 + \mu_2 & 0 & -\mu_1 \\
-\lambda_1 & 0 & s + \mu_1 + \lambda_2 & -\mu_2 \\
0 & -\lambda_1 & -\lambda_2 & s + \mu_1 + \mu_2
\end{bmatrix}
\tag{7.73}
$$

The desired solution to $\mathbf{C}^*\mathbf{Z} = [\,1, 0, 0, 0\,]^t$ is very onerous to obtain, involving as it does the expansion of one four-by-four and several three-by-three determinants. The development is left to readers who enjoy such activities.

In a simpler situation of similar type, reprised by Chatterjee (1971), there are two identical units in parallel and two equally competent repairmen; however, no repair is initiated until both units have failed. The four states, depicted in Figure 7.11, are

S_0: Both units in service.

S_1: One unit working, one failed but failure not noted.

S_2: Both units failed, both under repair.

S_3: One unit repaired and functioning, the other still under repair.

The exact transition matrix is

$$
\begin{array}{c}
\begin{array}{cccc} \quad 0 & \quad 1 & \quad 2 & \quad 3 \end{array} \\
\begin{array}{c} 0 \\ 1 \\ 2 \\ 3 \end{array}
\begin{bmatrix}
(1 - \lambda h)^2 & 2\lambda h(1 - \lambda h) & (\lambda h)^2 & 0 \\
0 & (1)(1 - \lambda h) & \lambda h(1) & 0 \\
(\mu h)^2 & 0 & (1 - \mu h)^2 & 2\mu h(1 - \mu h) \\
\mu h & 0 & \lambda h(1 - \mu h) & (1 - \lambda h)(1 - \mu h)
\end{bmatrix}
\end{array}
\tag{7.74}
$$

which leads, after the usual algebraic substitutions $\mathbf{C} = (\mathbf{T}^t - \mathbf{I})/h$ and $\mathbf{C}^* = s\mathbf{I} - \mathbf{C}$, to

$$
\mathbf{C}^* = \begin{bmatrix}
s + 2\lambda & 0 & 0 & -\mu \\
-2\lambda & s + \lambda & 0 & 0 \\
0 & -\lambda & s + 2\mu & -\lambda \\
0 & 0 & -2\mu & s + \lambda + \mu
\end{bmatrix}
\tag{7.75}
$$

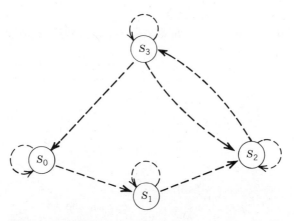

Figure 7.11   State diagram for the system in which repair awaits the failure of the second component.

The usual Cramer's rule solution is somewhat difficult to obtain for this situation, since the value of the denominator determinant is

$$\text{DNOM} = s^4 + s^3(3\mu + 4\lambda) + s^2(2\mu^2 + 9\lambda\mu + 5\lambda^2) \\ + 2\lambda s(3\mu^2 + 3\lambda\mu + \lambda^2) \tag{7.76}$$

This is factorable into $s(s - r_1)(s - r_2)(s - r_3)$ whose roots $\{r_i\}$ are not obtainable analytically. We do know that the solution form for $A(t) = 1 - P_2(t)$ is

$$A(t) = c_0 + c_1 e^{r_1 t} + c_2 e^{r_2 t} + c_3 e^{r_3 t} \tag{7.77}$$

which has the usual exponential convergence to the steady-state value $c_0 = A(\infty)$. For particular values of λ and μ an approximate numerical solution might be obtained.

The steady-state solution is not difficult to find. It comes from the usual procedure of setting $P'_i(t)$ equal to zero in the matrix equation

$$\mathbf{P}'(t) = \mathbf{C}\,\mathbf{P}(t)$$

The result, designating $P_i = \lim\limits_{t \to \infty} P_i(t)$, is

$$\begin{bmatrix} -2\lambda & 0 & 0 & \mu \\ 2\lambda & -\lambda & 0 & 0 \\ 0 & \lambda & -2\mu & \lambda \\ 0 & 0 & 2\mu & -(\lambda + \mu) \end{bmatrix} \begin{bmatrix} P_0 \\ P_1 \\ P_2 \\ P_3 \end{bmatrix} = \begin{bmatrix} 0 \\ 0 \\ 0 \\ 0 \end{bmatrix} \tag{7.78}$$

The requirement that $P_0 + P_1 + P_2 + P_3 = 1$ is used to replace any one of the redundant equations set corresponding to the expansion of (7.78). The steady-state availability is

$$A(\infty) = 1 - P_2 = \frac{2\lambda\mu + 3\mu^2}{\lambda^2 + 3\lambda\mu + 3\mu^2} \tag{7.79}$$

EXERCISES 7.4

1. Write out in detail the transition matrix for the situation presented in Figure 7.10. Verify that it leads to matrix 7.72.

2. If you are good at determinants, find the solutions to $\mathbf{C}*\mathbf{Z} = [\,1,\,0,\,0,\,0\,]^t$ with $\mathbf{C}*$ given in matrix 7.73.

3. Verify the correctness of the steady-state matrix 7.78.

4. Derive the steady-state availability $A(\infty)$ given by (7.79).

7.5 MULTIPLE UNITS AND REPAIRMEN

An interesting situation considered by Sandler (1963, page 128) is that of n units and r repairmen. For example, with three units and two repairmen the simplified transition matrix, with state k designating k failed units, is

$$
\mathbf{T} = \begin{array}{c} \\ 0 \\ 1 \\ 2 \\ 3 \end{array}
\begin{array}{cccc}
0 & 1 & 2 & 3 \\
\left[\begin{array}{cccc}
1 - 3\lambda h & 3\lambda h & 0 & 0 \\
\mu & 1 - (2\lambda + \mu)h & 2\lambda h & 0 \\
0 & 2\mu h & 1 - (\lambda + 2\mu)h & \lambda h \\
0 & 0 & 2\mu h & 1 - 2\mu h
\end{array}\right]
\end{array} \tag{7.80}
$$

The steady-state probabilities are found by solving the equations

$$
\left.\begin{array}{l}
-3\lambda P_0 \qquad\quad + \mu P_1 \qquad\qquad\qquad\qquad\qquad\quad = 0 \\
3\lambda P_0 - (2\lambda + \mu)P_1 \qquad\quad + 2\mu P_2 \qquad\qquad\quad = 0 \\
\qquad\qquad 2\lambda P_1 - (\lambda + 2\mu)P_2 + 2\mu P_3 = 0 \\
\qquad\qquad\qquad\qquad\qquad\quad \lambda P_2 - 2\mu P_3 = 0 \\
P_0 \qquad\quad + P_1 \qquad\qquad + P_2 \quad + P_3 = 1
\end{array}\right\} \tag{7.81}
$$

The resulting solution set is

$$
\left.\begin{array}{l}
\Delta = 4\mu^3 + 12\mu^2\lambda + 12\lambda^2\mu + 6\lambda^3 \\
P_0 = 4\mu^3/\Delta \\
P_1 = 12\lambda\mu^3/\Delta \\
P_2 = 12\lambda^2\mu/\Delta \\
P_3 = 6\lambda^3/\Delta
\end{array}\right\} \tag{7.82}
$$

The form of these equations is

$$
P_k = \frac{\text{term }(k)}{\sum_m \text{ term }(m)} \tag{7.83}
$$

which is encountered regularly in steady-state solution sets.

We turn now to the general case of n units with $r\,(< n)$ repairmen. The possible events are displayed in Table 7.3. The state index $k\;(k = 0, 1, \ldots, n)$ designates

TABLE 7.3 The Event Space for n Units, r Repairmen

Event	State at t	Net Activity, Coefficient	State at $t + h$
		Boundary Case	
$E\,1$:	0	No change, $M_{0,0}$	0
$E\,2$:	1	One repair, $M_{0,-1}$	0
		General Case	
$E\,3$:	$k - 1$	One failure, $M_{k,+1}$	k
$E\,4$:	k	No change, $M_{k,0}$	k
$E\,5$:	$k + 1$	One repair, $M_{k,-1}$	k

that there are k failed units. The procedure to be followed is a generalization of that used in Section 2.3 for the simple Poisson arrival process. To illustrate the technique, consider first the boundary condition related to events $E\,1$ and $E\,2$. We write

$$
\begin{aligned}
P_0 &= P(E\,1 \text{ or } E\,2) \\
&= P(E\,1) + P(E\,2) \\
&= P_0(t)\,M_{0,0} + P_1(t)\,M_{0,-1}
\end{aligned}
\tag{7.84}
$$

The coefficients $M_{k,j}$ indicate the terminal state k and the net change j in failed units during the infinitesimal interval of length h. The value $j = 1$ means that a failure has occurred in the interval h and the value $j = -1$ means that a repair has been made. Thus $M_{0,0}$ is the probability that the final failed-unit count at time $t + h$ is zero because there have been no failures and no repairs. It implies that the good-unit count at time t was n. Thus the value of $M_{0,0}$ is

$$
M_{0,0} = (1 - \lambda h)^n (1)
\tag{7.85}
$$

We repeatedly in this section invoke the approximation

$$
(1 - b)^m \simeq 1 - mb \qquad \text{for small } b
\tag{7.86}
$$

so that (7.85) reduces to

$$
M_{0,0} \simeq 1 - n\lambda h
\tag{7.87}
$$

Similarly, $M_{0,-1}$ is the probability that the final failed-unit count at time $t + h$ is zero because the single failed unit had been repaired at time t. Its value is

$$
\begin{aligned}
M_{0,-1} &= (1 - \lambda h)^{n-1}(\mu h) \\
&\simeq [1 - (n - 1)\lambda h]\mu h \\
&\simeq \mu h
\end{aligned}
\tag{7.88}
$$

Inserting (7.87) and (7.88) into (7.84) yields

$$
P_0(t + h) = (1 - n\lambda h)\,P_0(t) + \mu h\,P_1(t)
\tag{7.89}
$$

Transposing $P_O(t)$ to the left side, dividing by h, and taking $\lim_{h \to 0}$ gives the boundary differential equation

$$P_0'(t) = -n\lambda P_0(t) + \mu P_1(t) \tag{7.90}$$

The general case corresponding to events $E\,3$, $E\,4$, and $E\,5$ has the following structure:

$$
\begin{aligned}
P_k(t + h) &= P(E\,3 \text{ or } E\,4 \text{ or } E\,5) \\
&= P_{k-1}(t)\, M_{k,+1} + P_k(t)\, M_{k,0} + P_{k+1} M_{k,-1} \tag{7.91}
\end{aligned}
$$

$M_{k,+1} = P(1 \text{ failure, no repair, resulting in } k \text{ failed units})$

$M_{k,0} = P(\text{no failure, no repair, resulting in } k \text{ failed units})$

$M_{k,-1} = P(\text{no failure, 1 repair, resulting in } k \text{ failed units})$

The specific form for the M-coefficients is

$$
\begin{aligned}
M_{k,+1} &= P(1 \text{ failure among } n - k + 1 \text{ good units and} \\
&\qquad \text{no repairs among } k - 1 \text{ failed units}) \\
&= \binom{n-k+1}{1} \lambda h(1 - \lambda h)^{n-k}(1 - \mu h)^{k-1} \\
&\approx (n - k + 1)\lambda h[1 - (n - k)\lambda h][1 - (k - 1)\mu h] \\
&\approx (n - k + 1)\lambda h \tag{7.92}
\end{aligned}
$$

$$
\begin{aligned}
M_{k,0} &= P(\text{no failures among } n - k \text{ good units and} \\
&\qquad \text{no repairs among } k \text{ failed units}) \\
&= (1 - \lambda h)^{n-k}(1 - \mu h)^{k} \\
&\approx [1 - (n - k)\lambda h][1 - k\mu h] \\
&\approx 1 - [(n - k)\lambda + k\mu]h \qquad \text{for } k < r \tag{7.93a}
\end{aligned}
$$

$$
\begin{aligned}
M_{k,0} &= (1 - \lambda h)^{n-k}(1)^{k-r}(1 - \mu h)^{r} \\
&= 1 - [(n - k)\lambda + r\mu]h \qquad \text{for } k \ge r \tag{7.93b}
\end{aligned}
$$

$$
\begin{aligned}
M_{k,-1} &= P(\text{no failures among } n - k - 1 \text{ good units and} \\
&\qquad 1 \text{ repair among } (k + 1) \text{ failed units})
\end{aligned}
$$

The construction of $M_{k,-1}$ is difficult, so we analyze it in detail. In order to have k failed units at time $t + h$ as a result of the failure count decreasing by one unit during the interval h, there must have been $k + 1$ failed units at time t. We know that $(1 - \lambda h)^{n-k-1}$ is the probability of no failures in the interval h. For the failure count to decrease by one, one unit among the $k + 1$ failed units must be repaired, with a probability that depends on how many repairmen are available to work on the failed units. If $k < r$, there are enough repairmen to work on every unit, so the probability of a single repair is $\mu h(1 - \mu h)^{k}$ times $\binom{k+1}{1}$, the binomial coefficient for "selecting" one item to repair among $k + 1$ possibilities. On the other hand, if $k \ge r$, there are not enough repairmen to give every failed unit a chance to be repaired. A group of $k + 1 - r$ of them will remain unrepaired with absolute certainty, giving rise to the factor $(1)^{k+1-r}$; a group of r of them will be subject to the repair process, and a single one of these will actually be repaired with probability $\mu h(1 - \mu h)^{r-1}$ times $\binom{r}{1}$, the binomial coefficient for "selecting" one item to repair among r possibilities. In summary, there are two conditions to consider: $k < r$, enough repairmen, and $k \geq r$, not enough repairmen. The resulting equations are

$$
\begin{aligned}
M_{k,-1} &= (1 - \lambda h)^{n-k-1} \binom{k+1}{1} \mu h (1 - \mu h)^k \\
&\simeq [1 - (n - k - 1)\lambda h](k + 1)\mu h[1 - k\mu h] \\
&\simeq (k + 1)\mu h \qquad\qquad\qquad\qquad \text{for } k < r \qquad (7.94)
\end{aligned}
$$

and

$$
\begin{aligned}
M_{k,-1} &= (1 - \lambda h)^{n-k-1} \binom{r}{1} \mu h (1 - \mu h)^{r-1} (1)^{k+1-r} \\
&\simeq [1 - (n - k - 1)\lambda h] r \mu h[1 - (r - 1)\mu h] \\
&\simeq r \mu h \qquad\qquad\qquad\qquad\quad \text{for } k \geq r \qquad (7.95)
\end{aligned}
$$

The general case equations that result from inserting (7.93) through (7.95) into (7.91) have the form

$$
\begin{aligned}
P_k(t + h) = \; &P_{k-1}(t)(n - k + 1)\lambda h \\
&+ P_k(t)\{1 - [(n - k)\lambda + k\mu]h\} \\
&+ P_{k+1}(t)(k + 1)\mu h \qquad\qquad \text{for } k < r \qquad (7.96)
\end{aligned}
$$

$$
\begin{aligned}
P_k(t + h) = \; &P_{k-1}(t)\,(n - k + 1)\lambda h \\
&+ P_k(t)\,\{1 - [(n - k)\lambda + r\mu]h\} \\
&+ P_{k+1}(t)\,r\,\mu h \qquad\qquad\qquad \text{for } k \geq r \qquad (7.97)
\end{aligned}
$$

After the usual transposition of the term $P_k(t)$, dividing by h, and taking $\lim_{h\to 0}$, the resulting difference–differential set is obtained, to go with the boundary condition 7.90. This is straightforward and is left to the student as an exercise. Unfortunately, the solution of the resulting equations is beyond our competence. We can, however, deal easily with the steady-state equations that result from setting $P'_k(t)$ to zero. These are

$$
-n\lambda P_0 + \mu P_1 = 0 \qquad\qquad\qquad (7.98)
$$

$$
\begin{aligned}
(n - k + 1)\lambda P_{k-1} &- [(n - k)\lambda + k\mu]P_k \\
&+ (k + 1)\mu P_{k+1} = 0 \qquad \text{for } k < r \qquad (7.99)
\end{aligned}
$$

$$
\begin{aligned}
(n - k + 1)\lambda P_{k-1} &- [(n - k)\lambda + r\mu]P_k \\
&+ r\mu P_{k+1} = 0 \qquad \text{for } k \geq r \qquad (7.100)
\end{aligned}
$$

The solution is obtained by a domino process. Equation 7.98 is solved for P_1 in terms of P_0; the result is inserted in (7.99) with $k = 1$, which yields P_2. Setting $k = 2, 3, \ldots, n$ in turn in (7.98) allows determination of the P_k as multiples of P_0.

They are

$$\left.\begin{array}{l} P_1 = n\left(\dfrac{\lambda}{\mu}\right)P_0 \\[2mm] P_2 = \dfrac{n(n-1)}{2}\left(\dfrac{\lambda}{\mu}\right)^2 P_0 \\[2mm] \vdots \\[1mm] P_k = \dfrac{n!}{(n-k)!k!}\left(\dfrac{\lambda}{\mu}\right)^k P_0 \qquad \text{for } k < r \\[2mm] \vdots \\[1mm] P_r = \dfrac{n!}{(n-r)!r!}\left(\dfrac{\lambda}{\mu}\right)^r P_0 \\[2mm] P_{r+1} = \dfrac{n!}{(n-r)!r\,!r}\left(\dfrac{\lambda}{\mu}\right)^{r+1} P_0 \\[2mm] \vdots \\[1mm] P_k = \dfrac{n!}{(n-k)!r\,!r^{k-r}}\left(\dfrac{\lambda}{\mu}\right)^k P_0 \qquad \text{for } k > r \end{array}\right\} \qquad (7.101)$$

If this is written in the form $P_k = C_k P_0$, the value of P_0 comes from

$$P_0 = (C_0 + C_1 + \cdots + C_n)^{-1} \qquad (7.102)$$

Instead of the domino process just suggested, it is more convenient to develop the solution set 7.101 using a time-honored queueing theory trick, the rate diagram shown in Figure 7.12. The arcs across the top are labeled with the total failure rate for passing from one state to the next on the right; as units fail these values are $n\lambda$, $(n-1)\lambda, \ldots, \lambda$ in turn. The arcs along the bottom are labeled with the total repair rate for passing from one state to the next lower state. Notice that they increase with the number of failed units until the state r at which they level off to the maximum value $r\mu$, since there are only r repairmen. As soon as the repairmen are all busy, the repair rate cannot increase.

Now each P_k is calculated sequentially as the value of P_0 multiplied by the product of label ratios from each stage, going from S_0 up to S_k. Thus

$$P_1 = \frac{n\lambda}{\mu} P_0$$

$$P_2 = \frac{n\lambda}{\mu}\frac{(n-1)\lambda}{2\mu} P_0$$

$$\vdots$$

These formulae are familiar to students of queueing theory. Keep in mind that this "quick and dirty" method is not a proof or derivation; it is a *mnemonic* only, used to avoid the tedium of the step-by-step solution which leads from (7.98) through (7.100) to equation set 7.101. Recognize that unless $\lambda/\mu < 1$, the system will spend most of its time in the all-units-failed configuration.

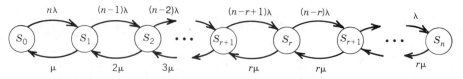

Figure 7.12   The rate diagram for the steady-state solution for n units with r repairmen.

Example 7.2

Consider a system with $n = 6$ units and $r = 3$ repairmen, for which the repair rate is ten times the failure rate, or $\lambda/\mu = 0.1$. Set up the state diagram and then compute the probabilities $P_0, \dots, P_6$.

Solution: The state diagram is shown in Figure 7.13. We see that

$$P_1 = 6(.1)P_0 = .6P_0$$

$$P_2 = \frac{(6)(5)}{(1)(2)}(.1)^2 P_0 = .15P_0$$

$$P_3 = \frac{(6)(5)(4)}{(1)(2)(3)}(.1)^3 P_0 = .02P_0$$

$$P_4 = \frac{(6)(5)(4)(3)}{(1)(2)(3)(3)}(.1)^4 P_0 = .002P_0$$

$$P_5 = \frac{(6)(5)(4)(3)(2)}{(1)(2)(3)(3)(3)}(.1)^5 P_0 = \left(\frac{40}{3}\right)(.1)^5 P_0 \simeq .0001333P_0$$

$$P_6 = \frac{(6)(5)(4)(3)(2)(1)}{(1)(2)(3)(3)(3)(3)}(.1)^6 P_0 = \frac{40}{9}(.1)^6 P_0 \simeq .00000444P_0$$

The value of P_0 is calculated from

$$P_0 = (1 + .6 + .15 + .02 + .002 + .000133 + .00000444)^{-1}$$
$$= (.1772138)^{-1} = .56429$$

The complete set of steady-state probabilities is

$$
\begin{array}{ll}
P_0 = .56429 & P_4 = .00113 \\
P_1 = .33857 & P_5 = .0000752 \\
P_2 = .08464 & P_6 = .00000251 \\
P_3 = .01129 &
\end{array}
$$

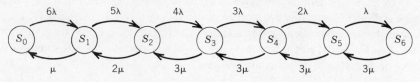

Figure 7.13   State diagram for Example 7.2.

The treatment of a number of other interesting multicomponent systems can be found in Dhillon and Singh (1981) and Henley and Kumamoto (1981).

EXERCISES 7.5

1. Verify that Equations 7.92 through 7.95 and 7.91 yield Equations 7.96 and 7.97.

2. Use (7.96) and (7.97) to develop the general difference–differential equation set.

3. Verify that your solution to Exercise 2 does lead to the steady-state solution set of (7.98) through (7.100).

4. For $n = 5$ and $r = 3$, use the step-by-step (domino) technique to derive the steady-state solution set from (7.98) through (7.99). Check it against the solution obtained by the quick method using a state diagram.

5. Use $\lambda = 1$ and $\mu = 5$ in Exercise 4 and find the steady-state probabilities for all states.

6. Use $\lambda = 1$ and $\mu = 5$ in Example 7.2 and recompute the P_i values.

7. Use $\lambda = \mu$ in Example 7.2 and recompute the P_i values.

7.6   PREVENTIVE MAINTENANCE

The previous section referred to systems whose repair was initiated only at the time of component failure. In the present section we deal briefly with the interesting problem of preventive maintenance. Does it pay to check a system at regular intervals and fix whatever might need fixing? The answer, in general, is a resounding "yes." The following development, laid out by von Alven (1964), is still the standard analysis based on the assumption of multiple CFR components, the easiest situation being that of two or more elements in parallel. We have seen in Section 3.3 that a parallel configuration of CFR components is not CFR.

A system with parallel redundancy can be functioning adequately even with failed components. It seems reasonable, however, that system reliability will be improved if we carry out inspection from time to time, replacing failed units by good ones.

Let T_0 be the fixed period between preventive maintenance actions, and let $R_M(t)$ be the reliability of the maintained system. It seems obvious that $R_M(t)$ will exceed the unmaintained reliability $R(t)$, but by how much?

The total elapsed time t can be written in the form

$$t = j\,T_0 + x \tag{7.103}$$

where j is the integer part of t/T_0 and x is the modulus (remainder) of t/T_0. Let A_i designate the event that the system functions from $t = (j - 1)T_0$ to $t = j\,T_0$ ($j = 1, 2, \dots$), and B the event that the system functions from $t = j\,T_0$ to $t = j\,T_0 + x$.

The value of $R_M(t)$ is

$$R_M(t) = P(A_1)\,P(A_2|A_1)\,P(A_3|A_2)\dots P(A_j|A_{j-1})\,P(B|A_j) \qquad (7.104)$$

But at each maintenance action all failed components are replaced, so the system starts "as good as new" at each time point $j\,T_0$. As a result, we see that

$$P(A_2|A_1) = P(A_2) = P(A_1) = R(T_0)$$
$$P(A_3|A_2) = P(A_3) = P(A_1) = R(T_0)$$
$$\vdots$$
$$P(B|A_j) = P(B) = R(x)$$

and hence

$$R_M(t) = [R(T_0)]^j\, R(x) \qquad (7.105)$$

The specific amount of increase of $R_M(t)$ over $R(t)$ depends on the length of the maintenance period T_0. The smaller the value of T_0, the more notable the improvement. Figure 7.14 shows the behavior of a system of two identical parallel units using $\lambda = 1$, or time measured in "standard units", and with $T_0 = 0.2, 0.5, 1$, compared with the unmaintained system ($T = \infty$). Figure 7.15 shows the same type of comparison for three units in parallel. Because it is difficult for the eye to compare the two systems on separate plots, we supply an additional comparison in Figure 7.16, where the two unmaintained systems are shown, along with their $T_0 = 1$ curves. Curves a and b show $R(t)$ and $R_M(t)$ for the 2-unit system, while curves c and d show them for the 3-unit system.

Notice that curve b lies above curve c after approximately two maintenance periods. Inexperienced readers may have a temporary feeling—because the compo-

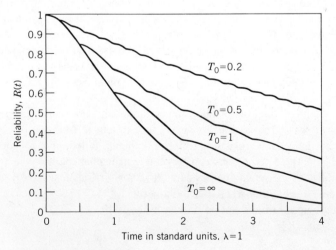

Figure 7.14 Reliability of a periodically maintained system of two parallel units.

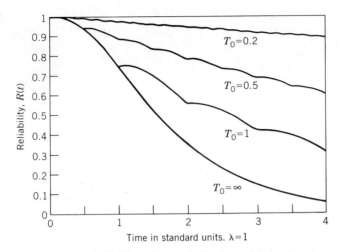

Figure 7.15 Reliability of a periodically maintained system of three parallel units.

nents start as good as new at each $j\,T_0$ time point—that the reliability should perhaps jump back to the value unity, which it had at time $t = 0$. It does not do so for the same reason that a single CFR component, one as good as new until it fails, shows a monotone decreasing reliability curve. Even with good-as-new units, successful long missions are less likely than successful short missions.

Since the reliability of the maintained systems is higher than that of unmaintained systems, it seems reasonable that the mean time to failure under preventive mainte-

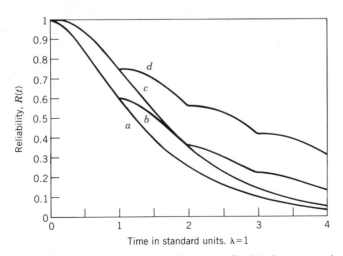

Figure 7.16 Comparison of $R(t)$ and $R_M(t)$ for two and three parallel units.

nance should also be higher. This is indeed the case. Let MTTF_{pm} designate the mean system life for preventatively maintained systems. Then from Theorem 2.1 we have

$$
\begin{aligned}
\text{MTTF}_{\text{pm}} &= \int_0^\infty R_M(t)\, dt \\
&= \int_0^{T_0} R_M(t)\, dt + \int_{T_0}^{2T_0} R_M(t)\, dt + \int_{2T_0}^{3T_0} R_M(t)\, dt + \cdots \\
&= \int_0^{T_0} R(x)\, dx + \int_0^{T_0} R(T_0)R(x)\, dx + \int_0^{T_0} [R(T_0)]^2 R(x)\, dx + \cdots \\
&= [1 + R(T_0) + R^2(T_0) + R^3(T_0) + \cdots] \int_0^{T_0} R(x)\, dx \\
&= \frac{\int_0^{T_0} R(x)\, dx}{1 - R(T_0)} = \frac{\int_0^{T_0} R(x)\, dx}{F(T_0)}
\end{aligned}
\tag{7.106}
$$

Example 7.3

Compute MTTF_{pm} for the two-component parallel system of Figure 7.14 using the five maintenance strategies $T_0 = 0.1, 0.2, 0.5, 1, \infty$ (unmaintained).

Solution: We know that for $n = 2$ units in parallel $R(t) = 1 - (1 - e^{-\lambda t})^2 = 2e^{-\lambda t} - e^{-2\lambda t}$, so that integration gives

$$
\int_0^{T_0} R(x)\, dx = \frac{1}{2\lambda}\left(3 - 4e^{-\lambda T_0} + e^{-2\lambda T_0}\right)
$$

In addition,

$$
F(T_0) = (1 - e^{-\lambda T_0})^2
$$

and the ratio of these quantities is

$$
\text{MTTF}_{\text{pm}} = \frac{3 - 4e^{-\lambda T_0} + e^{-2\lambda T_0}}{2\lambda(1 - e^{\lambda T_0})^2}
\tag{7.107}
$$

Using $\lambda = 1$ (standard units) we get

$$
\begin{aligned}
\text{MTTF}_{\text{pm}} &= 11.01 \quad \text{for } T_0 = 0.1 \\
&= 6.02 \quad \text{for } T_0 = 0.2 \\
&= 3.04 \quad \text{for } T_0 = 0.5 \\
&= 2.08 \quad \text{for } T_0 = 1 \\
&= 1.5 \quad \text{for } T_0 = \infty
\end{aligned}
$$

Example 7.4

Rework Example 7.3 for the three-component parallel system of Figure 7.15.

Solution: The numerator of $MTTF_{pm}$ is

$$\int_0^{T_0} R(x)\, dx = \int_0^{T_0} (3e^{-\lambda x} - 3e^{-2\lambda x} + e^{-3\lambda x})\, dx$$

$$= \frac{1}{6\lambda}(11 - 18e^{-\lambda T_0} + 9e^{-2\lambda T_0} - 2e^{-3\lambda T_0}) \qquad (7.108)$$

and the denominator is

$$F(T_0) = (1 - e^{-\lambda T_0})^3 \qquad (7.109)$$

Using $\lambda = 1$ we get

$$
\begin{aligned}
MTTF_{pm} &= 116.01 &&\text{for } T_0 = 0.1 \\
&= 33.53 &&\text{for } T_0 = 0.2 \\
&= 8.06 &&\text{for } T_0 = 0.5 \\
&= 3.63 &&\text{for } T_0 = 1 \\
&= 1.83 &&\text{for } T_0 = \infty
\end{aligned}
$$

EXERCISES 7.6

1. Master the details of proof leading to Equation 7.105.

2. Master the details of proof leading to Equation 7.106.

3. Verify the solution in Example 7.3.

4. Verify the solution in Example 7.4.

5. Show that for a single component the maintained system reduces to the unmaintained system in the sense that (7.106) leads to the usual value $MTTF = 1/\lambda$. Is there a nonmathematical way you can justify the equality of the two situations?

6. Use an electronic spreadsheet to reproduce Figure 7.14.

7. Use an electronic spreadsheet to reproduce Figure 7.15.

8. Repeat Example 7.4 for a four-component parallel system.

CHAPTER 8

Military Standards

For the theoretician, the military handbooks and manuals are sometimes perceived to be as interesting as a telephone book; like a telephone book, however, they contain information that can be found nowhere else and that is important to the practitioner. In this short chapter, we present an overview of some of the most frequently cited manuals published by the Department of Defense for the use of those involved at the applied level in various aspects of quality control and reliability. These publications are available upon request at little or no cost from

> Naval Publications and Forms Center
> 5801 Tabor Avenue
> Philadelphia, PA 19120-5099

The function of the Forms Center is described in an information sheet provided upon request; ask for leaflet NPFC-4120/3, entitled *DOD Single Stock Point for Specifications and Standards: A Guide for Private Industry*. Service is slow and stockouts are frequent, so readers are advised to anticipate their needs well in advance. Months were spent collecting the documents that are reviewed here; a number were ordered and have never arrived. In Appendix 8B we include a list of libraries that might be a source of the documents via interlibrary loan. But even though the listed libraries nominally participate in the federal depository program, I found that very few of the DOD documents were available from them.

The documents whose catalog number begins with AR (for Army Regulation) will not come from the Forms Center, which is an interservice organization, but from HDQA (DAMA-PPM) Washington, DC 20310.

There is a wide variation in the quantity and quality of information in the various documents. They range from very brief clarifications of procedures to fat tomes with much technical information. There is considerable redundancy among the nontechnical procedures manuals. Most manuals come in the form of loose 8 1/2 × 11 three-hole punched sheets, in plastic wrappers, ready to insert into snap-ring notebooks or folders. I have usually indicated the size of the document as a rough indication of how much information it might contain. When it is not convenient to give a page count—because the count begins again with each subsubsection—I have used thickness as a measure of bulk.

The publications are revised and updated from time to time by DOD; the last alphabetical digit indicates the version number. For example, MIL-STD-105A dated 1950 was replaced by MIL-STD-105B in 1958. The current version, released in 1963, is MIL-STD-105D. The purchaser will be sent the current edition even if no version (or an older version) is specified. Although the publications bear the DOD imprimatur, orders should be sent to the address listed.

There is a great deal of circularity of reference from the various documents to others; the reader who is consulting a single one may find it necessary to follow up with repeated requests to the Forms Center for other cited references. For example, on page 2 of MIL-HDBK-189 (*Reliability Growth Management*) we find that the user is assumed to have also—since they "form a part of the document"—the additional documents MIL-STD-721 (*Definitions of Effectiveness Terms for Reliability, Maintainability, Human Factors, and Safety*), MIL-STD-756 (*Reliability Prediction*), MIL-STD-785 (*Reliability Program for Systems and Equipment Development and Production*), MIL-STD-781 (*Reliability Design Qualification and Production Acceptance Tests: Exponential Distribution*), and MIL-STD-499 (*Engineering Management*).

Since reliability practitioners often have occasion to use attribute sampling, I have reviewed two documents relating to it, as well as those that deal with measurement or lifetime models.

MIL-STD-721: *Definitions of Terms for Reliability and Maintainability*
Version 721C, dated 12 June 1981, is a 16-page pamphlet which presents a glossary of terms used in reliability work. Examples of expressions defined are "burn-in," "dependability," "derating," "failure," "maintainability," and many, many more. Unfortunately, the acronyms used in the other military documents are not included; definitions for these are scattered throughout the other military documents. Thus the reader who needs to know that RD/GT means reliability development/growth testing will not find it in MIL-STD-721C. It can, however, be found on page 1 of MIL-STD-781D, as well as at its point of first use in other documents. The reader is advised to create his or her own glossary of acronyms, since the authors of the documents are somewhat stingy with their repetitions of definitions.

As a convenience to readers, my Appendix 8C lists a number of acronyms gleaned from the reviewed documents. Feel free to photocopy the list and keep it handy whenever military documents are referred to.

AR 702-3: *Product Assurance: Army Materiel Systems Reliability, Availability, and Maintainability* This 32-page army regulation pamphlet, whose current version is dated 15 June 1987, is an outline of broadly laid out reliability and maintainability procedures. It specifies the objectives, concepts, management, documentation, testing, and accounting steps to be followed in a RAM program. In each section are outlined the rationale, the responsible parties, the procedures to be used and where they are pertinent, and the military standard or handbook to be followed. The pamphlet provides a list of references, which we have included in Appendix 8A.

MIL-STD-105: *Sampling Procedures and Tables for Inspection by Attributes* Version 105D, dated 1963, is the current version. A 64-page monograph covering all facets of attribute sampling, it is familiar to students of traditional quality control or quality assurance. Attribute sampling corresponds to Bernoulli trials, that is, trials whose outcomes are dichotomous (defective or nondefective) with constant probability of success from trial to trial. The document contains sampling plan specifications (sample size, accept and reject numbers) for various desired outcomes expressed in terms of consumer's risk, producer's risk, average outgoing quality level (AOQL), and so on. Plans are given for single, double, and multiple sampling and a wide range of fraction-defective levels. Operating characteristic (OC) curves are given for all plans. The focal point of the plans is acceptable quality level (AQL), and they are indexed accordingly. The narrative introduction explains adequately how to select and use the appropriate plans based on lot size, degree of protection desired, and so on.

MIL-STD-1235: *Single and Multilevel Continuous Sampling Procedures and Table for Inspection by Attributes* Whereas the original version MIL-STD-1235-(ORD) dated 17 July 1962 was a 243-page document, the current MIL-STD-1235B, dated 10 December 1981, is a brief 53 pages. The original document contained 20 pages of definitions and instructions, followed by 20 pages of sampling plans, the balance of the book consisting of OC, AOQ, and AOQL curves. It is these 200-some pages of curves that have been omitted in the 1235B version; they are available in MIL-STD-1235A-1, *Functional Curves of the Continuous Sampling Plans*.

The 1235B document contains plans that were presented in handbooks H107 (now out of print) and H106 (also out of print) for single-level and multilevel attribute sampling, respectively. The plans are for units that arrive in a continuous stream— as from an assembly line or production process—rather than in discrete lots. This document is considered to be a supplement to MIL-STD-105.

Three kinds of inspection are used on the plans: 100 percent inspection (detailing), sampling inspection, and verifying inspection. Plans are indexed by AQL and sample-size code letter, as well as by AOQL. Inspection is assumed to be nondestructive, and the units are to be inspected in the order in which they were produced. A sample consists of one unit of product (except in CSP-F) selected from the production process.

1. *CSP-1* A single-level plan which alternates between sequences of 100 percent inspection and sampling inspection, with no limit on the number of such sequences.

When i (tabled value) consecutive good units are encountered (and are verified as good by the checking inspectors), then screening is terminated and samples are taken at frequency f (tabled value). When a defective unit is found, 100 percent inspection resumes. The value i is called the "clearance value."

2. *CSP-F* A single-level plan (a variation of CSP-1) wherein units are inspected N at a time (tabled value), rather than singly.

3. *CSP-2* A single-level plan with alternating sequences of 100 percent inspection and sampling inspection, with no limits on the numbers of such sequences. It does not require return to 100 percent inspection provided that i or more good units have been observed since the previous defective unit.

4. *CSP-T* A multilevel plan with alternating sequences as in the preceding plans, but this one provides for reduced inspection for observed superior product quality. The sampling frequency f (tabled value) can be reduced to $f/2$, then $f/4$ after a sufficiently long series of good units.

5. *CSP-V* A single-level plan with alternative sequences as in the preceding plans, but this one provides a reduced clearance number (the tabled value i) for observed superior product quality.

MIL-STD-690: *Failure Rate Sampling Plans and Procedures* Version 690B, dated 17 April 1968, is a 35-page pamphlet with two brief updates (12 August 1971 and 1 August 1974). It provides sampling plans for units that have exponential lifetimes. The language and terminology are more "Army" than "university." For instance, an acceptable failure rate (AFR) is described as "the failure rate specified for qualification." The term confidence level, often defined as the probability of rejecting the null hypothesis even though it is true, that is, of rejecting acceptable product, is defined as "the probability of disqualifying a product when the true failure rate is at the failure rate specified for qualification." Failure rate is defined as percent failures per thousand hours of test time, and tables are provided for values ranging from 1.0 to 0.0001 percent per thousand component hours, or $\lambda = 10^{-5}f$/hour to $10^{-9}f$/hour. Tables and figures are provided to show the relation between true failure rate and selected confidence levels.

The academically trained statistician will experience frustration with this standard, since some terms seem to be poorly defined or used in nonstandard ways. Examples are lacking, and there is no hint of how the tables were generated.

The "qualification sample" is to be carried out according to a specification table that designates the number of sample units to be inspected, the number of permissible failures ($c = 0, 1, \ldots, 5$), and the test duration. Two specification tables are provided, FRSP-60 (Table I) and FRSP-90 (Table II) for *failure rate sampling plans* with 60 percent confidence level and 90 percent confidence level. Table III provides two points on the corresponding OC curves. For example, suppose that the failure rate value is .001 percent per 1000 hours (code letter S) and the number of failures permitted is $c = 0$. Then the test period must be 230×10^6 unit hours. The anchor points of the OC curve are given in Table III (60 percent confidence level) as $\lambda_0 = .06 \times 10^{-5}$ with $\alpha = .05$ and $\lambda_1 = 2.51 \times 10^{-5}$ with $\beta = .10$. A nomograph is provided to estimate failure rate and confidence interval from experienced data. For example, consider a test experience of 10^8 unit hours with exactly one observed failure. Drawing a straight line on the nomograph provides the following combinations of failure rates and confidence levels:

Failure Rate	Confidence Level, percent
.005	95
.004	90
.002	60

Various OC curves and confidence tables are provided but are inadequately explained.

Readers should not attempt to use this document unless supplementary information can be unearthed that will explain and describe it better than its authors have done.

MIL-STD-781: *Reliability Testing for Engineering Development, Qualification, and Production*   Version 781D, dated 17 October 1986, is a 42-page pamphlet, organized on a "tailorable task" basis and coordinated with MIL-STD-785. The pamphlet consists of a list of definitions relating to hypothesis tests for life testing and several short sections on various test plan documents and procedures. The treatment is narrative; there are no tables, no sampling plans, no OC curves.

MIL-HDBK-217: *Reliability Prediction of Electronic Equipment*   Version 217E, dated 27 October 1986, is about $1^3/_8$ inches thick; it is basically a collection of tables and instructions for how to use them. Its purpose is to establish uniform methods of predicting reliability of electronic components and systems, based on the type of system and the operating environment. The book begins with a 30-page contents listing the types of components that are considered, the source documents, and the tables of parameters used. Parts categories include such things as microelectronics, lasers, tubes, switches, connectors, and fuses. Parameters are given for different environments such as ground, naval, and airborne. The failure rate λ_p for a typical part might be calculated from a formula such as

$$\lambda_p = \lambda_b(\Pi_E \Pi_A \Pi_{S2} \Pi_C)$$

where λ_p is the part failure rate,

   λ_b is the base failure rate under certain standard conditions,

   Π_E is a modifier whose value depends on the operating environment,

   Π_A is a modifier for the effect of the particular type of application,

   Π_{S2} is a modifier for the voltage stress,

   Π_C is a modifier for the system complexity.

This is only one example of a multitude of possible formulae; the possible types of Π-factors are listed in a table three and a half pages long. Tables of numerical values for the various parameters comprise the balance of the book. Two approaches are outlined for incorporating the given tabled values into a useful reliability estimate. They are the part stress analysis method, which can be used when most of the design is completed, and the parts count method, which is applicable in the early design phase and needs less detailed information.

MIL-HDBK-189: *Reliability Growth Management*   This 148-page monograph, the initial and still current version, is dated 13 February 1981. It is one of the most quantitative and theoretical of the various handbooks and standards. Its purpose is to guide the setting of priorities and allocation of resources in a reliability growth program. Designed to be used by both the manager and the analyst, the monograph gives guidelines and procedures to be used on systems during their development phase. The iterative process of design–test–fix is discussed at length with emphasis on two basic methods. The first is to look at the product, that is, do reliability testing; the second is to look at the process, to monitor the activities that are producing the product. There are different reliability growth curves over time, depending on when "fixes" are incorporated into the procedures. For instance, a "test–fix–test" procedure operates differently from a "test–find–test" procedure with delayed fixes.

In order for a planned growth curve to be developed, a starting point for the curve must be determined, based on experience with a predecessor system. The learning curve concept is incorporated, and the resulting mathematical model is fully developed, then illustrated with examples. "Appendix A: Engineering Analysis" is an overview of problems that arise in design changes. "Appendix B: Growth Models" is a mathematical section listing seventeen models that have been proposed in the reliability literature for the reliability function. "Appendix C: the AMSAA Reliability Growth Model" is dedicated to the army materiel system analysis activity (AMSAA) model, which uses the Weibull process to model reliability growth during the development test phase. The treatment is mathematical, with arithmetical examples and narrative discussion. A good 25-item bibliography of original sources is produced as Appendix D.

TR 4: *Sampling Procedures and Tables for Life and Reliability Testing Based on the Weibull Distribution (Hazard Rate Criterion)*   This 86-page pamphlet, dated 28 February 1962, is out of print and no longer available from the Naval Forms Center, but it might be found in libraries. It is one of the few manuals that bears the names of its authors, Henry P. Goode and John H. K. Kao. The assumption is that lots (batches) of units with Weibull-distributed lifetimes are submitted for inspection. The sampling procedures and plans are based on the evaluation of the hazard function. The two-parameter Weibull distribution is assumed; plans are provided for β-values (Weibull slope) of $\beta = 1/3$, $1/2$, $2/3$, 1, $4/3$, $5/3$, 2, $5/2$, $9/3$, 4, and 5. The acceptance procedure is of the following form. Select n items to test; carry out a life test for some period of time t; determine the failure count r; if $r \leq c$ (the accept number), accept the lot, otherwise reject. If β is known, the $P(\text{accept})$ is a simple function of the hazard rate $z(t)$ and β, and the OC curve can be constructed. Tables are provided for the product $tz(t)$ for various values of

$p' = 1 - R(t)$; these allow the inspector to construct sampling plans for a desired operating characteristic. Full mathematical details and a number of examples are supplied in introductory sections (43 pages).

Instructions are given (Section 5) for converting the sampling plans of MIL-STD-105D to Weibull use. No OC curves are provided, but the technique for constructing them is described. A 23-item bibliography is supplied.

TR 6: *Sampling Procedures and Tables for Life and Reliability Testing Based on the Weibull Distribution (Reliable Life Criteria)*   The 68-page pamphlet is dated 15 February 1963; written by the authors of TR 4, it covers some of the same ground and, like TR 4, is out of print. The sampling plans, however, have a different format. Here the acceptance procedure has the same form: select n items to test; carry out a life test for some period of time t; determine the failure count r; if $r \le c$ (the accept number), accept the lot, otherwise reject. The parameter β is assumed known (or approximately so). Instead of focusing on the hazard function, as TR 4 did, this document focuses on the rth quantile life ρ defined by $r = 1 - R(\rho)$, that is, $1 - r = e^{-(\rho/\eta)^\beta}$. The expression for the fraction p' failing by the truncation time is $p' = 1 - e^{-(t/\eta)^\beta}$, which can be rewritten using $t/\eta = (t/\rho)(\rho/\eta)$ so that p' depends only on t, ρ, and r. The OC-curve expression $1 - p'$ thus also depends only on t, ρ, and r (for any specified β) when the dimensionless unit of time t/ρ is one of the table entries. Acceptable quality level (AQL) and rejectable quality level (RQL), and their associated producer's risk and consumer's risk values, can be specified as two particular pairs of ρ, r values. A plan, a pair of c and n values, is established by referring to the appropriate tables.

An adequate explanatory section of 29 pages is provided, followed by 35 pages of tables. No OC curves are given, but the procedure for their calculation is indicated. A 17-item bibliography is provided.

H 108: *Sampling Procedures and Tables for Life and Reliability Testing (Based on Exponential Distribution)*   This 68-page pamphlet, dated 29 April 1960, is out of print but may be available in libraries. It contains tables and OC curves related to the life testing work of Epstein, which is covered in Chapter 5 of this primer on reliability, the one you are reading. No derivations are given, but adequate narrative is provided for cookbook usage of the tables. A number of OC curves are provided. The document provides plans for terminating testing at a predetermined destruction index r, or at a predetermined time t_0, as well as sequential plans. The plans may be used for sampling with or without replacement.

MIL-HDBK-251: *Reliability/Design Thermal Applications*   The current version, dated 19 January 1978, is a 700-page text and handbook on the thermal design of electronic equipment. The primary purposes are "To permit engineers and designers who are not heat transfer experts to design electronic equipment with adequate thermal performance . . . ; to assist heat transfer experts who are not electronic experts; to aid engineers in better understanding the thermal sections of DOD standards . . . " (page 1). It presents methods for analyzing stress on electronic parts, for selecting the maximum allowable temperature, and so on, in connection with shipboard, ground-based, and airborne electronics. Appropriate mathematical equations and computer information are included for the use of experts in heat transfer, but nomographs and curves are also given so that nonexperts can design adequate equipment. Clearly the book is technical but not in the mathematical sense.

MIL-STD-785: *Reliability Program for Systems and Equipment Development and Production* Version 785B, dated 15 September 1980, is a pamphlet less than a quarter inch thick. There is no mathematical or statistical discussion in this document; it contains only narrative material. Its purpose is to outline and describe the logical flow of tasks that must be performed when developing a reliability program plan. The needed program management tasks are identified and tied together. Instructions are given for describing tasks, designing schedules, recording data, managing staff, and so on. The instructions would be very useful to the supervisor who has been assigned the job of implementing a reliability program without any background in how to begin such a real-world assignment.

MIL-STD-2155(AS): *Failure Reporting, Analysis and Corrective Action System* The current version, dated 24 July 1985, is a pamphlet of ten pages describing the FRACAS procedure. The goal is to describe and define a "disciplined and aggressive closed loop" failure and reporting system. The standard establishes uniform requirements and criteria to implement the FRACAS requirement of MIL-STD-785.

Definitions are provided (one to three sentences each) for the following terms: acquiring activity, closed-loop failure-reporting system, contractor, corrective action effectivity, failure, failure analysis, failure cause, failure review board, failure symptom, fault, and laboratory analysis.

A short paragraph of discussion is supplied concerning each of the following requirements: contractor responsibility, FRACAS planning, failure review board, failure documentation, failure reporting, failure analysis, failure verification, corrective action, failure report closeout, and identification and control of failed items.

There is a short appendix (four pages), "Application and Tailoring Guide," with suggestions for implementation. (My copy had two pages of its four left unintentionally blank.)

DOD-HDBK-344(USAF): *Environmental Stress Screening (ESS) of Electronic Equipment* This 120-page booklet, dated 20 October 1986, provides techniques for designing ESS programs, using quantitative methods to plan and control both cost and effectiveness. It is intended to support the requirements of MIL-STD-785 and is for use by procuring agents and contractors during development and production. ESS is a process in which environmental stresses are applied to electronic items to precipitate early failures, with testing throughout to identify the resulting defects. The procedures given have the following titles: A, Part Fraction Defective—R&M 2000 Goals and Incoming Defect Density; B, Screen Selection and Placement; C, Failure-Free Acceptance Tests; D, Cost Effectiveness Analysis; and E, Monitoring, Evaluation, and Control. Task sequencing is thoroughly described, both narratively and with flow diagrams. Many tables are provided showing the fraction of parts that can be defective and the confidence limits for a variety of parts (connectors, relays, inductors, etc.) in various environments (ground benign, missile launch, etc.). The appendices (18 pages in all) give some algebraic formulae on which the procedures are based.

MIL-STD 810: *Environmental Test Methods and Engineering Guidelines*
Version 810D, dated 19 July 1983, is a one-inch-thick manual that comes with a 23-page update, NOTICE 1 (31 July 1986). Chapter 5 of this reliability primer speaks casually of life testing as though it were obvious to readers how to carry out such a procedure. MIL-STD-810D is the manual that may be more precious than rubies to the practitioner who designs and supervises actual testing. Its purpose is to provide guidelines for conducting individual tasks and tailoring environmental tests for specific items. The introductory 19 pages (Sections 1 through 4) contain a general discussion on testing procedures; then follows the lengthy Section 5, which comprises the bulk of the book. It is divided into 23 subsections, each subsection describing the test procedures appropriate for a particular environment condition. These include low pressure, high temperature, solar radiation, fungus, rain, salt fog, shock, and so on. A typical subsection discusses the purpose of the test, describes the environmental effects that nonconforming units might have, gives guidelines for test conditions (decompression rate, test duration, etc.), and refers to other DOD documents that might be pertinent or helpful. Most of the treatment is descriptive and narrative, although some tables and graphs are given.

MIL-STD-1543A (USAF): *Reliability Programs for Space and Missile Systems*   The purpose of this 49-page pamphlet, dated 25 June 1982, is to establish uniform reliability practices for space and missile systems. It applies to all prime and subtier contractors. The appendices, which provide guidance for implementing the standard in the most cost-effective way, make up most of the document.

MIL-STD-1629: *Procedures for Performing a Failure Mode, Effects and Criticality Analysis*   Version 1629A, dated 24 November 1980, is a quarter-inch-thick booklet, mostly narrative, which discusses the requirements and procedures for performing a FMECA to document the potential impact of each item failure on mission outcome. Each potential failure is ranked by the severity of its potential effect. There is outlined a system of tasks: Task 101, Failure Modes and Effects Analysis; Task 102, Criticality Analysis; Task 103, FMECA-Maintainability Information; Task 104, Damage Mode and Effects Analysis; Task 105, Failure Mode, Effects, and Criticality Analysis Plan. The appendix provides notes on contractual requirements in procurement, and an algebraic example is given for the calculation of the criticality number C_r from failure data. The criticality number is the number of failures of a specific type expected per million hours, owing to the item's modes of failure under a particular severity classification. Failure rates from MIL-HDBK-217 are used.

MIL-STD-2074: *Failure Classification for Reliability Testing*   This slender pamphlet of ten pages, dated 15 February 1978, is provided to establish criteria for classifying failures that occur during reliability tests. The categories defined are relevant versus nonrelevant. Relevant failures include design and workmanship failures, failures of component parts, parts that wear out, and intermittent failures. Nonrelevant failures are those caused by improper installation, operator error, failure of test equipment, and so on.

MIL-STD-2164 (EC): *Environmental Stress Screening Process for Electronic Equipment* This 42-page pamphlet, dated 5 April 1985, provides standards and guidelines for the environmental stress screening (ESS) process, similar to those already in existence for development and qualification testing. The goal is a test process that will permit latent defects to be located and eliminated before equipment is accepted. Specific criteria are given for the ambient temperature, humidity, and atmospheric pressure of the test situation. Vibration testing is described, as well as thermal cycling. The appendices discuss the ESS characteristic curve and the logistic curve for the hazard function. There is good discussion—with algebraic derivations—of how long the test time should be to provide given probability of acceptance and defect-free mission time.

MIL-STD-5662: *Calibration System Requirements* This five-page pamphlet, dated 10 June 1980, comes with update NOTICE 3 (14 December 1984), which replaces two of the five pages. NOTICE 1 and NOTICE 2 are missing. The pamphlet contains requirements for establishing and maintaining a calibration system for controlling the accuracy of measuring and test equipment. It describes the contractor's responsibilities with respect to maintaining the integrity of such pieces of equipment.

MIL-STD-790: *Reliability Assurance Program for Electronic Parts Specification* Version 790D, dated 30 May 1986, is a 20-page pamphlet. It establishes the criteria for the assurance program that are to be met by manufacturers who are qualifying parts to specifications. Distributors are categorized into three levels depending on their amount of involvement with supplied parts, and the testing requirements are given for each level. Training programs are mentioned, as are calibration, failure reporting, and analysis procedures. An appendix discusses self-audit and provides a checklist for performing it.

MIL-STD-1132: *Switches and Associated Hardware, Selection and Use of* Version 1132A, dated 23 March 1976, is a one-inch-thick book which comes with three thick sets of replacement pages labeled NOTICE 1 (20 June 1977), NOTICE 2 (6 June 1979), and NOTICE 3 (21 June 1983). It describes the standard switches and associated hardware that are mandatory for use in the design of military equipment. The document is technical, not in the engineering sense like MIL-STD-251 or like a math textbook, but in the way that a parts catalog is technical. It consists of page after page of blueprint-style drawings of all kinds of switches, along with accompanying tables of dimensions and part numbers. The book, although vital to the hardware designer, will be of little interest to most readers.

Other Government Reliability Documents

We list here a number of other DOD reliability documents, to show the types that are available. It is not feasible to provide a full list, which would require a volume the size of a big-city telephone book. Readers who need more information may obtain help from the technical documents librarian of a large library or from the Library of Congress in Washington, D.C.

MIL-STD-109	*Reliability and Maintainability Research and Development Procedures Manual*
MIL-STD-965	*Engineering Design Handbook, Environmental Factors*
MIL-STD-1670	*Electrostatic Discharge Control Handbook for Protection of Electrical and Electronic Parts, Assemblies, and Equipment (Excluding Electrically-Initiated Explosive Devices)*
DOD-HDBK-263	*Environmental Stress Screening (ESS) of Electronic Equipment*
MIL-STD-471	*Electrostatic Discharge Control Program for Protection of Electrical and Electronic Parts, Assemblies and Equipment (Excluding Electrically-Initiated Explosive Devices)*
MIL-STD-1388	*Clean Room and Work Station Requirements, Controlled Environment*
MIL-STD-2068	*Barrier Material, Flexible, Electrostatic-Free, Heat Sealable*
MIL-STD-1591	*Reliability Growth Management*
MIL-STD-882	*Reliability Prediction of Electronic Equipment*
MIL-STD-1547	*Reliability/Design Thermal Applications*
MIL-S-901	*Manual, Technical; Functionally Oriented Maintenance Manual for Systems and Equipment*
MIL-B-81705	*Microcircuits, General Specification for*
MIL-STD-167	*Standard Electronic Module Program: General Specification for*
MIL-STD-480	*Quality Program Requirements*
MIL-E-16400	*Reliability Index Determination for Avionic Equipment Models, General Specification for*
MIL-R-22973	*Sampling Procedures and Tables for Inspection by Attributes*

MIL-STD-1679	*Quality Assurance Terms and Definitions*
MIL-STD-701	*Switches and Associated Hardware, Selection and Use of*
AFSCM 80	*Glossary of Environmental Terms*
AMCP-706-116	*Single and Multilevel Continuous Sampling Procedures*
FED-STD-209	*Marking for Shipment and Storage*
MIL-STD-721	*Requirements for Employing Standard Electronic Modules*
MIL-S-19500	*Logistics Support Analysis*
MIL-M-24110	*Design Requirements for Standard Electronic Modules*
MIL-STD-2080	*Technical Reviews and Audits for Systems, Equipment, and Computer Programs*
RADC-TR-82-87	*Test Requirements for Space Vehicles*
MIL-STD-210	*Reliability Programs for Space and Missile Systems*
MIL-STD-1165	*Parts, Materials, and Processes for Space Launch Vehicles, Technical Requirements for*
MIL-STD-975	*Government/Industry Data Exchange Program*
MIL-STD-780	*On Aircraft, Fault Diagnosis, Subsystems, Analysis/Synthesis of*
MIL-STD-883	*Procedures for Performing a Failure Mode, Effects and Criticality Analysis*
MIL-Q-9858	*Mechanical Vibrations of Shipboard Equipment*
MIL-STD-280	*Environmental Criteria and Guidelines for Air-Launched Weapons*
RADC-TR-86-138	*Weapon System Software Development*
MIL-STD-454	*Reliability Development Tests*
AFWAL-TR-80-3086	*Survivability, Aircraft; Establishment and Conduct of Programs for*
MIL-STD-2072	*Failure Classification for Reliability Testing*
RADC-TR-86-149	*Maintenance Plan Analysis for Aircraft and Ground Support Equipments*
MIL-STD-470	*Climatic Extremes for Military Equipment*
MIL-STD-1387	*Failure Reporting, Analysis and Corrective Action System*
NAVMAT P-9492	*Environmental Stress Screening Process for Electronic Equipment*
MIL-STD-1389	*Definitions of Item Levels, Item Exchangability, Models and Related Terms*
MIL-M-28787	*Standard General Requirements for Electronic Equipment*
MIL-STD-1540	*Calibration System Requirements*
MIL-STD-129	*Maintainability Program Requirements (for Systems and Equipment)*
NAVORD OSTD 80	*Maintainability Demonstration*

MIL-M-38510	*Configuration Control—Engineering Changes, Deviations and Waivers*
MIL-STD-1521	*Lists of Standard Semiconductor Devices*
MIL-STD-756	*Definitions of Effectiveness Terms for Reliability, Maintainability, Human Factors, and Safety*
MIL-STD-1556	*Definitions of Terms for Reliability and Maintainability*
MIL-STD-499	*Engineering Management*
AR 70-1	*Army Research, Development, and Acquisition*
AR 70-10	*Test and Evaluation During Development and Acquisition of Materiel*
AR 70-15	*Product Improvement of Materiel*
AR 71-9	*Materiel Objectives and Requirements*
AR 700-127	*Integrated Logistic Support*
AR 750-37	*Sample Data Collection: The Army Maintenance Management System*
AR 750-43	*Test, Measurement, and Diagnostic Equipment (Including Prognostic Equipment and Calibration Test/Measurement Equipment)*
AR 1000-1	*Basic Policies for Systems Acquisition*
DA Pam 11-4	*Operating and Support Cost Guide for Army Materiel Systems*
DA Pam 11-25	*Life Cycle System Management Model for Army Systems*
DA Pam 70-21	*The Coordinated Test Program*
AR 5-4	*Department of the Army Productivity Improvement Program*
AR 10-5	*Department of the Army*
AR 11-14	*Logistic Readiness*
AR 15-14	*System Acquisition Review Council Procedures*
AR 70-17	*System/Program/Project/Product Management*
AR 70-37	*Configuration Management*
AR 70-38	*Research, Development, Test, and Evaluation of Materiel for Extreme Climatic Conditions*
AR 70-61	*Type Classification of Army Materiel*
AR 71-3	*User Testing*
AR 310-25	*Dictionary of United States Army Terms*
AR 310-50	*Catalog of Abbreviations and Brevity Codes*
AR 350-35	*New Equipment Training*
AR 385-16	*System Safety Engineering and Management*
AR 602-1	*Human Factors Engineering Program*
AR 700-18	*Provisioning of US Army Equipment*

AR 700-47	*Defense Standardization and Specification Program*
AR 700-90	*Army Industrial Preparedness Program*
AR 702-6	*Ammunition Stockpile Reliability Program*
AR 702-9	*Production Testing of Army Materiel*
AR 702-10	*Post-Production Testing of Army Materiel*
AR 702-11	*Army Quality Program*
AR 702-13	*Army Warranty Program*
AR 715-6	*Proposal Evaluation and Source Selection*
AR 750-1	*Army Materiel Maintenance Concepts and Policies*

APPENDIX 8B

Federal Depository Library Program

The Federal Depository Library Program provides government publications to designated libraries throughout the United States. The Regional Depository Libraries listed allegedly receive and retain at least one copy of nearly every federal government publication, in either printed or microfilm form for use by the general public. *These libraries provide reference services and interlibrary loans; however, they are not sales outlets. Send no checks or orders to these libraries.* You may wish to ask your local library to contact a Regional Depository to help you locate specific publications, or you may contact the Regional Depository yourself.

Auburn University at Montgomery Library
Documents Dept.
Montgomery, AL   36193
(205) 279-9110, ext. 253

University of Alabama Library
Documents Dept., Box S
University, AL   35486
(205) 348-7369

Arkansas State Library
One Capitol Mall
Little Rock, AR   72201
(501) 371-2326

Dept. of Library Archives and Public Records
Third Floor—State Capitol
1700 West Washington
Phoenix, AZ   85007
(602) 255-4121

University of Arizona Library
Government Documents Dept.
Tucson, AZ   85721
(602) 626-5233

California State Library
Government Publications Section
P.O. Box 2037
Sacramento, CA   95809
(916) 322-4572

University of Colorado Libraries
Government Publications Division
Campus Box 184
Boulder, CO   80309
(303) 492-8834

Denver Public Library
Government Publications Dept.
1357 Broadway
Denver, CO   80203
(303) 571-2131

Connecticut State Library
Government Documents Unit
231 Capitol Avenue
Hartford, CT   06115
(203) 566-4971

University of Florida Libraries
Library West
Documents Dept.
Gainesville, FL   32601
(904) 392-0367

University of Georgia Libraries
Government Reference Dept.
Athens, GA   30602
(404) 542-8951

University of Hawaii Library
Government Documents Collection
2550 The Mall
Honolulu, HI   96822
(808) 948-8230

University of Idaho Library
Documents Section
Moscow, ID   83843
(208) 885-6344

Illinois State Library
Information Services Branch
Centennial Building
Springfield, IL   62706
(217) 782-5185

Indiana State Library
Serials and Documents Section
140 North Senate Avenue
Indianapolis, IN   46204
(317) 232-3686

University of Iowa Libraries
Government Publication Dept.
Iowa City, IA   52242
(319) 353-3318

University of Kansas
Documents Collection, Spencer Library
Lawrence, KS   66045
(913) 864-4662

University of Kentucky Libraries
Government Publications Department
Lexington, KY   40506
(606) 258-8686

Louisiana State University Library
BA/Documents Dept.
Middleton Library
Baton Rouge, LA   70803
(504) 388-2570

Louisiana Technical University Library
Documents Dept.
Ruston, LA   71272
(318) 257-4962

University of Maine
Raymond H. Fogler Library
Documents Depository
Orono, ME   04469
(207) 581-7178

University of Maryland
McKeldin Library, Documents Division
College Park, MD   20742
(301) 454-3034

Boston Public Library
Government Documents Dept.
Boston, MA   02117
(617) 536-5400, ext. 226

Detroit Public Library
Sociology Dept.
5201 Woodward Avenue
Detroit, MI   48202
(313) 833-1409

Michigan State Library
P.O. Box 30007
Lansing, MI   48909
(517) 373-0640

University of Minnesota
Government Publications Division
400 Wilson Library
309 19th Avenue South
Minneapolis, MN   55455
(612) 373-7813

University of Mississippi Library
Documents Dept.
University, MS   38677
(601) 232-7091, ext. 7

University of Montana
Mansfield Library
Documents Division
Missoula, MT   59812
(406) 243-6700

Nebraska Library Community
Federal Documents
1420 P Street
Lincoln, NE   68508
(402) 471-2045

University of Nebraska at Lincoln
D. L. Love Memorial Library
Lincoln, NE   68588
(402) 472-2562

University of Nevada Library
Government Publications Dept.
Reno, NV   89557
(702) 784-6579

Newark Public Library
5 Washington Street
Newark, NJ   07101
(201) 733-7812

University of New Mexico
Zimmerman Library
Government Publications Dept.
Albuquerque, NM   87131
(505) 277-5441

New Mexico State Library
Reference Dept.
325 Don Gaspar Avenue
Santa Fe, NM   87503
(505) 827-2033

New York State Library
Empire State Plaza
Albany, NY   12230
(518) 474-5563

**University of North Carolina
at Chapel Hill Library**
BA/SS Division Documents
Chapel Hill, NC   27514
(919) 962-1151

University of North Dakota
Chester Fritz Library
Documents Dept.
Grand Forks, ND   58202
(701) 777-2617
(in cooperation with) **North Dakota
State University Library**

State Library of Ohio
Documents Dept.
65 South Front Street
Columbus, OH   43215
(614) 462-7051

Oklahoma Dept. of Libraries
Government Documents
200 NE 18th Street
Oklahoma City, OK   73105
(405) 521-2502

Oklahoma State University Library
Documents Dept.
Stillwater, OK   74078
(405) 624-6546

Portland State University Library
Documents Dept.
P.O. Box 1151
Portland, OR   97207
(503) 229-3673

State Library of Pennsylvania
Government Publications Section
P.O. Box 1601
Harrisburg, PA   17105
(717) 787-3752

Texas State Library
Public Service Dept.
P.O. Box 12927—Cap. Sta.
Austin, TX   78711
(512) 475-2996

Texas Tech University Library
Government Documents Dept.
Lubbock, TX   79409
(806) 742-2268

Utah State University
Merril Library, U.M.C. 30
Logan, UT   84322
(801) 750-2682

University of Virginia
Alderman Library, Public Documents
Charlottesville, VA   22901
(804) 924-3133

Washington State Library
Documents Section
Olympia, WA   98501
(206) 753-6525

West Virginia University Library
Documents Dept.
Morgantown, WV   26506
(304) 293-3640

Milwaukee Public Library
814 West Wisconsin Avenue
Milwaukee, WI   53233
(414) 278-3065

State Historical Library of Wisconsin
Government Publications Section
816 State Street
Madison, WI   53706
(608) 262-4347

Wyoming State Library
Supreme Court/Library Building
Cheyenne, WY   82002
(307) 777-6344

APPENDIX 8C

Department of Defense Acronyms

Readers are free to photocopy this list for their personal use. Clearly it is not exhaustive, containing only terms that were encountered during the preparation of this chapter. More extensive lists can be found in the Army publications AR 310-25, *Dictionary of United States Army Terms* and AR 310-50, *Catalog of Abbreviations and Brevity Codes*.

ACSI	assistant chief of staff for intelligence
ADPE	automatic data processing equipment
AHR	acceptable hazard rate
AIB	airborne inhabited bomber
AIC	airborne inhabited cargo
AIF	airborne inhabited fighter
AIT	airborne inhabited trainer
AMETA	army management engineering training agency
AMSAA	army materiel system analysis activity
AOQ	average outgoing quality
AOQL	average outgoing quality limit

AQL	acceptable quality level
AR	army regulation
ARW	airborne rotary wing
ASARC	army systems acquisition review council
AUA	airborne uninhabited attack
AUB	airborne uninhabited bomber
AUC	airborne uninhabited cargo
AUF	airborne uninhabited fighter
AUT	airborne uninhabited trainer
BIT	built-in test
BITE	built-in test equipment
CA	criticality analysis
CDE	chance defective exponential
CDR	critical design review
CDRL	contract data requirements list
CFE	contractor-furnished equipment
CG	commanding general
CI	configuration item
CL	cannon launch
CND	cannot duplicate
COE	chief of engineers
COEA	cost and operational effectiveness analysis
COMSEC	communication security
CPE	contract proposal evaluation
CT/ME	calibration test–measurement equipment
CTP	coordinated test program
DA	department of the army
DAMRIP	department of the army productivity improvement program
DARCOM	development and readiness command
DCP	decision coordinating paper
DCSLOG	deputy chief of staff for logistics
DCSOPS	deputy chief of staff for operations and plans
DCSPER	deputy chief of staff for personnel
DCSRDA	chief of staff for research, development, and acquisition
DD	defect density
DE	detection efficiency (probability of detection)
DF	detectable failure
DI	data item
DID	data item description
DMEA	damage mode and effects analysis
DMMH/MA	direct maintenance man hours per maintenance action
DMMH/ME	direct maintenance man hours per maintenance event
DMWR	depot maintenance work requirements
DSARC	defense systems acquisition review council
DT	development testing

DT&E	development test and evaluation
DT/OT	development test/operation test
EDT	engineering development test
ER	established reliability
ESD	electrostatic discharge
ESD/EOS	electrostatic discharge/electrical overstress
ESDS	ESD-sensitive
ESS	environmental stress screening
FBT	functional board tester
FDSC	failure definition and scoring criteria
FD/SC	failure definition and scoring criteria
FFAT	failure-free acceptance test
FFP	failure-free period
FL	fault location
FMEA	failure mode and effects analysis
FMECA	failure modes, effects, and criticality analysis
FR	failure rate
FRACAS	failure reporting, analysis, and corrective action system
FRB	failure review board
FRSP	failure rate sampling plan
FSED	full-scale engineering development
GB	ground benign
GF	ground fixed
GFE	government-furnished equipment
GFP	government-furnished property
GIDEP	government–industry data exchange program
GM	ground mobile
GPR	government plant representative(s)
HI	high impact
IC	integrated circuit
ICA	in-circuit analyzer
ICT	in-circuit tester
IEP	independent evaluation plan
IER	independent evaluation report
ILS	integrated logistic support
IPR	in-process review
IPS	integrated program summary
JTG	joint test group
LBS	loaded board shorts
LOA	letter(s) of agreement
LR	letter requirement
LRM	line-replaceable model
LRU	line-replaceable unit
LSAP	logistic support analysis program
LSAR	logistic support analysis records
LSI	large-scale integration

LTPD	lot tolerance percent defective
MACOM	major army commands
MCBF	mean cycles between failures
MCSP	mission completion success probability
MFF	missile free flight
ML	missile launch
MLE	maximum likelihood estimate
MMBF	mean miles between failures
MMD	mean mission duration
MOR	materiel objectives and requirements
MP	mission profiles
MP	manpack
MPA	maintenance plan analysis
MR	maintenance ratio
MSI	medium-scale integration
MTBC	mean time between critical failures
MTBDE	mean time between downing events
MTBF	mean time between failures
MTBMA	mean time between maintenance actions
MTBOMF	mean time between operational mission failures
MTBR	mean time between removals
MTBUMA	mean time between unscheduled maintenance actions
NASC	naval air systems command
NFF	no fault found
NH	naval hydrofoil
NS	naval sheltered
NSB	naval submarine
NU	naval unsheltered
NUU	naval undersea unsheltered
O&S	operating and support
OC	operating characteristic
OEM	original equipment manufacturer
OMS	operational mode summary
OT	operational testing
OT&E	operational test and evaluation
OTEA	operational test and evaluation agency
PA	procuring activity (including program–project offices)
PCA	physical configuration audit
PCB	parts control board
PDF	pre-defect-free [test]
PDR	preliminary design review
PEP	production engineering phase
PIM	product improvement of materiel
PM	performance monitoring
PM	project manager
PMO	program manager's office

PMP	parts, material, and processes
PMP	program management plan
PPB	parts per billion
PPM	parts per million
PPSL	program parts selection list
PRAT	production reliability acceptance test
PRST	probability ratio sequential test
PWA	printed wiring assembly
QRC	quick reaction capability
RA	rationale annex
RAM	reliability and maintainability
R&M	reliability and maintainability
RCM	reliability-centered maintenance
RDGT	reliability development/growth test(ing)
RD/GT	reliability development/growth test(ing)
RFP	request for proposal
RHR	rejectable hazard rate
RIW	reliability improvement warranty
ROC	required operational capability
RQL	rejectable quality level
RQT	reliability qualification test(ing)
RTOK	retest OK
SCA	sneak circuit analysis
SF	space flight
SOW	statement of work
SPFM	single-point failure mode
SQL	specified quality level
SRO	system readiness objective
SRU	shop replaceable unit
SS	screening strength
SV	specified value
TAAF	test, analyze, and fix
TAMMS	the army maintenance management system
TDLOA	training device letters of agreement
TDLR	training device letter requirement
TDP	technical data package
TDR	training device requirement
TEMP	test and evaluation master plan
TMDE	test, measurement, and diagnostic equipment
TR	test report
TRADOC	training and doctrine command
TS	test strength
TSG	the surgeon general
TTR	time to repair
USALEA	U.S. army logistics evaluation agency
USL	undersea launch

Tables

INTRODUCTION

The tables presented here represent a combination of sources; insofar as possible I computed them myself, partly to make them more complete than those generally available and partly as an intellectual challenge. The software available to me was Statistical Analysis System (SAS), International Mathematical and Statistical Library (IMSL), and LOTUS 1-2-3.

As a rule the SAS functions seem to be more accurate than the IMSL functions (ten decimal places versus six decimal places in some cases). However, the formatting of output is so much easier and flexible in FORTRAN than in SAS that I used the IMSL functions whenever the requisite accuracy could be obtained, that is, when the output agreed with published tables.

Table T1, the standard normal critical points, was computed with the IMSL routine MDNRIS.

Table T2, the standard normal CDF, was computed using the SAS function PROBNORM. The first attempt was based on the IMSL subroutine MDNOR and was found to be insufficiently accurate on the high tail for the desired number of decimal places.

Table T3, the χ^2 critical points, was computed using the SAS function CINV.

Table T4, the Student's t critical points, was computed from the IMSL routine MDSTI.

Table T5, the gamma function for noninteger arguments, was calculated in FORTRAN by my student Vic Simonis using a formula from Abramowitz and Stegun (1968).

321

Table T6, the incomplete gamma function, was computed using the IMSL routine MDGAM.

Table T7, Snedacor's F distribution, is a combination of two sources. The tables for $\alpha = .005, .025, .05, .10, .25$ are copied with permission from the standard Biometrika tables. The numerical values agree with those I obtained, but the typesetting plates were already available from other Wiley publications and it seemed sensible to utilize them. The tables for the values $\alpha = .001, .20, .30, .40$ were calculated using the IMSL routine MDFI. Their format differs slightly from the Biometrika tables because IMSL balks at the computation of a critical point for $\nu_2 = \infty$, so I settled for $\nu_2 = 1000$ in the last column. In addition, a number of rows for larger ν_1 values were included.

Table T8, the cumulative Poisson probabilities, was calculated using a LOTUS 1-2-3 spreadsheet with columns based on the recursion formula $f(x) = (\lambda/x)f(x-1)$.

Table T9, the DUD(0,9) random digits, was created with the RAND function of LOTUS 1-2-3. The table has not received the intense statistical scrutiny accorded the famous Rand Corporation random number tables, but it should be adequate for the classroom and homework use it is intended for.

Table T10, the Kolmogorov–Smirnov critical points, it copied with permission from the cited article in *Journal of the American Statistical Association*.

Table T11, the random standard normal deviates, was computed in LOTUS 1-2-3 using the RAND function inserted into the Box–Mueller formulae 2.133a and b.

TABLE T1 Standard Normal Critical Points

Cumulative Probability	z-value	Cumulative Probability	z-value	Cumulative Probability	z-value
.0001	−3.719	.33	−0.440	.69	0.496
.005	−2.576	.34	−0.412	.70	0.524
.001	−3.090	.35	−0.385	.71	0.553
.01	−2.326	.36	−0.358	.72	0.583
.02	−2.054	.37	−0.332	.73	0.613
.025	−1.960	.38	−0.305	.74	0.643
.03	−1.881	.39	−0.279	.75	0.674
.04	−1.751	.40	−0.253	.76	0.706
.05	−1.645	.41	−0.228	.77	0.739
.06	−1.555	.42	−0.202	.78	0.772
.07	−1.476	.43	−0.176	.79	0.806
.08	−1.405	.44	−0.151	.80	0.842
.09	−1.341	.45	−0.126	.81	0.878
.10	−1.282	.46	−0.100	.82	0.915
.11	−1.227	.47	−0.075	.83	0.954
.12	−1.175	.48	−0.050	.84	0.994
.13	−1.126	.49	−0.025	.85	1.036
.14	−1.080	.50	0.000	.86	1.080
.15	−1.036	.51	0.025	.87	1.126
.16	−0.994	.52	0.050	.88	1.175
.17	−0.954	.53	0.075	.89	1.227
.18	−0.915	.54	0.100	.90	1.282
.19	−0.878	.55	0.126	.91	1.341
.20	−0.842	.56	0.151	.92	1.405
.21	−0.806	.57	0.176	.93	1.476
.22	−0.772	.58	0.202	.94	1.555
.23	−0.739	.59	0.228	.95	1.645
.24	−0.706	.60	0.253	.96	1.751
.25	−0.674	.61	0.279	.97	1.881
.26	−0.643	.62	0.305	.975	1.960
.27	−0.613	.63	0.332	.98	2.054
.28	−0.583	.64	0.358	.99	2.326
.29	−0.553	.65	0.385	.999	3.090
.30	−0.524	.66	0.412	.9995	3.290
.31	−0.496	.67	0.440	.9999	3.719
.32	−0.468	.68	0.468		

TABLE T2 Standard Normal Distribution Function

$$\Phi(z) = \frac{1}{\sqrt{2\pi}} \int_{-\infty}^{z} e^{-x^2/2}\,dx \text{ for } 0.00 \leqslant z \leqslant 4.99.$$

z	.00	.01	.02	.03	.04	.05	.06	.07	.08	.09
.0	.5000	.5040	.5080	.5120	.5160	.5199	.5239	.5279	.5319	.5359
.1	.5398	.5438	.5478	.5517	.5557	.5596	.5636	.5675	.5714	.5753
.2	.5793	.5832	.5871	.5910	.5948	.5987	.6026	.6064	.6103	.6141
.3	.6179	.6217	.6255	.6293	.6331	.6368	.6406	.6443	.6480	.6517
.4	.6554	.6591	.6628	.6664	.6700	.6736	.6772	.6808	.6844	.6879
.5	.6915	.6950	.6985	.7019	.7054	.7088	.7123	.7157	.7190	.7224
.6	.7257	.7291	.7324	.7357	.7389	.7422	.7454	.7486	.7517	.7549
.7	.7580	.7611	.7642	.7673	.7703	.7734	.7764	.7794	.7823	.7852
.8	.7881	.7910	.7939	.7967	.7995	.8023	.8051	.8078	.8106	.8133
.9	.8159	.8186	.8212	.8238	.8264	.8289	.8315	.8340	.8365	.8389
1.0	.8413	.8438	.8461	.8485	.8508	.8531	.8554	.8577	.8599	.8621
1.1	.8643	.8665	.8686	.8708	.8729	.8749	.8770	.8790	.8810	.8830
1.2	.8849	.8869	.8888	.8907	.8925	.8944	.8962	.8980	.8997	.90147
1.3	.90320	.90490	.90658	.90824	.90988	.91149	.91309	.91466	.91621	.91774
1.4	.91924	.92073	.92220	.92364	.92507	.92647	.92785	.92922	.93056	.93189
1.5	.93319	.93448	.93574	.93699	.93822	.93943	.94062	.94179	.94295	.94408
1.6	.94520	.94630	.94738	.94845	.94950	.95053	.95154	.95254	.95352	.95449
1.7	.95543	.95637	.95728	.95818	.95907	.95994	.96080	.96164	.96246	.96327
1.8	.96407	.96485	.96562	.96638	.96712	.96784	.96856	.96926	.96995	.97062
1.9	.97128	.97193	.97257	.97320	.97381	.97441	.97500	.97558	.97615	.97670
2.0	.97725	.97778	.97831	.97882	.97932	.97982	.98030	.98077	.98124	.98169
2.1	.98214	.98257	.98300	.98341	.98382	.98422	.98461	.98500	.98537	.98574
2.2	.98610	.98645	.98679	.98713	.98745	.98778	.98809	.98840	.98870	.98899
2.3	.98928	.98956	.98983	$.9^2 0097$	$.9^2 0358$	$.9^2 0613$	$.9^2 0863$	$.9^2 1106$	$.9^2 1344$	$.9^2 1576$
2.4	$.9^2 1802$	$.9^2 2024$	$.9^2 2240$	$.9^2 2451$	$.9^2 2656$	$.9^2 2857$	$.9^2 3053$	$.9^2 3244$	$.9^2 3431$	$.9^2 3613$

For negative arguments $\Phi(z) = 1 - \Phi(z)$

TABLE T2 Standard Normal Distribution Function (Continued)

$$\Phi(z) = \frac{1}{\sqrt{2\pi}} \int_{-\infty}^{z} e^{-x^2/2}\, dx \text{ for } 0.00 \leqslant z \leqslant 4.99.$$

z	.00	.01	.02	.03	.04	.05	.06	.07	.08	.09
2.5	$.9^23790$	$.9^23963$	$.9^24132$	$.9^24297$	$.9^24457$	$.9^24614$	$.9^24766$	$.9^24915$	$.9^25060$	$.9^25201$
2.6	$.9^25339$	$.9^25473$	$.9^25604$	$.9^25731$	$.9^25855$	$.9^25975$	$.9^26093$	$.9^26207$	$.9^26319$	$.9^26427$
2.7	$.9^26533$	$.9^26636$	$.9^26736$	$.9^26833$	$.9^26928$	$.9^27020$	$.9^27110$	$.9^27197$	$.9^27282$	$.9^27365$
2.8	$.9^27445$	$.9^27523$	$.9^27599$	$.9^27673$	$.9^27744$	$.9^27814$	$.9^27882$	$.9^27948$	$.9^28012$	$.9^28074$
2.9	$.9^28134$	$.9^28193$	$.9^28250$	$.9^28305$	$.9^28359$	$.9^28411$	$.9^28462$	$.9^28511$	$.9^28559$	$.9^28605$
3.0	$.9^28650$	$.9^28694$	$.9^28736$	$.9^28777$	$.9^28817$	$.9^28856$	$.9^28893$	$.9^28930$	$.9^28965$	$.9^28999$
3.1	$.9^30324$	$.9^30646$	$.9^30957$	$.9^31260$	$.9^31553$	$.9^31836$	$.9^32112$	$.9^32378$	$.9^32636$	$.9^32886$
3.2	$.9^33129$	$.9^33363$	$.9^33590$	$.9^33810$	$.9^34024$	$.9^34230$	$.9^34429$	$.6^34623$	$.9^34810$	$.9^34991$
3.3	$.9^35166$	$.9^35335$	$.9^35499$	$.9^35658$	$.9^35811$	$.9^35959$	$.9^36103$	$.9^36242$	$.9^36376$	$.9^36505$
3.4	$.9^36631$	$.9^36752$	$.9^36869$	$.9^36982$	$.9^37091$	$.9^37197$	$.9^37299$	$.9^37398$	$.9^37493$	$.9^37585$
3.5	$.9^37674$	$.9^37759$	$.9^37842$	$.9^37922$	$.9^37999$	$.9^38074$	$.9^38146$	$.9^38215$	$.9^38282$	$.9^38347$
3.6	$.9^38409$	$.9^38469$	$.9^38527$	$.9^38583$	$.9^38637$	$.9^38689$	$.9^38739$	$.9^38787$	$.9^38834$	$.9^38879$
3.7	$.9^38922$	$.9^38964$	$.9^40039$	$.9^40426$	$.9^40799$	$.9^41158$	$.9^41504$	$.9^41838$	$.9^42159$	$.9^42568$
3.8	$.9^42765$	$.9^43052$	$.9^43327$	$.9^43593$	$.9^43848$	$.9^44094$	$.9^44331$	$.9^44558$	$.9^44777$	$.9^44988$
3.9	$.9^45190$	$.9^45385$	$.9^45573$	$.9^45753$	$.9^45926$	$.9^46092$	$.9^46253$	$.9^46406$	$.9^46554$	$.9^46696$
4.0	$.9^46833$	$.9^46964$	$.9^47090$	$.9^47211$	$.9^47327$	$.9^47439$	$.9^47546$	$.9^47649$	$.9^47748$	$.9^47843$
4.1	$.9^47934$	$.9^48022$	$.9^48106$	$.9^48186$	$.9^48263$	$.9^48338$	$.9^48409$	$.9^48477$	$.9^48542$	$.9^48605$
4.2	$.9^48665$	$.9^48723$	$.9^48778$	$.9^48832$	$.9^48882$	$.9^48931$	$.9^48978$	$.9^50226$	$.9^50655$	$.9^51066$
4.3	$.9^51460$	$.9^51837$	$.9^52199$	$.9^52545$	$.9^52876$	$.9^53193$	$.9^53497$	$.9^53788$	$.9^54066$	$.9^54332$
4.4	$.9^54587$	$.9^54831$	$.9^55065$	$.9^55288$	$.9^55502$	$.9^55706$	$.9^55902$	$.9^56089$	$.9^56268$	$.9^56439$
4.5	$.9^56602$	$.9^56759$	$.9^56908$	$.9^57051$	$.9^57187$	$.9^57318$	$.9^57442$	$.9^57561$	$.9^57675$	$.9^57784$
4.6	$.9^57888$	$.9^57987$	$.9^58081$	$.9^58172$	$.9^58258$	$.9^58340$	$.9^58419$	$.9^58494$	$.9^58566$	$.9^58634$
4.7	$.9^58699$	$.9^58761$	$.9^58821$	$.9^58877$	$.9^58931$	$.9^58983$	$.9^60320$	$.9^60789$	$.9^61235$	$.9^61661$
4.8	$.9^62067$	$.9^62453$	$.9^62822$	$.9^63173$	$.9^63508$	$.9^63827$	$.9^64131$	$.9^64420$	$.9^64696$	$.9^64958$
4.9	$.9^65208$	$.9^65446$	$.9^65673$	$.9^65889$	$.9^66094$	$.9^66289$	$.9^66475$	$.9^66652$	$.9^66821$	$.9^66981$

Example: $\Phi(3.39) = 0.9996505$

For negative arguments $\Phi(z) = 1 - \Phi(z)$

TABLE T3 Critical Points of the Chi-Square Distribution

Values of χ^2_α for ν degrees of freedom

ν	$\chi^2_{.999}$	$\chi^2_{.995}$	$\chi^2_{.99}$	$\chi^2_{.975}$	$\chi^2_{.95}$	$\chi^2_{.90}$	$\chi^2_{.75}$	$\chi^2_{.50}$	$\chi^2_{.25}$	$\chi^2_{.10}$	$\chi^2_{.05}$	$\chi^2_{.025}$	$\chi^2_{.01}$	$\chi^2_{.005}$	$\chi^2_{.001}$
1	0.000	0.000	0.000	0.001	0.004	0.016	0.102	0.455	1.323	2.706	3.841	5.024	6.635	7.879	10.828
2	0.002	0.010	0.020	0.051	0.103	0.211	0.575	1.386	2.773	4.605	5.991	7.378	9.210	10.597	13.815
3	0.024	0.072	0.115	0.216	0.352	0.584	1.213	2.366	4.108	6.251	7.815	9.348	11.345	12.838	16.266
4	0.091	0.207	0.297	0.484	0.711	1.064	1.923	3.357	5.385	7.779	9.488	11.143	13.277	14.860	18.467
5	0.210	0.412	0.554	0.831	1.145	1.610	2.675	4.351	6.626	9.236	11.070	12.833	15.086	16.750	20.515
6	0.381	0.676	0.872	1.237	1.635	2.204	3.455	5.348	7.841	10.645	12.592	14.449	16.812	18.548	22.458
7	0.598	0.989	1.239	1.690	2.167	2.833	4.255	6.346	9.037	12.017	14.067	16.013	18.475	20.278	24.322
8	0.857	1.344	1.646	2.180	2.733	3.490	5.071	7.344	10.219	13.362	15.507	17.535	20.090	21.955	26.124
9	1.152	1.735	2.088	2.700	3.325	4.168	5.899	8.343	11.389	14.684	16.919	19.023	21.666	23.589	27.877
10	1.479	2.156	2.558	3.247	3.940	4.865	6.737	9.342	12.549	15.987	18.307	20.483	23.209	25.188	29.588
11	1.834	2.603	3.053	3.816	4.575	5.578	7.584	10.341	13.701	17.275	19.675	21.920	24.725	26.757	31.264
12	2.214	3.074	3.571	4.404	5.226	6.304	8.438	11.340	14.845	18.549	21.026	23.337	26.217	28.300	32.909
13	2.617	3.565	4.107	5.009	5.892	7.042	9.299	12.340	15.984	19.812	22.362	24.736	27.688	29.819	34.528
14	3.041	4.075	4.660	5.629	6.571	7.790	10.165	13.339	17.117	21.064	23.685	26.119	29.141	31.319	36.123
15	3.483	4.601	5.229	6.262	7.261	8.547	11.037	14.339	18.245	22.307	24.996	27.488	30.578	32.801	37.697
16	3.942	5.142	5.812	6.908	7.962	9.312	11.912	15.338	19.369	23.542	26.296	28.845	32.000	34.267	39.252
17	4.416	5.697	6.408	7.564	8.672	10.085	12.792	16.338	20.489	24.769	27.587	30.191	33.409	35.718	40.790
18	4.905	6.265	7.015	8.231	9.390	10.865	13.675	17.338	21.605	25.989	28.869	31.526	34.805	37.156	42.312
19	5.407	6.844	7.633	8.907	10.117	11.651	14.562	18.338	22.718	27.204	30.144	32.852	36.191	38.582	43.820
20	5.921	7.434	8.260	9.591	10.851	12.443	15.452	19.337	23.828	28.412	31.410	34.170	37.566	39.997	45.315
21	6.447	8.034	8.897	10.283	11.591	13.240	16.344	20.337	24.935	29.615	32.671	35.479	38.932	41.401	46.797
22	6.983	8.643	9.542	10.982	12.338	14.041	17.240	21.337	26.039	30.813	33.924	36.781	40.289	42.796	48.268
23	7.529	9.260	10.196	11.689	13.091	14.848	18.137	22.337	27.141	32.007	35.172	38.076	41.638	44.181	49.728
24	8.085	9.886	10.856	12.401	13.848	15.659	19.037	23.337	28.241	33.196	36.415	39.364	42.980	45.559	51.179
25	8.649	10.520	11.524	13.120	14.611	16.473	19.939	24.337	29.339	34.382	37.652	40.646	44.314	46.928	52.620
26	9.222	11.160	12.198	13.844	15.379	17.292	20.843	25.336	30.435	35.563	38.885	41.923	45.642	48.290	54.052
27	9.803	11.808	12.879	14.573	16.151	18.114	21.749	26.336	31.528	36.741	40.113	43.195	46.963	49.645	55.476

TABLE T3 Critical Points of the Chi-Square Distribution (Continued)

Values of χ^2_α for ν degrees of freedom

ν	$\chi^2_{.999}$	$\chi^2_{.995}$	$\chi^2_{.99}$	$\chi^2_{.975}$	$\chi^2_{.95}$	$\chi^2_{.90}$	$\chi^2_{.75}$	$\chi^2_{.50}$	$\chi^2_{.25}$	$\chi^2_{.10}$	$\chi^2_{.05}$	$\chi^2_{.025}$	$\chi^2_{.01}$	$\chi^2_{.005}$	$\chi^2_{.001}$
28	10.391	12.461	13.565	15.308	16.928	18.939	22.657	27.336	32.620	37.916	41.337	44.461	48.278	50.993	56.892
29	10.986	13.121	14.256	16.047	17.708	19.768	23.567	28.336	33.711	39.087	42.557	45.722	49.588	52.336	58.301
30	11.588	13.787	14.953	16.791	18.493	20.599	24.478	29.336	34.800	40.256	43.773	46.979	50.892	53.672	59.703
35	14.688	17.192	18.509	20.569	22.465	24.797	29.054	34.336	40.223	46.059	49.802	53.203	57.342	60.275	66.619
40	17.916	20.707	22.164	24.433	26.509	29.051	33.660	39.335	45.616	51.805	55.758	59.342	63.691	66.766	73.402
45	21.251	24.311	25.901	28.366	30.612	33.350	38.291	44.335	50.985	57.505	61.656	65.410	69.957	73.166	80.077
50	24.674	27.991	29.707	32.357	34.764	37.689	42.942	49.335	56.334	63.167	67.505	71.420	76.154	79.490	86.661
55	28.173	31.735	33.570	36.398	38.958	42.060	47.610	54.335	61.665	68.796	73.311	77.380	82.292	85.749	93.168
60	31.738	35.534	37.485	40.482	43.188	46.459	52.294	59.335	66.981	74.397	79.082	83.298	88.379	91.952	99.607
65	35.362	39.383	41.444	44.603	47.450	50.883	56.990	64.335	72.285	79.973	84.821	89.177	94.422	98.105	105.99
70	39.036	43.275	45.442	48.758	51.739	55.329	61.698	69.334	77.577	85.527	90.531	95.023	100.43	104.21	112.32
75	42.757	47.206	49.475	52.942	56.054	59.795	66.417	74.334	82.858	91.061	96.217	100.84	106.39	110.29	118.60
80	46.520	51.172	53.540	57.153	60.391	64.278	71.145	79.334	88.130	96.578	101.88	106.63	112.33	116.32	124.84
85	50.320	55.170	57.634	61.389	64.749	68.777	75.881	84.334	93.394	102.08	107.52	112.39	118.24	122.32	131.04
90	54.155	59.196	61.754	65.647	69.126	73.291	80.625	89.334	98.650	107.57	113.15	118.14	124.12	128.30	137.21
95	58.022	63.250	65.898	69.925	73.520	77.818	85.376	94.334	103.90	113.04	118.75	123.86	129.97	134.25	143.34
100	61.918	67.328	70.065	74.222	77.929	82.358	90.133	99.334	109.14	118.50	124.34	129.56	135.81	140.17	149.45
200	143.84	152.24	156.43	162.73	168.28	174.84	186.17	199.33	213.10	226.02	233.99	241.06	249.45	255.26	267.54
300	229.96	240.66	245.97	253.91	260.88	269.07	283.14	299.33	316.14	331.79	341.40	349.87	359.91	366.84	381.43
400	318.26	330.90	337.16	346.48	354.64	364.21	380.58	399.33	418.70	436.65	447.63	457.31	468.72	476.61	493.13
500	407.95	422.30	429.39	439.94	449.15	459.93	478.32	499.33	520.95	540.93	553.13	563.85	576.49	585.21	603.45
600	498.62	514.53	522.37	534.02	544.18	556.06	576.29	599.33	622.99	644.80	658.09	669.77	683.52	692.98	712.77
700	590.05	607.38	615.91	628.58	639.61	652.50	674.41	699.33	724.86	748.36	762.66	775.21	789.97	800.13	821.35
800	682.07	700.73	709.90	723.51	735.36	749.19	772.67	799.33	826.60	851.67	866.91	880.28	895.98	906.79	929.33
900	774.57	794.48	804.25	818.76	831.37	846.08	871.03	899.33	928.24	954.78	970.90	985.03	1001.6	1013.0	1036.8
1000	867.48	888.56	898.91	914.26	927.59	943.13	969.48	999.33	1029.8	1057.7	1074.7	1089.5	1107.0	1119.0	1143.9

TABLE T4 Student's t Critical Points

Values of t_α for ν degrees of freedom

α-Level

ν	.45	.40	.35	.30	.25	.20	.15	.10	.05	.025	.01	.005	.001	.0001
1	0.158	0.325	0.510	0.727	1.000	1.376	1.963	3.078	6.314	12.706	31.821	63.657	318.31	3183.1
2	0.142	0.289	0.445	0.617	0.816	1.061	1.386	1.886	2.920	4.303	6.965	9.925	22.327	70.700
3	0.137	0.277	0.424	0.584	0.765	0.978	1.250	1.638	2.353	3.182	4.541	5.841	10.215	22.204
4	0.134	0.271	0.414	0.569	0.741	0.941	1.190	1.533	2.132	2.776	3.747	4.604	7.173	13.034
5	0.132	0.267	0.408	0.559	0.727	0.920	1.156	1.476	2.015	2.571	3.365	4.032	5.893	9.678
6	0.131	0.265	0.404	0.553	0.718	0.906	1.134	1.440	1.943	2.447	3.143	3.707	5.208	8.025
7	0.130	0.263	0.402	0.549	0.711	0.896	1.119	1.415	1.895	2.365	2.998	3.499	4.785	7.063
8	0.130	0.262	0.399	0.546	0.706	0.889	1.108	1.397	1.860	2.306	2.896	3.355	4.501	6.442
9	0.129	0.261	0.398	0.543	0.703	0.883	1.100	1.383	1.833	2.262	2.821	3.250	4.297	6.010
10	0.129	0.260	0.397	0.542	0.700	0.879	1.093	1.372	1.812	2.228	2.764	3.169	4.144	5.694
11	0.129	0.260	0.396	0.540	0.697	0.876	1.088	1.363	1.796	2.201	2.718	3.106	4.025	5.453
12	0.128	0.259	0.395	0.539	0.695	0.873	1.083	1.356	1.782	2.179	2.681	3.055	3.930	5.263
13	0.128	0.259	0.394	0.537	0.694	0.870	1.079	1.350	1.771	2.160	2.650	3.012	3.852	5.111
14	0.128	0.258	0.393	0.537	0.692	0.868	1.076	1.345	1.761	2.145	2.624	2.977	3.787	4.985
15	0.128	0.258	0.393	0.536	0.691	0.866	1.074	1.341	1.753	2.131	2.602	2.947	3.733	4.880
16	0.128	0.258	0.392	0.535	0.690	0.865	1.071	1.337	1.746	2.120	2.583	2.921	3.686	4.791
17	0.128	0.257	0.392	0.534	0.689	0.863	1.069	1.333	1.740	2.110	2.567	2.898	3.646	4.714
18	0.127	0.257	0.392	0.534	0.688	0.862	1.067	1.330	1.734	2.101	2.552	2.878	3.610	4.648
19	0.127	0.257	0.391	0.533	0.688	0.861	1.066	1.328	1.729	2.093	2.539	2.861	3.579	4.590
20	0.127	0.257	0.391	0.533	0.687	0.860	1.064	1.325	1.725	2.086	2.528	2.845	3.552	4.539
21	0.127	0.257	0.391	0.532	0.686	0.859	1.063	1.323	1.721	2.080	2.518	2.831	3.527	4.493

TABLE T4 Student's *t* Critical Points (Continued)

Values of t_α for ν degrees of freedom

ν	α-Level													
	.45	.40	.35	.30	.25	.20	.15	.10	.05	.025	.01	.005	.001	.0001
22	0.127	0.256	0.390	0.532	0.686	0.858	1.061	1.321	1.717	2.074	2.508	2.819	3.505	4.452
23	0.127	0.256	0.390	0.532	0.685	0.858	1.060	1.319	1.714	2.069	2.500	2.807	3.485	4.415
24	0.127	0.256	0.390	0.531	0.685	0.857	1.059	1.318	1.711	2.064	2.492	2.797	3.467	4.382
25	0.127	0.256	0.390	0.531	0.684	0.856	1.058	1.316	1.708	2.060	2.485	2.787	3.450	4.352
26	0.127	0.256	0.390	0.531	0.684	0.856	1.058	1.315	1.706	2.056	2.479	2.779	3.435	4.324
27	0.127	0.256	0.389	0.531	0.684	0.855	1.057	1.314	1.703	2.052	2.473	2.771	3.421	4.299
28	0.127	0.256	0.389	0.530	0.683	0.855	1.056	1.313	1.701	2.048	2.467	2.763	3.408	4.275
29	0.127	0.256	0.389	0.530	0.683	0.854	1.055	1.311	1.699	2.045	2.462	2.756	3.396	4.254
30	0.127	0.256	0.389	0.530	0.683	0.854	1.055	1.310	1.697	2.042	2.457	2.750	3.385	4.234
35	0.127	0.255	0.389	0.529	0.682	0.852	1.052	1.306	1.690	2.030	2.438	2.724	3.340	4.153
40	0.126	0.255	0.388	0.529	0.681	0.851	1.050	1.303	1.684	2.021	2.423	2.704	3.307	4.094
45	0.126	0.255	0.388	0.528	0.680	0.850	1.049	1.301	1.679	2.014	2.412	2.690	3.281	4.049
50	0.126	0.255	0.388	0.528	0.679	0.849	1.047	1.299	1.676	2.009	2.403	2.678	3.261	4.014
60	0.126	0.254	0.387	0.527	0.679	0.848	1.045	1.296	1.671	2.000	2.390	2.660	3.232	3.962
70	0.126	0.254	0.387	0.527	0.678	0.847	1.044	1.294	1.667	1.994	2.381	2.648	3.211	3.926
80	0.126	0.254	0.387	0.526	0.678	0.846	1.043	1.292	1.664	1.990	2.374	2.639	3.195	3.899
90	0.126	0.254	0.387	0.526	0.677	0.846	1.042	1.291	1.662	1.987	2.369	2.632	3.183	3.878
100	0.126	0.254	0.386	0.526	0.677	0.845	1.042	1.290	1.660	1.984	2.364	2.626	3.174	3.862
150	0.126	0.254	0.386	0.526	0.676	0.844	1.040	1.287	1.655	1.976	2.351	2.609	3.145	3.813
200	0.126	0.254	0.386	0.525	0.676	0.843	1.039	1.286	1.652	1.972	2.345	2.601	3.131	3.789
300	0.126	0.254	0.386	0.525	0.675	0.843	1.038	1.284	1.650	1.968	2.339	2.592	3.118	3.766
400	0.126	0.254	0.386	0.525	0.675	0.843	1.038	1.284	1.649	1.966	2.336	2.588	3.111	3.754
inf	0.126	0.253	0.385	0.524	0.674	0.842	1.036	1.282	1.645	1.960	2.326	2.576	3.090	3.719

TABLE T5 Gamma Functions for Noninteger Arguments

α	.00	.01	.02	.03	.04	.05	.06	.07	.08	.09
					$\Gamma(\alpha)$					
1.0	1.0000	.9943	.9888	.9836	.9784	.9735	.9687	.9642	.9597	.9555
1.1	.9514	.9474	.9435	.9399	.9364	.9330	.9298	.9267	.9237	.9209
1.2	.9182	.9156	.9131	.9108	.9085	.9064	.9045	.9025	.9007	.8990
1.3	.8974	.8960	.8946	.8934	.8922	.8912	.8902	.8893	.8885	.8879
1.4	.8873	.8868	.8864	.8860	.8858	.8857	.8856	.8856	.8857	.8859
1.5	.8862	.8866	.8870	.8876	.8882	.8889	.8896	.8905	.8914	.8924
1.6	.8935	.8947	.8959	.8972	.8986	.9001	.9017	.9033	.9050	.9068
1.7	.9086	.9106	.9126	.9147	.9168	.9191	.9214	.9238	.9262	.9288
1.8	.9314	.9341	.9368	.9397	.9426	.9456	.9487	.9518	.9551	.9584
1.9	.9618	.9652	.9688	.9724	.9761	.9799	.9867	.9877	.9917	.9958

Calculated by Vic Simonis. Permission granted. Formula from Abramowitz and Stegun (1968).

TABLE T6 Incomplete Gamma Function

$$F(x;\alpha) = \frac{1}{\Gamma(\alpha)} \int_0^x v^{\alpha-1} e^{-v} dv$$

x \ α	0	1	2	3	4	5	6	7	8	9
0.2	.1813	.0175	.0011	.0001	.0000	.0000	.0000	.0000	.0000	.0000
0.4	.3297	.0616	.0079	.0008	.0001	.0000	.0000	.0000	.0000	.0000
0.6	.4512	.1219	.0231	.0034	.0004	.0000	.0000	.0000	.0000	.0000
0.8	.5507	.1912	.0474	.0091	.0014	.0002	.0000	.0000	.0000	.0000
1.0	.6321	.2642	.0803	.0190	.0037	.0006	.0001	.0000	.0000	.0000
1.2	.6988	.3374	.1205	.0338	.0077	.0015	.0003	.0000	.0000	.0000
1.4	.7534	.4082	.1665	.0537	.0143	.0032	.0006	.0001	.0000	.0000
1.6	.7981	.4751	.2166	.0788	.0237	.0060	.0013	.0003	.0000	.0000
1.8	.8347	.5372	.2694	.1087	.0364	.0104	.0026	.0006	.0001	.0000
2.0	.8647	.5940	.3233	.1429	.0527	.0166	.0045	.0011	.0002	.0000
2.2	.8892	.6454	.3773	.1806	.0725	.0249	.0075	.0020	.0005	.0001
2.4	.9093	.6916	.4303	.2213	.0959	.0357	.0116	.0033	.0009	.0002
2.6	.9257	.7326	.4816	.2640	.1226	.0490	.0172	.0053	.0015	.0004
2.8	.9392	.7689	.5305	.3081	.1523	.0651	.0244	.0081	.0024	.0007
3.0	.9502	.8009	.5768	.3528	.1847	.0839	.0335	.0119	.0038	.0011

TABLE T6 Incomplete Gamma Function (Continued)

$$F(x; \alpha) = \frac{1}{\Gamma(\alpha)} \int_0^x v^{\alpha-1} e^{-v} dv$$

x \ α	0	1	2	3	4	5	6	7	8	9
3.2	.9592	.8288	.6201	.3975	.2194	.1054	.0446	.0168	.0057	.0018
3.4	.9666	.8532	.6603	.4416	.2558	.1295	.0579	.0231	.0083	.0027
3.6	.9727	.8743	.6973	.4848	.2936	.1559	.0733	.0308	.0117	.0040
3.8	.9776	.8926	.7311	.5265	.3322	.1844	.0909	.0401	.0160	.0058
4.0	.9817	.9084	.7619	.5665	.3712	.2149	.1107	.0511	.0214	.0081
4.2	.9850	.9220	.7898	.6046	.4102	.2469	.1325	.0639	.0279	.0111
4.4	.9877	.9337	.8149	.6406	.4488	.2801	.1564	.0786	.0358	.0149
4.6	.9899	.9437	.8374	.6743	.4868	.3142	.1820	.0951	.0451	.0195
4.8	.9918	.9523	.8575	.7058	.5237	.3490	.2092	.1133	.0558	.0251
5.0	.9933	.9596	.8753	.7350	.5595	.3840	.2378	.1334	.0681	.0318
5.2	.9945	.9658	.8912	.7619	.5939	.4191	.2676	.1551	.0819	.0397
5.4	.9955	.9711	.9052	.7867	.6267	.4539	.2983	.1783	.0974	.0488
5.6	.9963	.9756	.9176	.8094	.6579	.4881	.3297	.2030	.1143	.0591
5.8	.9970	.9794	.9285	.8300	.6873	.5217	.3616	.2290	.1328	.0708
6.0	.9975	.9826	.9380	.8488	.7149	.5543	.3937	.2560	.1528	.0839
6.2	.9980	.9854	.9464	.8658	.7408	.5859	.4258	.2840	.1741	.0984
6.4	.9983	.9877	.9537	.8811	.7649	.6163	.4577	.3127	.1967	.1142
6.6	.9986	.9897	.9600	.8948	.7873	.6453	.4892	.3419	.2204	.1314
6.8	.9989	.9913	.9656	.9072	.8080	.6730	.5201	.3715	.2452	.1498
7.0	.9991	.9927	.9704	.9182	.8270	.6993	.5503	.4013	.2709	.1695
7.2	.9993	.9939	.9745	.9281	.8445	.7241	.5796	.4311	.2973	.1903
7.4	.9994	.9949	.9781	.9368	.8605	.7474	.6080	.4607	.3243	.2123
7.6	.9995	.9957	.9812	.9446	.8751	.7693	.6354	.4900	.3518	.2351
7.8	.9996	.9964	.9839	.9515	.8883	.7897	.6616	.5188	.3796	.2589
8.0	.9997	.9970	.9862	.9576	.9004	.8088	.6866	.5470	.4074	.2834
8.2	.9997	.9975	.9882	.9630	.9113	.8264	.7104	.5746	.4353	.3085
8.4	.9998	.9979	.9900	.9677	.9211	.8427	.7330	.6013	.4631	.3341
8.6	.9998	.9982	.9914	.9719	.9299	.8578	.7543	.6272	.4906	.3600
8.8	.9998	.9985	.9927	.9756	.9379	.8716	.7744	.6522	.5177	.3863
9.0	.9999	.9988	.9938	.9788	.9450	.8843	.7932	.6761	.5443	.4126
9.2	.9999	.9990	.9947	.9816	.9514	.8959	.8108	.6990	.5704	.4389
9.4	.9999	.9991	.9955	.9840	.9571	.9065	.8273	.7208	.5958	.4651
9.6	.9999	.9993	.9962	.9862	.9622	.9162	.8426	.7416	.6204	.4911
9.8	.9999	.9994	.9967	.9880	.9667	.9250	.8567	.7612	.6442	.5168
10.0	1.000	.9995	.9972	.9897	.9707	.9329	.8699	.7798	.6672	.5421

TABLE T7 Critical Points of the F Distribution

Upper 0.1 Percent Points

$\nu_2 \backslash \nu_1$	1	2	3	4	5	6	7	8	9
1	405295.	50014.	540393.	562516.	576422.	585952.	592877.	598163.	602295.
2	998.52	999.02	999.18	999.26	999.31	999.34	999.37	999.39	999.40
3	167.03	148.50	141.01	137.12	134.76	132.97	131.74	130.70	130.09
4	74.14	61.25	56.17	53.45	51.71	50.52	49.66	48.96	48.47
5	47.18	37.12	33.20	31.08	29.75	28.84	28.15	27.65	27.25
6	35.51	27.00	23.70	21.93	20.80	20.03	19.46	19.03	18.69
7	29.25	21.69	18.77	17.20	16.21	15.52	15.02	14.63	14.33
8	25.41	18.49	15.83	14.39	13.48	12.86	12.40	12.05	11.77
9	22.86	16.39	13.90	12.56	11.71	11.13	10.70	10.37	10.11
10	21.04	14.91	12.55	11.28	10.48	9.93	9.52	9.20	8.96
11	19.69	13.81	11.56	10.35	9.58	9.05	8.65	8.35	8.12
12	18.64	12.97	10.80	9.63	8.89	8.38	8.00	7.71	7.48
13	17.82	12.31	10.21	9.07	8.35	7.86	7.49	7.21	6.98
14	17.14	11.78	9.73	8.62	7.92	7.44	7.08	6.80	6.58
15	16.59	11.34	9.34	8.25	7.57	7.09	6.74	6.47	6.26
16	16.12	10.97	9.01	7.94	7.27	6.80	6.46	6.19	5.98
17	15.72	10.66	8.73	7.68	7.02	6.56	6.22	5.96	5.75
18	15.38	10.39	8.49	7.46	6.81	6.35	6.02	5.76	5.56
19	15.08	10.16	8.28	7.27	6.62	6.18	5.84	5.59	5.39
20	14.82	9.95	8.10	7.10	6.46	6.02	5.69	5.44	5.24
21	14.59	9.77	7.94	6.95	6.32	5.88	5.56	5.31	5.11
22	14.38	9.61	7.80	6.81	6.19	5.76	5.44	5.19	4.99
23	14.20	9.47	7.67	6.70	6.08	5.65	5.33	5.09	4.89
24	14.03	9.34	7.55	6.59	5.98	5.55	5.24	4.99	4.80
25	13.88	9.22	7.45	6.49	5.89	5.46	5.15	4.91	4.71
26	13.74	9.12	7.36	6.41	5.80	5.38	5.07	4.83	4.64
27	13.61	9.02	7.27	6.33	5.73	5.31	5.00	4.76	4.57
28	13.50	8.93	7.19	6.25	5.66	5.24	4.93	4.69	4.50
29	13.39	8.85	7.12	6.19	5.59	5.18	4.87	4.64	4.45
30	13.29	8.77	7.05	6.12	5.53	5.12	4.82	4.58	4.39
40	12.61	8.25	6.59	5.70	5.13	4.73	4.44	4.21	4.02
60	11.97	7.77	6.17	5.31	4.76	4.37	4.09	3.86	3.69
100	11.50	7.41	5.86	5.02	4.48	4.11	3.83	3.61	3.44
500	10.96	7.00	5.51	4.69	4.18	3.81	3.54	3.33	3.16
10000	10.81	6.91	5.43	4.62	4.11	3.75	3.48	3.27	3.10

10	12	16	20	24	30	40	60	100	1000
605634.	610688.	617083.	620929.	623524.	626086.	628753.	631359.	633395.	636274.
999.41	999.43	999.47	999.47	999.47	999.46	999.49	999.49	999.47	999.60
129.47	128.25	127.15	126.69	126.88	125.45	124.96	124.47	124.07	123.53
48.01	47.42	46.63	46.03	45.82	45.50	45.09	44.75	44.47	44.10
26.91	26.42	25.79	25.41	25.15	24.86	24.59	24.33	24.12	23.82
18.41	17.99	17.45	17.12	16.90	16.68	16.44	16.21	16.03	15.77
14.08	13.71	13.23	12.93	12.73	12.53	12.33	12.13	11.95	11.72
11.54	11.19	10.75	10.48	10.29	10.11	9.92	9.72	9.57	9.36
9.89	9.57	9.15	8.90	8.73	8.55	8.37	8.18	8.04	7.84
8.75	8.45	8.05	7.80	7.64	7.47	7.30	7.12	6.98	6.78
7.92	7.63	7.24	7.01	6.85	6.68	6.52	6.35	6.21	6.02
7.29	7.00	6.63	6.40	6.25	6.09	5.93	5.76	5.63	5.44
6.80	6.52	6.16	5.93	5.78	5.63	5.47	5.31	5.17	4.99
6.40	6.13	5.78	5.56	5.41	5.25	5.10	4.94	4.81	4.62
6.08	5.81	5.46	5.25	5.10	4.95	4.80	4.64	4.51	4.33
5.81	5.55	5.20	4.99	4.85	4.70	4.54	4.39	4.26	4.08
5.58	5.32	4.99	4.78	4.63	4.48	4.33	4.18	4.05	3.87
5.39	5.13	4.80	4.59	4.45	4.30	4.15	4.00	3.87	3.69
5.22	4.97	4.64	4.43	4.29	4.14	3.99	3.84	3.71	3.53
5.08	4.82	4.49	4.29	4.15	4.00	3.86	3.70	3.58	3.40
4.95	4.70	4.37	4.17	4.03	3.88	3.74	3.58	3.46	3.28
4.83	4.58	4.26	4.06	3.92	3.78	3.63	3.48	3.35	3.17
4.73	4.48	4.16	3.96	3.82	3.68	3.53	3.38	3.25	3.08
4.64	4.39	4.07	3.87	3.74	3.59	3.45	3.29	3.17	2.99
4.56	4.31	3.99	3.79	3.66	3.52	3.37	3.22	3.09	2.91
4.48	4.24	3.92	3.72	3.59	3.44	3.30	3.15	3.02	2.84
4.41	4.17	3.86	3.66	3.52	3.38	3.23	3.08	2.96	2.78
4.35	4.11	3.80	3.60	3.46	3.32	3.18	3.02	2.90	2.72
4.29	4.05	3.74	3.54	3.41	3.27	3.12	2.97	2.84	2.66
4.24	4.00	3.69	3.49	3.36	3.22	3.07	2.92	2.79	2.61
3.87	3.64	3.34	3.15	3.01	2.87	2.73	2.57	2.44	2.25
3.54	3.32	3.02	2.83	2.69	2.55	2.41	2.25	2.12	1.92
3.30	3.07	2.78	2.59	2.46	2.32	2.17	2.01	1.87	1.64
3.02	2.81	2.52	2.33	2.20	2.05	1.90	1.73	1.57	1.28
2.96	2.75	2.46	2.27	2.14	1.99	1.84	1.66	1.50	1.15

TABLE T7   Critical Points of the F Distribution (Continued)

Upper 0.5 Percent Points

ν_2 \ ν_1	1	2	3	4	5	6	7	8	9
1	16211.	20000.	21615.	22500.	23056.	23437.	23715.	23925.	24091.
2	198.5	199.0	199.2	199.2	199.3	199.3	199.4	199.4	199.4
3	55.55	49.80	47.47	46.19	45.39	44.84	44.43	44.13	43.88
4	31.33	26.28	24.26	23.15	22.46	21.97	21.62	21.35	21.14
5	22.78	18.31	16.53	15.56	14.94	14.51	14.20	13.96	13.77
6	18.63	14.54	12.92	12.03	11.46	11.07	10.79	10.57	10.39
7	16.24	12.40	10.88	10.05	9.52	9.16	8.89	8.68	8.51
8	14.69	11.04	9.60	8.81	8.30	7.95	7.69	7.50	7.34
9	13.61	10.11	8.72	7.96	7.47	7.13	6.88	6.69	6.54
10	12.83	9.43	8.08	7.34	6.87	6.54	6.30	6.12	5.97
11	12.23	8.91	7.60	6.88	6.42	6.10	5.86	5.68	5.54
12	11.75	8.51	7.23	6.52	6.07	5.76	5.52	5.35	5.20
13	11.37	8.19	6.93	6.23	5.79	5.48	5.25	5.08	4.94
14	11.06	7.92	6.68	6.00	5.56	5.26	5.03	4.86	4.72
15	10.80	7.70	6.48	5.80	5.37	5.07	4.85	4.67	4.54
16	10.58	7.51	6.30	5.64	5.21	4.91	4.69	4.52	4.38
17	10.38	7.35	6.16	5.50	5.07	4.78	4.56	4.39	4.25
18	10.22	7.21	6.03	5.37	4.96	4.66	4.44	4.28	4.14
19	10.07	7.09	5.92	5.27	4.85	4.56	4.34	4.18	4.04
20	9.94	6.99	5.82	5.17	4.76	4.47	4.26	4.09	3.96
21	9.83	6.89	5.73	5.09	4.68	4.39	4.18	4.01	3.88
22	9.73	6.81	5.65	5.02	4.61	4.32	4.11	3.94	3.81
23	9.63	6.73	5.58	4.95	4.54	4.26	4.05	3.88	3.75
24	9.55	6.66	5.52	4.89	4.49	4.20	3.99	3.83	3.69
25	9.48	6.60	5.46	4.84	4.43	4.15	3.94	3.78	3.64
26	9.41	6.54	5.41	4.79	4.38	4.10	3.89	3.73	3.60
27	9.34	6.49	5.36	4.74	4.34	4.06	3.85	3.69	3.56
28	9.28	6.44	5.32	4.70	4.30	4.02	3.81	3.65	3.52
29	9.23	6.40	5.28	4.66	4.26	3.98	3.77	3.61	3.48
30	9.18	6.35	5.24	4.62	4.23	3.95	3.74	3.58	3.45
40	8.83	6.07	4.98	4.37	3.99	3.71	3.51	3.35	3.22
60	8.49	5.79	4.73	4.14	3.76	3.49	3.29	3.13	3.01
120	8.18	5.54	4.50	3.92	3.55	3.28	3.09	2.93	2.81
∞	7.88	5.30	4.28	3.72	3.35	3.09	2.90	2.74	2.62

$F_{0.005;\ \nu_1,\nu_2} = 1/F_{0.995,\nu_2,\nu_1}$

10	12	15	20	24	30	40	60	120	∞
24224.	24426.	24630.	24836.	24940.	25044.	25148.	25253.	25359.	25465.
199.4	199.4	199.4	199.4	199.5	199.5	199.5	199.5	199.5	199.5
43.69	43.39	43.08	42.78	42.62	42.47	42.31	42.15	41.99	41.83
20.97	20.70	20.44	20.17	20.03	19.89	19.75	19.61	19.47	19.32
13.62	13.38	13.15	12.90	12.78	12.66	12.53	12.40	12.27	12.14
10.25	10.03	9.81	9.59	9.47	9.36	9.24	9.12	9.00	8.88
8.38	8.18	7.97	7.75	7.65	7.53	7.42	7.31	7.10	7.08
7.21	7.01	6.81	6.61	6.50	6.40	6.29	6.18	6.06	5.95
6.42	6.23	6.03	5.83	5.73	5.62	5.52	5.41	5.30	5.19
5.85	5.66	5.47	5.27	5.17	5.07	4.97	4.86	4.75	4.61
5.42	5.24	5.05	4.86	4.76	4.65	4.55	4.44	4.34	4.23
5.09	4.91	4.72	4.53	4.43	4.33	4.23	4.12	4.01	3.90
4.82	4.64	4.46	4.27	4.17	4.07	3.97	3.87	3.76	3.65
4.60	4.43	4.25	4.06	3.96	3.86	3.76	3.66	3.55	3.44
4.42	4.25	4.07	3.88	3.79	3.69	3.58	3.48	3.37	3.26
4.27	4.10	3.92	3.73	3.64	3.54	3.44	3.33	3.22	3.11
4.14	3.97	3.79	3.61	3.51	3.41	3.31	3.21	3.10	2.98
4.03	3.86	3.68	3.50	3.40	3.30	3.20	3.10	2.09	2.87
3.93	3.76	3.59	3.40	3.31	3.21	3.11	3.00	2.89	2.78
3.85	3.68	3.50	3.32	3.22	3.12	3.02	2.92	2.81	2.69
3.77	3.60	3.43	3.24	3.15	3.05	2.95	2.84	2.73	2.61
3.70	3.54	3.36	3.18	3.08	2.98	2.88	2.77	2.66	2.55
3.64	3.47	3.30	3.12	3.02	2.92	2.82	2.71	2.60	2.48
3.59	3.42	3.25	3.06	2.97	2.87	2.77	2.66	2.55	2.43
3.54	3.37	3.20	3.01	2.92	2.82	2.72	2.61	2.50	2.38
3.49	3.33	3.15	2.97	2.87	2.77	2.67	2.56	2.45	2.33
3.45	3.28	3.11	2.93	2.83	2.73	2.63	2.52	2.41	2.29
3.41	3.25	3.07	2.89	2.79	2.69	2.59	2.48	2.37	2.25
3.38	3.21	3.04	2.86	2.76	2.66	2.56	2.45	2.33	2.21
3.34	3.18	3.01	2.82	2.73	2.63	2.52	2.42	2.30	2.18
3.12	2.95	2.78	2.60	2.50	2.40	2.30	2.18	2.06	1.93
2.90	2.74	2.57	2.39	2.29	2.19	2.08	1.96	1.83	1.69
2.71	2.54	2.37	2.19	2.09	1.98	1.87	1.75	1.61	1.43
2.52	2.36	2.19	2.00	1.90	1.79	1.67	1.53	1.36	1.00

TABLE T7 Critical Points of the F Distribution (Continued)

Upper 1 Percent Points

ν_2 \ ν_1	1	2	3	4	5	6	7	8	9
1	4052.	4999.5	5403.	5625.	5764.	5859.	5928.	5982.	6022.
2	98.50	99.00	99.17	99.25	99.30	99.33	99.36	99.37	99.39
3	34.12	30.82	29.46	28.71	28.24	27.91	27.67	27.49	27.35
4	21.20	18.00	16.69	15.98	15.52	15.21	14.98	14.80	14.66
5	16.26	13.27	12.06	11.39	10.97	10.67	10.46	10.29	10.16
6	13.75	10.92	9.78	9.15	8.75	8.47	8.26	8.10	7.98
7	12.25	9.55	8.45	7.85	7.46	7.19	6.99	6.84	6.72
8	11.26	8.65	7.59	7.01	6.63	6.37	6.18	6.03	5.91
9	10.56	8.02	6.99	6.42	6.06	5.80	5.61	5.47	5.35
10	10.04	7.56	6.55	5.99	5.64	5.39	5.20	5.06	4.94
11	9.65	7.21	6.22	5.67	5.32	5.07	4.80	4.74	4.63
12	9.33	6.93	5.95	5.41	5.06	4.82	4.64	4.50	4.39
13	9.07	6.70	5.74	5.21	4.86	4.62	4.44	4.30	4.19
14	8.86	6.51	5.56	5.04	4.69	4.46	4.28	4.14	4.03
15	8.68	6.36	5.42	4.89	4.56	4.32	4.14	4.00	3.89
16	8.53	6.23	5.29	4.77	4.44	4.20	4.03	3.89	3.78
17	8.40	6.11	5.18	4.67	4.34	4.10	3.93	3.79	3.68
18	8.29	6.01	5.09	4.58	4.25	4.01	3.84	3.71	3.60
19	8.18	5.93	5.01	4.50	4.17	3.94	3.77	3.63	3.52
20	8.10	5.85	4.94	4.43	4.10	3.87	3.70	3.56	3.46
21	8.02	5.78	4.87	4.37	4.04	3.81	3.64	3.51	3.40
22	7.95	5.72	4.82	4.31	3.99	3.76	3.59	3.45	3.35
23	7.88	5.66	4.76	4.26	3.94	3.71	3.54	3.41	3.30
24	7.82	5.61	4.72	4.22	3.90	3.67	3.50	3.36	3.26
25	7.77	5.57	4.68	4.18	3.85	3.63	3.46	3.32	3.22
26	7.72	5.53	4.64	4.14	3.82	3.59	3.42	3.29	3.18
27	7.68	5.49	4.60	4.11	3.78	3.56	3.39	3.26	3.15
28	7.64	5.45	4.57	4.07	3.75	3.53	3.30	3.23	3.12
29	7.60	5.42	4.54	4.04	3.73	3.50	3.33	3.20	3.09
30	7.56	5.39	4.51	4.02	3.70	3.47	3.30	3.17	3.07
40	7.31	5.18	4.31	3.83	3.51	3.29	3.12	2.99	2.89
60	7.08	4.98	4.13	3.65	3.34	3.12	2.95	2.82	2.72
120	6.85	4.79	3.95	3.48	3.17	2.96	2.79	2.66	2.56
∞	6.63	4.61	3.78	3.32	3.02	2.80	2.64	2.51	2.41

10	12	15	20	24	30	40	60	120	∞
6056.	6106.	6157.	6209.	6235.	6261.	6287.	6313.	6339.	6366.
99.40	99.42	99.43	99.45	99.46	99.47	99.47	99.78	99.49	99.50
27.23	27.05	26.87	26.69	26.60	26.50	26.41	26.32	26.22	26.13
14.55	14.37	14.20	14.02	13.93	13.84	13.75	13.65	13.56	13.46
10.05	9.89	9.72	9.55	9.47	9.38	9.29	9.20	9.11	9.02
7.87	7.72	7.56	7.40	7.31	7.23	7.14	7.06	6.97	6.88
6.62	6.47	6.31	6.16	6.07	5.99	5.91	5.82	5.74	5.65
5.81	5.67	5.52	5.36	5.28	5.20	5.12	5.03	4.95	4.86
5.26	5.11	4.96	4.81	4.73	4.65	4.57	4.48	4.40	4.31
4.85	4.71	4.56	4.41	4.33	4.25	4.17	4.08	4.00	3.91
4.54	4.40	4.25	4.10	4.02	3.94	3.86	3.78	3.69	3.60
4.30	4.16	4.01	3.86	3.78	3.70	3.62	3.54	3.45	3.36
4.10	3.96	3.82	3.66	3.59	3.51	3.43	3.34	3.25	3.17
3.94	3.80	3.66	3.51	3.43	3.35	3.27	3.18	3.09	3.00
3.80	3.67	3.52	3.37	3.29	3.21	3.13	3.05	2.96	2.87
3.69	3.55	3.41	3.26	3.18	3.10	3.02	2.93	2.84	2.76
3.59	3.46	3.31	3.16	3.08	3.00	2.92	2.83	2.75	2.65
3.51	3.37	3.23	3.08	3.00	2.92	2.84	2.75	2.66	2.57
3.43	3.30	3.15	3.00	2.92	2.84	2.76	2.67	2.58	2.49
3.37	3.23	3.09	2.94	2.86	2.78	2.69	2.61	2.52	2.42
3.31	3.17	3.03	2.88	2.80	2.72	2.64	2.55	2.46	2.36
3.26	3.12	2.98	2.83	2.75	2.67	2.58	2.50	2.40	2.31
3.21	3.07	2.93	2.78	2.70	2.62	2.54	2.45	2.35	2.26
3.17	3.03	2.89	2.74	2.66	2.58	2.49	2.40	2.31	2.21
3.13	2.99	2.85	2.70	2.62	2.54	2.45	2.36	2.27	2.17
3.09	2.96	2.81	2.66	2.58	2.50	2.42	2.33	2.23	2.13
3.06	2.93	2.78	2.63	2.55	2.47	2.38	2.29	2.20	2.10
3.03	2.90	2.75	2.60	2.52	2.44	2.35	2.26	2.17	2.06
3.00	2.87	2.73	2.57	2.49	2.41	2.33	2.23	2.14	2.03
2.98	2.84	2.70	2.55	2.47	2.39	2.30	2.21	2.11	2.01
2.80	2.66	2.52	2.37	2.29	2.20	2.11	2.02	1.92	1.80
2.63	2.50	2.35	2.20	2.12	2.03	1.94	1.84	1.73	1.60
2.47	2.34	2.19	2.03	1.95	1.86	1.76	1.66	1.53	1.38
2.32	2.18	2.04	1.88	1.79	1.70	1.59	1.47	1.32	1.00

TABLE T7 Critical Points of the F Distribution (Continued)

Upper 2.5 Percent Points

ν_2 \ ν_1	1	2	3	4	5	6	7	8	9
1	647.8	799.5	864.2	899.6	921.8	937.1	948.2	956.7	963.3
2	38.51	39.00	39.17	39.25	39.30	39.33	39.36	39.37	39.39
3	17.44	16.04	15.44	15.10	14.88	14.73	14.62	14.54	14.47
4	12.22	10.65	9.98	9.60	9.36	9.20	9.07	8.98	8.90
5	10.01	8.43	7.76	7.39	7.15	6.98	6.85	6.76	6.68
6	8.81	7.26	6.60	6.23	5.99	5.82	5.70	5.60	5.52
7	8.07	6.54	5.89	5.52	5.29	5.12	4.99	4.90	4.82
8	7.57	6.06	5.42	5.05	4.82	4.65	4.53	4.43	4.36
9	7.21	5.71	5.08	4.72	4.48	4.32	4.20	4.10	4.03
10	6.94	5.46	4.83	4.47	4.24	4.07	3.95	3.85	3.78
11	6.72	5.26	4.63	4.28	4.04	3.88	3.76	3.66	3.59
12	6.55	5.10	4.47	4.12	3.89	3.73	3.61	3.51	3.44
13	6.41	4.97	4.35	4.00	3.77	3.60	3.48	3.39	3.31
14	6.30	4.86	4.24	3.89	3.66	3.50	3.38	3.29	3.21
15	6.20	4.77	4.15	3.80	3.58	3.41	3.29	3.20	3.12
16	6.12	4.69	4.08	3.73	3.50	3.34	3.22	3.12	3.05
17	6.04	4.62	4.01	3.66	3.44	3.28	3.16	3.06	2.98
18	5.98	4.56	3.95	3.61	3.38	3.22	3.10	3.01	2.93
19	5.92	4.51	3.90	3.56	3.33	3.17	3.05	2.96	2.88
20	5.87	4.46	3.86	3.51	3.29	3.13	3.01	2.91	2.84
21	5.83	4.42	3.82	3.48	3.25	3.09	2.97	2.87	2.80
22	5.79	4.38	3.78	3.44	3.22	3.05	2.93	2.84	2.76
23	5.75	4.35	3.75	3.41	3.18	3.02	2.90	2.81	2.73
24	5.72	4.32	3.72	3.38	3.15	2.99	2.87	2.78	2.70
25	5.69	4.29	3.69	3.35	3.13	2.97	2.85	2.75	2.68
26	5.66	4.27	3.67	3.33	3.10	2.94	2.82	2.73	2.65
27	5.63	4.24	3.65	3.31	3.08	2.92	2.80	2.71	2.63
28	5.61	4.22	3.63	3.29	3.06	2.90	2.78	2.69	2.61
29	5.59	4.20	3.61	3.27	3.04	2.88	2.76	2.67	2.59
30	5.57	4.18	3.59	3.25	3.03	2.87	2.75	2.65	2.57
40	5.42	4.05	3.46	3.13	2.90	2.74	2.62	2.53	2.45
60	5.29	3.93	3.34	3.01	2.79	2.63	2.51	2.41	2.33
120	5.15	3.80	3.23	2.89	2.67	2.52	2.39	2.30	2.22
∞	5.02	3.69	3.12	2.79	2.57	2.41	2.29	2.19	2.11

10	12	15	20	24	30	40	60	120	∞
968.6	976.7	984.9	993.1	997.2	1001.	1006.	1010.	1014.	1018.
39.40	39.41	39.43	39.45	39.46	39.46	39.47	39.48	39.49	39.50
14.42	14.34	14.25	14.17	14.12	14.08	14.04	13.99	13.95	13.90
8.84	8.75	8.66	8.56	8.51	8.46	8.41	8.36	8.31	8.20
6.62	6.52	6.43	6.33	6.28	6.23	6.18	6.12	6.07	6.02
5.46	5.37	5.27	5.17	5.12	5.07	5.01	4.96	4.90	4.85
4.76	4.67	4.57	4.47	4.42	4.36	4.31	4.25	4.20	4.14
4.30	4.20	4.10	4.00	3.95	3.89	3.84	3.78	3.73	3.67
3.96	3.87	3.77	3.67	3.61	3.56	3.51	3.45	3.39	3.33
3.72	3.62	3.52	3.42	3.37	3.31	3.26	3.20	3.14	3.08
3.53	3.43	3.33	3.23	3.17	3.12	3.06	3.00	2.94	2.88
3.37	3.28	3.18	3.07	3.02	2.96	2.91	2.85	2.79	2.72
3.25	3.15	3.05	2.95	2.89	2.84	2.78	2.72	2.66	2.60
3.15	3.05	2.95	2.84	2.79	2.73	2.67	2.61	2.55	2.49
3.06	2.96	2.86	2.76	2.70	2.64	2.59	2.52	2.46	2.40
2.99	2.89	2.79	2.68	2.63	2.57	2.51	2.45	2.38	2.32
2.92	2.82	2.72	2.62	2.56	2.50	2.44	2.38	2.32	2.25
2.87	2.77	2.67	2.56	2.50	2.44	2.38	2.32	2.26	2.19
2.82	2.72	2.62	2.51	2.45	2.39	2.33	2.27	2.20	2.13
2.77	2.68	2.57	2.46	2.41	2.35	2.29	2.22	2.16	2.09
2.73	2.64	2.53	2.42	2.37	2.31	2.25	2.18	2.11	2.04
2.70	2.60	2.50	2.39	2.33	2.27	2.21	2.14	2.08	2.00
2.67	2.57	2.47	2.36	2.30	2.24	2.18	2.11	2.04	1.97
2.64	2.54	2.44	2.33	2.27	2.21	2.15	2.08	2.01	1.94
2.61	2.51	2.41	2.30	2.24	2.18	2.12	2.05	1.98	1.91
2.59	2.49	2.39	2.28	2.22	2.16	2.09	2.03	1.95	1.88
2.57	2.47	2.36	2.25	2.19	2.13	2.07	2.00	1.93	1.85
2.55	2.45	2.34	2.23	2.17	2.11	2.05	1.98	1.91	1.83
2.53	2.43	2.32	2.21	2.15	2.09	2.03	1.96	1.89	1.81
2.51	2.41	2.31	2.20	2.14	2.07	2.01	1.94	1.87	1.79
2.39	2.29	2.18	2.07	2.01	1.94	1.88	1.80	1.72	1.64
2.27	2.17	2.06	1.94	1.88	1.82	1.74	1.67	1.58	1.48
2.16	2.05	1.94	1.82	1.76	1.69	1.61	1.53	1.43	1.31
2.05	1.94	1.83	1.71	1.64	1.57	1.48	1.39	1.27	1.00

TABLE T7 Critical Points of the F Distribution (Continued)

Upper 5 Percent Points

ν_2 \ ν_1	1	2	3	4	5	6	7	8	9
1	161.4	199.5	215.7	224.6	230.2	234.0	236.8	238.9	240.5
2	18.51	19.00	19.16	19.25	19.30	19.33	19.35	19.37	19.38
3	10.13	9.55	9.28	9.12	9.01	8.94	8.89	8.85	8.81
4	7.71	6.94	6.59	6.39	6.26	6.16	6.09	6.04	6.00
5	6.61	5.79	5.41	5.19	5.05	4.95	4.88	4.82	4.77
6	5.99	5.14	4.76	4.53	4.39	4.28	4.21	4.15	4.10
7	5.59	4.74	4.35	4.12	3.97	3.87	3.79	3.73	3.68
8	5.32	4.46	4.07	3.84	3.69	3.58	3.50	3.44	3.39
9	5.12	4.26	3.86	3.63	3.48	3.37	3.29	3.23	3.18
10	4.96	4.10	3.71	3.48	3.33	3.22	3.14	3.07	3.02
11	4.84	3.98	3.59	3.36	3.20	3.09	3.01	2.95	2.90
12	4.75	3.89	3.49	3.26	3.11	3.00	2.91	2.85	2.80
13	4.67	3.81	3.41	3.18	3.03	2.92	2.83	2.77	2.71
14	4.60	3.74	3.34	3.11	2.96	2.85	2.76	2.70	2.65
15	4.54	3.68	3.29	3.06	2.90	2.79	2.71	2.64	2.59
16	4.49	3.63	3.24	3.01	2.85	2.74	2.66	2.59	2.54
17	4.45	3.59	3.20	2.96	2.81	2.70	2.61	2.55	2.49
18	4.41	3.55	3.16	2.93	2.77	2.66	2.58	2.51	2.46
19	4.38	3.52	3.13	2.90	2.74	2.63	2.54	2.48	2.42
20	4.35	3.49	3.10	2.87	2.71	2.60	2.51	2.45	2.39
21	4.32	3.47	3.07	2.84	2.68	2.57	2.49	2.42	2.37
22	4.30	3.44	3.05	2.82	2.66	2.55	2.46	2.40	2.34
23	4.28	3.42	3.03	2.80	2.64	2.53	2.44	2.37	2.32
24	4.26	3.40	3.01	2.78	2.62	2.51	2.42	2.36	2.30
25	4.24	3.39	2.99	2.76	2.60	2.49	2.40	2.34	2.28
26	4.23	3.37	2.98	2.74	2.59	2.47	2.39	2.32	2.27
27	4.21	3.35	2.96	2.73	2.57	2.46	2.37	2.31	2.25
28	4.20	3.34	2.95	2.71	2.56	2.45	2.36	2.29	2.24
29	4.18	3.33	2.93	2.70	2.55	2.43	2.35	2.28	2.22
30	4.17	3.32	2.92	2.69	2.53	2.42	2.33	2.27	2.21
40	4.08	3.23	2.84	2.61	2.45	2.34	2.25	2.18	2.12
60	4.00	3.15	2.76	2.53	2.37	2.25	2.17	2.10	2.04
120	3.92	3.07	2.68	2.45	2.29	2.17	2.09	2.02	1.96
∞	3.84	3.00	2.60	2.37	2.21	2.10	2.01	1.94	1.88

10	12	15	20	24	30	40	60	120	∞
241.9	243.9	245.9	248.0	249.1	250.1	251.1	252.2	253.3	254.3
19.40	19.41	19.43	19.45	19.45	19.46	19.47	19.48	19.49	19.50
8.79	8.74	8.70	8.66	8.64	8.62	8.59	8.57	8.55	8.53
5.96	5.91	5.86	5.80	5.77	5.75	5.72	5.69	5.66	5.63
4.74	4.68	4.62	4.56	4.53	4.50	4.46	4.43	4.40	4.36
4.06	4.00	3.94	3.87	3.84	3.81	3.77	3.74	3.70	3.67
3.64	3.57	3.51	3.44	3.41	3.38	3.34	3.30	3.27	3.23
3.35	3.28	3.22	3.15	3.12	3.08	3.04	3.01	2.97	2.93
3.14	3.07	3.01	2.94	2.90	2.86	2.83	2.79	2.75	2.71
2.98	2.91	2.85	2.77	2.74	2.70	2.66	2.62	2.58	2.54
2.85	2.79	2.72	2.65	2.61	2.57	2.53	2.49	2.45	2.40
2.75	2.69	2.62	2.54	2.51	2.47	2.43	2.38	2.34	2.30
2.67	2.60	2.53	2.46	2.42	2.38	2.34	2.30	2.25	2.21
2.60	2.53	2.46	2.39	2.35	2.31	2.27	2.22	2.18	2.13
2.54	2.48	2.40	2.33	2.29	2.25	2.20	2.16	2.11	2.07
2.49	2.42	2.35	2.28	2.24	2.19	2.15	2.11	2.06	2.01
2.45	2.38	2.31	2.23	2.19	2.15	2.10	2.06	2.01	1.96
2.41	2.34	2.27	2.19	2.15	2.11	2.06	2.02	1.97	1.92
2.38	2.31	2.23	2.16	2.11	2.07	2.03	1.98	1.93	1.88
2.35	2.28	2.20	2.12	2.08	2.04	1.99	1.95	1.90	1.84
2.32	2.25	2.18	2.10	2.05	2.01	1.96	1.92	1.87	1.81
2.30	2.23	2.15	2.07	2.03	1.98	1.94	1.89	1.84	1.78
2.27	2.20	2.13	2.05	2.01	1.96	1.91	1.86	1.81	1.76
2.25	2.18	2.11	2.03	1.98	1.94	1.89	1.84	1.79	1.73
2.24	2.16	2.09	2.01	1.96	1.92	1.87	1.82	1.77	1.71
2.22	2.15	2.07	1.99	1.95	1.90	1.85	1.80	1.75	1.69
2.20	2.13	2.06	1.97	1.93	1.88	1.84	1.79	1.73	1.67
2.19	2.12	2.04	1.96	1.91	1.87	1.82	1.77	1.71	1.65
2.18	2.10	2.03	1.94	1.90	1.85	1.81	1.75	1.70	1.64
2.16	2.09	2.01	1.93	1.89	1.84	1.79	1.74	1.68	1.62
2.08	2.00	1.92	1.84	1.79	1.74	1.69	1.64	1.58	1.51
1.99	1.92	1.84	1.75	1.70	1.65	1.59	1.53	1.47	1.39
1.91	1.83	1.75	1.66	1.61	1.55	1.55	1.43	1.35	1.25
1.83	1.75	1.67	1.57	1.52	1.46	1.39	1.32	1.22	1.00

TABLE T7   Critical Points of the F Distribution (Continued)

Upper 10 Percent Points

ν_2 \ ν_1	1	2	3	4	5	6	7	8	9
1	39.86	49.50	53.59	55.83	57.24	58.20	58.91	59.44	59.86
2	8.53	9.00	9.16	9.24	9.29	9.33	9.35	9.37	9.38
3	5.54	5.46	5.39	5.34	5.31	5.28	5.27	5.25	5.24
4	4.54	4.32	4.19	4.11	4.05	4.01	3.98	3.95	3.94
5	4.06	3.78	3.62	3.52	3.45	3.40	3.37	3.34	3.32
6	3.78	3.46	3.29	3.18	3.11	3.05	3.01	2.98	2.96
7	3.59	3.26	3.07	2.96	2.88	2.83	2.78	2.75	2.72
8	3.46	3.11	2.92	2.81	2.73	2.67	2.62	2.59	2.56
9	3.36	3.01	2.81	2.69	2.61	2.55	2.51	2.47	2.44
10	3.29	2.92	2.73	2.61	2.52	2.46	2.41	2.38	2.35
11	3.23	2.86	2.66	2.54	2.45	2.39	2.34	2.30	2.27
12	3.18	2.81	2.61	2.48	2.39	2.33	2.28	2.24	2.21
13	3.14	2.76	2.56	2.43	2.35	2.28	2.23	2.20	2.16
14	3.10	2.73	2.52	2.39	2.31	2.24	2.19	2.15	2.12
15	3.07	2.70	2.49	2.36	2.27	2.21	2.16	2.12	2.09
16	3.05	2.67	2.46	2.33	2.24	2.18	2.13	2.09	2.06
17	3.03	2.64	2.44	2.31	2.22	2.15	2.10	2.06	2.03
18	3.01	2.62	2.42	2.29	2.20	2.13	2.08	2.04	2.00
19	2.99	2.61	2.40	2.27	2.18	2.11	2.06	2.02	1.98
20	2.97	2.59	2.38	2.25	2.16	2.09	2.04	2.00	1.96
21	2.96	2.57	2.36	2.23	2.14	2.08	2.02	1.98	1.95
22	2.95	2.56	2.35	2.22	2.13	2.06	2.01	1.97	1.93
23	2.94	2.55	2.34	2.21	2.11	2.05	1.99	1.95	1.92
24	2.93	2.54	2.33	2.19	2.10	2.04	1.98	1.94	1.91
25	2.92	2.53	2.32	2.18	2.09	2.02	1.97	1.93	1.89
26	2.91	2.52	2.31	2.17	2.08	2.01	1.96	1.92	1.88
27	2.90	2.51	2.30	2.17	2.07	2.00	1.95	1.91	1.87
28	2.89	2.50	2.29	2.16	2.06	2.00	1.94	1.90	1.87
29	2.89	2.50	2.28	2.15	2.06	1.99	1.93	1.89	1.86
30	2.88	2.49	2.28	2.14	2.05	1.98	1.93	1.88	1.85
40	2.84	2.44	2.23	2.09	2.00	1.93	1.87	1.83	1.79
60	2.79	2.39	2.18	2.04	1.95	1.87	1.82	1.77	1.74
120	2.75	2.35	2.13	1.99	1.90	1.82	1.77	1.72	1.68
∞	2.71	2.30	2.08	1.94	1.85	1.77	1.72	1.67	1.63

10	12	15	20	24	30	40	60	120	∞
60.19	60.71	61.22	61.74	62.00	62.26	62.53	62.79	63.06	63.33
9.39	9.41	9.42	9.44	9.45	9.46	9.47	9.47	9.48	9.49
5.23	5.22	5.20	5.18	5.18	5.17	5.16	5.15	5.14	5.13
3.92	3.90	3.87	3.84	3.83	3.82	3.80	3.79	3.78	3.76
3.30	3.27	3.24	3.21	3.19	3.17	3.16	3.14	3.12	3.10
2.94	2.90	2.87	2.84	2.82	2.80	2.78	2.76	2.74	2.72
2.70	2.67	2.63	2.59	2.58	2.56	2.54	2.51	2.49	2.47
2.54	2.50	2.46	2.42	2.40	2.38	2.36	2.34	2.32	2.29
2.42	2.38	2.34	2.30	2.28	2.25	2.23	2.21	2.18	2.16
2.32	2.28	2.24	2.20	2.18	2.16	2.13	2.11	2.08	2.06
2.25	2.21	2.17	2.12	2.10	2.08	2.05	2.03	2.00	1.97
2.19	2.15	2.10	2.06	2.04	2.01	1.99	1.96	1.93	1.90
2.14	2.10	2.05	2.01	1.98	1.96	1.93	1.90	1.88	1.85
2.10	2.05	2.01	1.96	1.94	1.91	1.89	1.86	1.83	1.80
2.06	2.02	1.97	1.92	1.90	1.87	1.85	1.82	1.79	1.76
2.03	1.99	1.94	1.89	1.87	1.84	1.81	1.78	1.75	1.72
2.00	1.96	1.91	1.86	1.84	1.81	1.78	1.75	1.72	1.69
1.98	1.93	1.89	1.84	1.81	1.78	1.75	1.72	1.69	1.66
1.96	1.91	1.86	1.81	1.79	1.76	1.73	1.70	1.67	1.63
1.94	1.89	1.84	1.79	1.77	1.74	1.71	1.68	1.64	1.61
1.92	1.87	1.83	1.78	1.75	1.72	1.69	1.66	1.62	1.59
1.90	1.86	1.81	1.76	1.73	1.70	1.67	1.64	1.60	1.57
1.89	1.84	1.80	1.74	1.72	1.69	1.66	1.62	1.59	1.55
1.88	1.83	1.78	1.73	1.70	1.67	1.64	1.61	1.57	1.53
1.87	1.82	1.77	1.72	1.69	1.66	1.63	1.59	1.56	1.52
1.86	1.81	1.76	1.71	1.68	1.65	1.61	1.58	1.54	1.50
1.85	1.80	1.75	1.70	1.67	1.64	1.60	1.57	1.53	1.49
1.84	1.79	1.74	1.69	1.66	1.63	1.59	1.56	1.52	1.48
1.83	1.78	1.73	1.68	1.65	1.62	1.58	1.55	1.51	1.47
1.82	1.77	1.72	1.67	1.64	1.61	1.57	1.54	1.50	1.46
1.76	1.71	1.66	1.61	1.57	1.54	1.51	1.47	1.42	1.38
1.71	1.66	1.60	1.54	1.51	1.48	1.44	1.40	1.35	1.29
1.65	1.60	1.55	1.48	1.45	1.41	1.37	1.32	1.26	1.19
1.60	1.55	1.49	1.42	1.38	1.34	1.30	1.24	1.17	1.00

TABLE T7 Critical Points of the F Distribution (Continued)

Upper 20 Percent Points

ν_2 \ ν_1	1	2	3	4	5	6	7	8	9
1	9.47	12.00	13.06	13.64	14.01	14.26	14.44	14.58	14.68
2	3.56	4.00	4.16	4.24	4.28	4.32	4.34	4.36	4.37
3	2.68	2.89	2.94	2.96	2.97	2.97	2.97	2.98	2.98
4	2.35	2.47	2.48	2.48	2.48	2.47	2.47	2.47	2.46
5	2.18	2.26	2.25	2.24	2.23	2.22	2.21	2.20	2.20
6	2.07	2.13	2.11	2.09	2.08	2.06	2.05	2.04	2.03
7	2.00	2.04	2.02	1.99	1.97	1.96	1.94	1.93	1.93
8	1.95	1.98	1.95	1.92	1.90	1.88	1.87	1.86	1.85
9	1.91	1.93	1.90	1.87	1.85	1.83	1.81	1.80	1.79
10	1.88	1.90	1.86	1.83	1.80	1.78	1.77	1.75	1.74
11	1.86	1.87	1.83	1.80	1.77	1.75	1.73	1.72	1.70
12	1.84	1.85	1.80	1.77	1.74	1.72	1.70	1.69	1.67
13	1.82	1.83	1.78	1.75	1.72	1.69	1.68	1.66	1.65
14	1.81	1.81	1.76	1.73	1.70	1.67	1.65	1.64	1.63
15	1.80	1.80	1.75	1.71	1.68	1.66	1.64	1.62	1.61
16	1.79	1.78	1.74	1.70	1.67	1.64	1.62	1.61	1.59
17	1.78	1.77	1.72	1.68	1.65	1.63	1.61	1.59	1.58
18	1.77	1.76	1.71	1.67	1.64	1.62	1.60	1.58	1.56
19	1.76	1.75	1.70	1.66	1.63	1.61	1.58	1.57	1.55
20	1.76	1.75	1.70	1.65	1.62	1.60	1.58	1.56	1.54
21	1.75	1.74	1.69	1.65	1.61	1.59	1.57	1.55	1.53
22	1.75	1.73	1.68	1.64	1.61	1.58	1.56	1.54	1.53
23	1.74	1.73	1.68	1.63	1.60	1.57	1.55	1.53	1.52
24	1.74	1.72	1.67	1.63	1.59	1.57	1.55	1.53	1.51
25	1.73	1.72	1.66	1.62	1.59	1.56	1.54	1.52	1.51
26	1.73	1.71	1.66	1.62	1.58	1.56	1.53	1.52	1.50
27	1.73	1.71	1.66	1.61	1.58	1.55	1.53	1.51	1.49
28	1.72	1.71	1.65	1.61	1.57	1.55	1.52	1.51	1.49
29	1.72	1.70	1.65	1.60	1.57	1.54	1.52	1.50	1.49
30	1.72	1.70	1.64	1.60	1.57	1.54	1.52	1.50	1.48
40	1.70	1.68	1.62	1.57	1.54	1.51	1.49	1.47	1.45
60	1.68	1.65	1.60	1.55	1.51	1.48	1.46	1.44	1.42
100	1.66	1.64	1.58	1.53	1.49	1.46	1.43	1.41	1.40
500	1.65	1.61	1.55	1.50	1.46	1.43	1.41	1.39	1.37
10000	1.63	1.61	1.55	1.50	1.46	1.43	1.40	1.38	1.36

10	12	16	20	24	30	40	60	100	1000
14.77	14.90	15.07	15.17	15.24	15.31	15.37	15.44	15.50	15.57
4.38	4.40	4.42	4.43	4.44	4.45	4.46	4.46	4.47	4.48
2.98	2.98	2.98	2.98	2.98	2.98	2.98	2.98	2.98	2.98
2.46	2.46	2.45	2.44	2.44	2.44	2.44	2.43	2.43	2.43
2.19	2.18	2.17	2.17	2.16	2.16	2.15	2.15	2.14	2.13
2.03	2.02	2.00	2.00	1.99	1.98	1.98	1.97	1.96	1.96
1.92	1.91	1.89	1.88	1.87	1.86	1.86	1.85	1.84	1.83
1.84	1.83	1.81	1.80	1.79	1.78	1.77	1.76	1.75	1.74
1.78	1.76	1.74	1.73	1.72	1.71	1.70	1.69	1.69	1.67
1.73	1.72	1.70	1.68	1.67	1.66	1.65	1.64	1.63	1.62
1.69	1.68	1.66	1.64	1.63	1.62	1.61	1.60	1.59	1.58
1.66	1.65	1.62	1.61	1.60	1.59	1.58	1.56	1.55	1.54
1.64	1.62	1.60	1.58	1.57	1.56	1.55	1.53	1.52	1.51
1.62	1.60	1.57	1.56	1.55	1.53	1.52	1.51	1.50	1.48
1.60	1.58	1.55	1.54	1.52	1.51	1.50	1.49	1.47	1.46
1.58	1.56	1.54	1.52	1.51	1.49	1.48	1.47	1.45	1.44
1.57	1.55	1.52	1.50	1.49	1.48	1.46	1.45	1.44	1.42
1.55	1.53	1.51	1.49	1.48	1.46	1.45	1.43	1.42	1.40
1.54	1.52	1.49	1.48	1.46	1.45	1.44	1.42	1.41	1.39
1.53	1.51	1.48	1.47	1.45	1.44	1.42	1.41	1.39	1.37
1.52	1.50	1.47	1.46	1.44	1.43	1.41	1.40	1.38	1.36
1.51	1.49	1.47	1.45	1.43	1.42	1.40	1.39	1.37	1.35
1.51	1.49	1.46	1.44	1.42	1.41	1.39	1.38	1.36	1.34
1.50	1.48	1.45	1.43	1.42	1.40	1.39	1.37	1.35	1.33
1.49	1.47	1.44	1.42	1.41	1.39	1.38	1.36	1.35	1.32
1.49	1.47	1.44	1.42	1.40	1.39	1.37	1.35	1.34	1.31
1.48	1.46	1.43	1.41	1.40	1.38	1.36	1.35	1.33	1.31
1.48	1.46	1.43	1.41	1.39	1.37	1.36	1.34	1.32	1.30
1.47	1.45	1.42	1.40	1.39	1.37	1.35	1.33	1.32	1.29
1.47	1.45	1.42	1.39	1.38	1.36	1.35	1.33	1.31	1.29
1.44	1.41	1.38	1.36	1.34	1.33	1.31	1.29	1.27	1.24
1.41	1.38	1.35	1.32	1.31	1.29	1.27	1.24	1.22	1.19
1.38	1.36	1.32	1.30	1.28	1.26	1.23	1.21	1.18	1.14
1.35	1.33	1.29	1.26	1.24	1.22	1.19	1.16	1.13	1.07
1.34	1.32	1.28	1.25	1.23	1.21	1.18	1.15	1.12	1.04

TABLE T7 Critical Points of the F Distribution (Continued)

Upper 25 Percent Points

v_2 \ v_1	1	2	3	4	5	6	7	8	9
1	5.83	7.50	8.20	8.58	8.82	8.98	9.10	9.19	9.26
2	2.57	3.00	3.15	3.23	3.28	3.31	3.34	3.35	3.37
3	2.02	2.28	2.36	2.39	2.41	2.42	2.43	2.44	2.44
4	1.81	2.00	2.05	2.06	2.07	2.08	2.08	2.08	2.08
5	1.69	1.85	1.88	1.89	1.89	1.89	1.89	1.89	1.89
6	1.62	1.76	1.78	1.79	1.79	1.78	1.78	1.78	1.77
7	1.57	1.70	1.72	1.72	1.71	1.71	1.70	1.70	1.70
8	1.54	1.66	1.67	1.66	1.66	1.65	1.64	1.64	1.63
9	1.51	1.62	1.63	1.63	1.62	1.61	1.60	1.60	1.59
10	1.49	1.60	1.60	1.59	1.59	1.58	1.57	1.56	1.56
11	1.47	1.58	1.58	1.57	1.56	1.55	1.54	1.53	1.53
12	1.46	1.56	1.56	1.55	1.54	1.53	1.52	1.51	1.51
13	1.45	1.55	1.55	1.53	1.52	1.51	1.50	1.49	1.49
14	1.44	1.53	1.53	1.52	1.51	1.50	1.49	1.48	1.47
15	1.43	1.52	1.52	1.51	1.49	1.48	1.47	1.46	1.46
16	1.42	1.51	1.51	1.50	1.48	1.47	1.46	1.45	1.44
17	1.42	1.51	1.50	1.49	1.47	1.46	1.45	1.44	1.43
18	1.41	1.50	1.49	1.48	1.46	1.45	1.44	1.43	1.42
19	1.41	1.49	1.49	1.47	1.46	1.44	1.43	1.42	1.41
20	1.40	1.49	1.48	1.47	1.45	1.44	1.43	1.42	1.41
21	1.40	1.48	1.48	1.46	1.44	1.43	1.42	1.41	1.40
22	1.40	1.48	1.47	1.45	1.44	1.42	1.41	1.40	1.39
23	1.39	1.47	1.47	1.45	1.43	1.42	1.41	1.40	1.39
24	1.39	1.47	1.46	1.44	1.43	1.41	1.40	1.39	1.38
25	1.39	1.47	1.46	1.44	1.42	1.41	1.40	1.39	1.38
26	1.38	1.46	1.45	1.44	1.42	1.41	1.39	1.38	1.37
27	1.38	1.46	1.45	1.43	1.42	1.40	1.39	1.38	1.37
28	1.38	1.46	1.45	1.43	1.41	1.40	1.39	1.38	1.37
29	1.38	1.45	1.45	1.43	1.41	1.40	1.38	1.37	1.36
30	1.38	1.45	1.44	1.42	1.41	1.39	1.38	1.37	1.36
40	1.36	1.44	1.42	1.40	1.39	1.37	1.36	1.35	1.34
60	1.35	1.42	1.41	1.38	1.37	1.35	1.33	1.32	1.31
120	1.34	1.40	1.39	1.37	1.35	1.33	1.31	1.30	1.29
∞	1.32	1.39	1.37	1.35	1.33	1.31	1.29	1.28	1.27

[a]This table is reproduced from Table 18 of *Biometrika Tables for Statisticians*, Volume I, Second

10	12	15	20	24	30	40	60	120	∞
9.32	9.41	9.49	9.58	9.63	9.67	9.71	9.76	9.80	9.85
3.38	3.39	3.41	3.43	3.43	3.44	3.45	3.46	3.47	3.48
2.44	2.45	2.46	2.46	2.46	2.47	2.47	2.47	2.47	2.47
2.08	2.08	2.08	2.08	2.08	2.08	2.08	2.08	2.08	2.08
1.89	1.89	1.89	1.88	1.88	1.88	1.88	1.87	1.87	1.87
1.77	1.77	1.76	1.76	1.75	1.75	1.75	1.74	1.74	1.74
1.69	1.68	1.68	1.67	1.67	1.66	1.66	1.65	1.65	1.65
1.63	1.62	1.62	1.61	1.60	1.60	1.59	1.59	1.58	1.58
1.59	1.58	1.57	1.56	1.56	1.55	1.54	1.54	1.53	1.53
1.55	1.54	1.53	1.52	1.52	1.51	1.51	1.50	1.49	1.48
1.52	1.51	1.50	1.49	1.49	1.48	1.47	1.47	1.46	1.45
1.50	1.49	1.48	1.47	1.46	1.45	1.45	1.44	1.43	1.42
1.48	1.47	1.46	1.45	1.44	1.43	1.42	1.42	1.41	1.40
1.46	1.45	1.44	1.43	1.42	1.41	1.41	1.40	1.39	1.38
1.45	1.44	1.43	1.41	1.41	1.40	1.39	1.38	1.37	1.36
1.44	1.43	1.41	1.40	1.39	1.38	1.37	1.36	1.35	1.34
1.43	1.41	1.40	1.39	1.38	1.37	1.36	1.35	1.34	1.33
1.42	1.40	1.39	1.38	1.37	1.36	1.35	1.34	1.33	1.32
1.41	1.40	1.38	1.37	1.36	1.35	1.34	1.33	1.32	1.30
1.40	1.39	1.37	1.36	1.35	1.34	1.33	1.32	1.31	1.29
1.39	1.38	1.37	1.35	1.34	1.33	1.32	1.31	1.30	1.28
1.39	1.37	1.36	1.34	1.33	1.32	1.31	1.30	1.29	1.28
1.38	1.37	1.35	1.34	1.33	1.32	1.31	1.30	1.28	1.27
1.38	1.36	1.35	1.33	1.32	1.31	1.30	1.29	1.28	1.26
1.37	1.36	1.34	1.33	1.32	1.31	1.29	1.28	1.27	1.25
1.37	1.35	1.34	1.32	1.31	1.30	1.29	1.28	1.26	1.25
1.36	1.35	1.33	1.32	1.31	1.30	1.28	1.27	1.26	1.24
1.36	1.34	1.33	1.31	1.30	1.29	1.28	1.27	1.25	1.24
1.35	1.34	1.32	1.31	1.30	1.29	1.27	1.26	1.25	1.23
1.35	1.34	1.32	1.30	1.29	1.28	1.27	1.26	1.24	1.23
1.33	1.31	1.30	1.28	1.26	1.25	1.24	1.22	1.21	1.19
1.30	1.29	1.27	1.25	1.24	1.22	1.21	1.19	1.17	1.15
1.28	1.26	1.24	1.22	1.21	1.19	1.18	1.16	1.13	1.10
1.25	1.24	1.22	1.19	1.18	1.16	1.14	1.12	1.08	1.00

TABLE T7 Critical Points of the F Distribution (Continued)

Upper 30 Percent Points

ν_2 \ ν_1	1	2	3	4	5	6	7	8	9
1	3.85	5.06	5.56	5.83	6.00	6.12	6.20	6.27	6.32
2	1.92	2.33	2.48	2.56	2.61	2.64	2.66	2.68	2.69
3	1.56	1.85	1.94	1.98	2.01	2.03	2.04	2.05	2.06
4	1.42	1.65	1.72	1.75	1.77	1.78	1.79	1.79	1.80
5	1.34	1.55	1.60	1.63	1.64	1.65	1.65	1.66	1.66
6	1.29	1.48	1.53	1.55	1.56	1.57	1.57	1.57	1.57
7	1.25	1.44	1.48	1.50	1.51	1.51	1.51	1.51	1.51
8	1.23	1.40	1.45	1.46	1.47	1.47	1.47	1.47	1.47
9	1.21	1.38	1.42	1.43	1.44	1.44	1.44	1.43	1.43
10	1.19	1.36	1.40	1.41	1.41	1.41	1.41	1.41	1.41
11	1.18	1.35	1.38	1.39	1.39	1.39	1.39	1.39	1.39
12	1.17	1.33	1.37	1.37	1.38	1.38	1.37	1.37	1.37
13	1.17	1.32	1.35	1.36	1.36	1.36	1.36	1.36	1.35
14	1.16	1.31	1.34	1.35	1.35	1.35	1.35	1.34	1.34
15	1.15	1.31	1.34	1.34	1.34	1.34	1.34	1.33	1.33
16	1.15	1.30	1.33	1.33	1.33	1.33	1.33	1.32	1.32
17	1.14	1.29	1.32	1.33	1.33	1.32	1.32	1.32	1.31
18	1.14	1.29	1.32	1.32	1.32	1.32	1.31	1.31	1.31
19	1.14	1.28	1.31	1.32	1.31	1.31	1.31	1.30	1.30
20	1.13	1.28	1.31	1.31	1.31	1.31	1.30	1.30	1.29
21	1.13	1.28	1.30	1.31	1.30	1.30	1.30	1.29	1.29
22	1.13	1.27	1.30	1.30	1.30	1.30	1.29	1.29	1.28
23	1.12	1.27	1.29	1.30	1.30	1.29	1.29	1.28	1.28
24	1.12	1.27	1.29	1.29	1.29	1.29	1.28	1.28	1.28
25	1.12	1.26	1.29	1.29	1.29	1.28	1.28	1.28	1.27
26	1.12	1.26	1.29	1.29	1.29	1.28	1.28	1.27	1.27
27	1.12	1.26	1.28	1.29	1.28	1.28	1.27	1.27	1.26
28	1.12	1.26	1.28	1.28	1.28	1.28	1.27	1.27	1.26
29	1.11	1.26	1.28	1.28	1.28	1.27	1.27	1.26	1.26
30	1.11	1.25	1.28	1.28	1.28	1.27	1.27	1.26	1.26
40	1.10	1.24	1.26	1.26	1.26	1.25	1.25	1.24	1.24
60	1.09	1.23	1.25	1.25	1.24	1.24	1.23	1.23	1.22
100	1.09	1.22	1.24	1.24	1.23	1.22	1.22	1.21	1.21
500	1.08	1.21	1.22	1.22	1.22	1.21	1.20	1.19	1.19
10000	1.06	1.20	1.22	1.22	1.21	1.21	1.20	1.19	1.18

10	12	16	20	24	30	40	60	100	1000
6.36	6.42	6.50	6.54	6.58	6.61	6.64	6.67	6.70	6.73
2.70	2.72	2.74	2.75	2.76	2.77	2.78	2.79	2.79	2.80
2.06	2.07	2.08	2.08	2.09	2.09	2.10	2.10	2.10	2.11
1.80	1.80	1.81	1.81	1.81	1.82	1.82	1.82	1.82	1.82
1.66	1.66	1.66	1.66	1.67	1.67	1.67	1.67	1.67	1.67
1.57	1.57	1.57	1.57	1.57	1.57	1.57	1.57	1.57	1.57
1.51	1.51	1.51	1.51	1.51	1.51	1.50	1.50	1.50	1.50
1.47	1.46	1.46	1.46	1.46	1.46	1.45	1.45	1.45	1.45
1.43	1.43	1.43	1.42	1.42	1.42	1.42	1.41	1.41	1.41
1.41	1.40	1.40	1.40	1.39	1.39	1.39	1.38	1.38	1.38
1.38	1.38	1.38	1.37	1.37	1.37	1.36	1.36	1.36	1.35
1.37	1.36	1.36	1.35	1.35	1.35	1.34	1.34	1.33	1.33
1.35	1.35	1.34	1.34	1.33	1.33	1.33	1.32	1.32	1.31
1.34	1.33	1.33	1.32	1.32	1.31	1.31	1.31	1.30	1.29
1.33	1.32	1.32	1.31	1.31	1.30	1.30	1.29	1.29	1.28
1.32	1.31	1.31	1.30	1.30	1.29	1.29	1.28	1.28	1.27
1.31	1.30	1.30	1.29	1.29	1.28	1.28	1.27	1.27	1.26
1.30	1.30	1.29	1.28	1.28	1.27	1.27	1.26	1.26	1.25
1.30	1.29	1.28	1.28	1.27	1.27	1.26	1.25	1.25	1.24
1.29	1.28	1.28	1.27	1.26	1.26	1.25	1.25	1.24	1.23
1.28	1.28	1.27	1.26	1.26	1.25	1.25	1.24	1.23	1.22
1.28	1.27	1.26	1.26	1.25	1.25	1.24	1.23	1.23	1.22
1.28	1.27	1.26	1.25	1.25	1.24	1.23	1.23	1.22	1.21
1.27	1.26	1.25	1.25	1.24	1.24	1.23	1.22	1.21	1.20
1.27	1.26	1.25	1.24	1.24	1.23	1.22	1.22	1.21	1.20
1.26	1.26	1.25	1.24	1.23	1.23	1.22	1.21	1.21	1.19
1.26	1.25	1.24	1.24	1.23	1.22	1.22	1.21	1.20	1.19
1.26	1.25	1.24	1.23	1.23	1.22	1.21	1.20	1.20	1.19
1.26	1.25	1.24	1.23	1.22	1.22	1.21	1.20	1.19	1.18
1.25	1.25	1.23	1.23	1.22	1.21	1.21	1.20	1.19	1.18
1.23	1.23	1.21	1.20	1.20	1.19	1.18	1.17	1.16	1.15
1.22	1.21	1.19	1.18	1.18	1.17	1.16	1.15	1.14	1.12
1.20	1.19	1.18	1.17	1.16	1.15	1.14	1.12	1.11	1.09
1.18	1.17	1.16	1.14	1.13	1.12	1.11	1.09	1.08	1.04
1.18	1.17	1.15	1.14	1.13	1.12	1.10	1.09	1.07	1.02

TABLE T7 Critical Points of the F Distribution (Continued)

Upper 40 Percent Points

ν_2 \ ν_1	1	2	3	4	5	6	7	8	9
1	1.89	2.63	2.93	3.09	3.20	3.27	3.32	3.36	3.39
2	1.13	1.50	1.64	1.72	1.76	1.80	1.82	1.84	1.85
3	0.96	1.26	1.37	1.43	1.47	1.49	1.51	1.52	1.53
4	0.89	1.16	1.26	1.31	1.34	1.36	1.37	1.38	1.39
5	0.85	1.11	1.20	1.24	1.27	1.29	1.30	1.31	1.31
6	0.82	1.07	1.16	1.20	1.22	1.24	1.25	1.26	1.27
7	0.80	1.05	1.13	1.17	1.19	1.21	1.22	1.23	1.23
8	0.79	1.03	1.11	1.15	1.17	1.19	1.20	1.20	1.21
9	0.78	1.02	1.10	1.13	1.15	1.17	1.18	1.18	1.19
10	0.77	1.01	1.08	1.12	1.14	1.15	1.16	1.17	1.17
11	0.77	1.00	1.07	1.11	1.13	1.14	1.15	1.16	1.16
12	0.76	0.99	1.07	1.10	1.12	1.13	1.14	1.15	1.15
13	0.76	0.98	1.06	1.09	1.11	1.13	1.13	1.14	1.14
14	0.75	0.98	1.05	1.09	1.11	1.12	1.13	1.13	1.14
15	0.75	0.97	1.05	1.08	1.10	1.11	1.12	1.13	1.13
16	0.75	0.97	1.04	1.08	1.10	1.11	1.12	1.12	1.13
17	0.75	0.97	1.04	1.07	1.09	1.10	1.11	1.12	1.12
18	0.74	0.96	1.04	1.07	1.09	1.10	1.11	1.11	1.12
19	0.74	0.96	1.03	1.07	1.09	1.10	1.10	1.11	1.11
20	0.74	0.96	1.03	1.06	1.08	1.09	1.10	1.11	1.11
21	0.74	0.96	1.03	1.06	1.08	1.09	1.10	1.10	1.11
22	0.74	0.96	1.03	1.06	1.08	1.09	1.10	1.10	1.10
23	0.74	0.95	1.02	1.06	1.07	1.09	1.09	1.10	1.10
24	0.73	0.95	1.02	1.06	1.07	1.08	1.09	1.10	1.10
25	0.73	0.95	1.02	1.05	1.07	1.08	1.09	1.09	1.10
26	0.73	0.95	1.02	1.05	1.07	1.08	1.09	1.09	1.09
27	0.73	0.95	1.02	1.05	1.07	1.08	1.08	1.09	1.09
28	0.73	0.95	1.02	1.05	1.07	1.08	1.08	1.09	1.09
29	0.73	0.95	1.02	1.05	1.06	1.08	1.08	1.09	1.09
30	0.73	0.94	1.01	1.05	1.06	1.07	1.08	1.08	1.09
40	0.72	0.94	1.01	1.04	1.05	1.06	1.07	1.07	1.08
60	0.72	0.93	1.00	1.03	1.04	1.05	1.06	1.06	1.07
100	0.71	0.92	0.99	1.02	1.04	1.05	1.05	1.06	1.06
500	0.71	0.92	0.98	1.01	1.03	1.04	1.04	1.05	1.05
10000	0.71	0.92	0.98	1.01	1.03	1.03	1.04	1.04	1.04

10	12	16	20	24	30	40	60	100	1000
3.41	3.45	3.49	3.52	3.54	3.56	3.58	3.60	3.61	3.63
1.86	1.88	1.90	1.91	1.92	1.92	1.93	1.94	1.95	1.96
1.54	1.55	1.56	1.57	1.58	1.58	1.59	1.59	1.60	1.60
1.40	1.41	1.42	1.43	1.43	1.43	1.44	1.44	1.45	1.45
1.32	1.33	1.34	1.34	1.35	1.35	1.36	1.36	1.36	1.37
1.27	1.28	1.29	1.29	1.30	1.30	1.30	1.31	1.31	1.31
1.24	1.24	1.25	1.26	1.26	1.26	1.27	1.27	1.27	1.27
1.21	1.22	1.23	1.23	1.23	1.24	1.24	1.24	1.24	1.25
1.19	1.20	1.21	1.21	1.21	1.21	1.22	1.22	1.22	1.22
1.18	1.18	1.19	1.19	1.20	1.20	1.20	1.20	1.20	1.21
1.17	1.17	1.18	1.18	1.18	1.18	1.19	1.19	1.19	1.19
1.16	1.16	1.17	1.17	1.17	1.17	1.17	1.18	1.18	1.18
1.15	1.15	1.16	1.16	1.16	1.16	1.17	1.17	1.17	1.17
1.14	1.14	1.15	1.15	1.15	1.16	1.16	1.16	1.16	1.16
1.13	1.14	1.14	1.15	1.15	1.15	1.15	1.15	1.15	1.15
1.13	1.13	1.14	1.14	1.14	1.14	1.14	1.14	1.14	1.14
1.12	1.13	1.13	1.13	1.14	1.14	1.14	1.14	1.14	1.14
1.12	1.12	1.13	1.13	1.13	1.13	1.13	1.13	1.13	1.13
1.12	1.12	1.12	1.13	1.13	1.13	1.13	1.13	1.13	1.13
1.11	1.12	1.12	1.12	1.12	1.12	1.12	1.12	1.12	1.12
1.11	1.11	1.12	1.12	1.12	1.12	1.12	1.12	1.12	1.12
1.11	1.11	1.11	1.12	1.12	1.12	1.12	1.12	1.12	1.12
1.10	1.11	1.11	1.11	1.11	1.11	1.11	1.11	1.11	1.11
1.10	1.10	1.11	1.11	1.11	1.11	1.11	1.11	1.11	1.11
1.10	1.10	1.11	1.11	1.11	1.11	1.11	1.11	1.11	1.11
1.10	1.10	1.10	1.10	1.11	1.11	1.11	1.11	1.10	1.10
1.10	1.10	1.10	1.10	1.10	1.10	1.10	1.10	1.10	1.10
1.09	1.10	1.10	1.10	1.10	1.10	1.10	1.10	1.10	1.10
1.09	1.10	1.10	1.10	1.10	1.10	1.10	1.10	1.10	1.10
1.09	1.09	1.10	1.10	1.10	1.10	1.10	1.10	1.10	1.09
1.08	1.08	1.08	1.09	1.09	1.08	1.08	1.08	1.08	1.08
1.07	1.07	1.07	1.07	1.07	1.07	1.07	1.07	1.07	1.06
1.06	1.06	1.06	1.06	1.06	1.06	1.06	1.06	1.05	1.04
1.05	1.05	1.05	1.05	1.05	1.05	1.04	1.04	1.03	1.02
1.05	1.05	1.05	1.05	1.05	1.04	1.04	1.04	1.03	1.01

TABLE T8 Cumulative Poisson Probabilities

$$F(x:\lambda) = \sum_{j=0}^{x} \frac{e^{-\lambda}\lambda^j}{j!}$$

λ \ x	0	1	2	3	4	5	6
0.02	0.9802	0.9998	1.0000				
0.04	0.9608	0.9992	1.0000				
0.06	0.9418	0.9983	1.0000				
0.08	0.9231	0.9970	0.9999	1.0000			
0.10	0.9048	0.9953	0.9998	1.0000			
0.15	0.8607	0.9898	0.9995	1.0000			
0.20	0.8187	0.9825	0.9989	0.9999	1.0000		
0.25	0.7788	0.9735	0.9978	0.9999	1.0000		
0.30	0.7408	0.9631	0.9964	0.9997	1.0000		
0.35	0.7047	0.9513	0.9945	0.9995	1.0000		
0.40	0.6703	0.9384	0.9921	0.9992	0.9999	1.0000	
0.45	0.6376	0.9246	0.9891	0.9988	0.9999	1.0000	
0.50	0.6065	0.9098	0.9856	0.9982	0.9998	1.0000	
0.55	0.5769	0.8943	0.9815	0.9975	0.9997	1.0000	
0.60	0.5488	0.8781	0.9769	0.9966	0.9996	1.0000	
0.65	0.5220	0.8614	0.9717	0.9956	0.9994	0.9999	1.0000
0.70	0.4966	0.8442	0.9659	0.9942	0.9992	0.9999	1.0000
0.75	0.4724	0.8266	0.9595	0.9927	0.9989	0.9999	1.0000
0.80	0.4493	0.8088	0.9526	0.9909	0.9986	0.9998	1.0000
0.85	0.4274	0.7907	0.9451	0.9889	0.9982	0.9997	1.0000
0.90	0.4066	0.7725	0.9371	0.9865	0.9977	0.9997	1.0000
0.95	0.3867	0.7541	0.9287	0.9839	0.9971	0.9995	0.9999
1.0	0.3679	0.7358	0.9197	0.9810	0.9963	0.9994	0.9999
1.1	0.3329	0.6990	0.9004	0.9743	0.9946	0.9990	0.9999
1.2	0.3012	0.6626	0.8795	0.9662	0.9923	0.9985	0.9997
1.3	0.2725	0.6268	0.8571	0.9569	0.9893	0.9978	0.9996
1.4	0.2466	0.5918	0.8335	0.9463	0.9857	0.9968	0.9994
1.5	0.2231	0.5578	0.8088	0.9344	0.9814	0.9955	0.9991
1.6	0.2019	0.5249	0.7834	0.9212	0.9763	0.9940	0.9987
1.7	0.1827	0.4932	0.7572	0.9068	0.9704	0.9920	0.9981
1.8	0.1653	0.4628	0.7306	0.8913	0.9636	0.9896	0.9974
1.9	0.1496	0.4337	0.7037	0.8747	0.9559	0.9868	0.9966
2.0	0.1353	0.4060	0.6767	0.8571	0.9473	0.9834	0.9955
2.2	0.1108	0.3546	0.6227	0.8194	0.9275	0.9751	0.9925
2.4	0.0907	0.3084	0.5697	0.7787	0.9041	0.9643	0.9884
2.6	0.0743	0.2674	0.5184	0.7360	0.8774	0.9510	0.9828
2.8	0.0608	0.2311	0.4695	0.6919	0.8477	0.9349	0.9756
3.0	0.0498	0.1991	0.4232	0.6472	0.8153	0.9161	0.9665
3.2	0.0408	0.1712	0.3799	0.6025	0.7806	0.8946	0.9554
3.4	0.0334	0.1468	0.3397	0.5584	0.7442	0.8705	0.9421
3.6	0.0273	0.1257	0.3027	0.5152	0.7064	0.8441	0.9267

TABLE T8 Cumulative Poisson Probabilities (Continued)

λ \ x	0	1	2	3	4	5	6
3.8	0.0224	0.1074	0.2689	0.4735	0.6678	0.8156	0.9091
4.0	0.0183	0.0916	0.2381	0.4335	0.6288	0.7851	0.8893
4.2	0.0150	0.0780	0.2102	0.3954	0.5898	0.7531	0.8675
4.4	0.0123	0.0663	0.1851	0.3594	0.5512	0.7199	0.8436
4.6	0.0101	0.0563	0.1626	0.3257	0.5132	0.6858	0.8180
4.8	0.0082	0.0477	0.1425	0.2942	0.4763	0.6510	0.7908
5.0	0.0067	0.0404	0.1247	0.2650	0.4405	0.6160	0.7622
5.2	0.0055	0.0342	0.1088	0.2381	0.4061	0.5809	0.7324
5.4	0.0045	0.0289	0.0948	0.2133	0.3733	0.5461	0.7017
5.6	0.0037	0.0244	0.0824	0.1906	0.3422	0.5119	0.6703
5.8	0.0030	0.0206	0.0715	0.1700	0.3127	0.4783	0.6384
6.0	0.0025	0.0174	0.0620	0.1512	0.2851	0.4457	0.6063
6.2	0.0020	0.0146	0.0536	0.1342	0.2592	0.4141	0.5742
6.4	0.0017	0.0123	0.0463	0.1189	0.2351	0.3837	0.5423
6.6	0.0014	0.0103	0.0400	0.1052	0.2127	0.3547	0.5108
6.8	0.0011	0.0087	0.0344	0.0928	0.1920	0.3270	0.4799
7.0	0.0009	0.0073	0.0296	0.0818	0.1730	0.3007	0.4497
7.2	0.0007	0.0061	0.0255	0.0719	0.1555	0.2759	0.4204
7.4	0.0006	0.0051	0.0219	0.0632	0.1395	0.2526	0.3920
7.6	0.0005	0.0043	0.0188	0.0554	0.1249	0.2307	0.3646
7.8	0.0004	0.0036	0.0161	0.0485	0.1117	0.2103	0.3384
8.0	0.0003	0.0030	0.0138	0.0424	0.0996	0.1912	0.3134
8.2	0.0003	0.0025	0.0118	0.0370	0.0887	0.1736	0.2896
8.4	0.0002	0.0021	0.0100	0.0323	0.0789	0.1573	0.2670
8.6	0.0002	0.0018	0.0086	0.0281	0.0701	0.1422	0.2457
8.8	0.0002	0.0015	0.0073	0.0244	0.0621	0.1284	0.2256
9.0	0.0001	0.0012	0.0062	0.0212	0.0550	0.1157	0.2068
9.2	0.0001	0.0010	0.0053	0.0184	0.0486	0.1041	0.1892
9.4	0.0001	0.0009	0.0045	0.0160	0.0429	0.0935	0.1727
9.6	0.0001	0.0007	0.0038	0.0138	0.0378	0.0838	0.1574
9.8	0.0001	0.0006	0.0033	0.0120	0.0333	0.0750	0.1433
10.0	0.0000	0.0005	0.0028	0.0103	0.0293	0.0671	0.1301
10.5	0.0000	0.0003	0.0018	0.0071	0.0211	0.0504	0.1016
11.0	0.0000	0.0002	0.0012	0.0049	0.0151	0.0375	0.0786
11.5	0.0000	0.0001	0.0008	0.0034	0.0107	0.0277	0.0603
12.0	0.0000	0.0001	0.0005	0.0023	0.0076	0.0203	0.0458
12.5	0.0000	0.0001	0.0003	0.0016	0.0053	0.0148	0.0346
13.0	0.0000	0.0000	0.0002	0.0011	0.0037	0.0107	0.0259
13.5	0.0000	0.0000	0.0001	0.0007	0.0026	0.0077	0.0193
14.0	0.0000	0.0000	0.0001	0.0005	0.0018	0.0055	0.0142
14.5	0.0000	0.0000	0.0001	0.0003	0.0012	0.0039	0.0105
15.0	0.0000	0.0000	0.0000	0.0002	0.0009	0.0028	0.0076
15.5	0.0000	0.0000	0.0000	0.0001	0.0006	0.0020	0.0055
16.0	0.0000	0.0000	0.0000	0.0001	0.0004	0.0014	0.0040

TABLE T8 Cumulative Poisson Probabilities (Continued)

λ＼x	0	1	2	3	4	5	6
16.5	0.0000	0.0000	0.0000	0.0001	0.0003	0.0010	0.0029
17.0	0.0000	0.0000	0.0000	0.0000	0.0002	0.0007	0.0021
17.5	0.0000	0.0000	0.0000	0.0000	0.0001	0.0005	0.0015
18.0	0.0000	0.0000	0.0000	0.0000	0.0001	0.0003	0.0010
18.5	0.0000	0.0000	0.0000	0.0000	0.0001	0.0002	0.0007
19.0	0.0000	0.0000	0.0000	0.0000	0.0000	0.0002	0.0005
19.5	0.0000	0.0000	0.0000	0.0000	0.0000	0.0001	0.0004
20.0	0.0000	0.0000	0.0000	0.0000	0.0000	0.0001	0.0003

λ＼x	7	8	9	10	11	12	13
0.95	1.0000						
1.0	1.0000						
1.2	1.0000						
1.3	0.9999	1.0000					
1.4	0.9999	1.0000					
1.5	0.9998	1.0000					
1.6	0.9997	1.0000					
1.7	0.9996	0.9999	1.0000				
1.8	0.9994	0.9999	1.0000				
1.9	0.9992	0.9998	1.0000				
2.0	0.9989	0.9998	1.0000				
2.2	0.9980	0.9995	0.9999	1.0000			
2.4	0.9967	0.9991	0.9998	1.0000			
2.6	0.9947	0.9985	0.9996	0.9999	1.0000		
2.8	0.9919	0.9976	0.9993	0.9998	1.0000		
3.0	0.9881	0.9962	0.9989	0.9997	0.9999	1.0000	
3.2	0.9832	0.9943	0.9982	0.9995	0.9999	1.0000	
3.4	0.9769	0.9917	0.9973	0.9992	0.9998	0.9999	1.0000
3.6	0.9692	0.9883	0.9960	0.9987	0.9996	0.9999	1.0000
3.8	0.9599	0.9840	0.9942	0.9981	0.9994	0.9998	1.0000
4.0	0.9489	0.9678	0.9919	0.9972	0.9991	0.9997	0.9999
4.2	0.9361	0.9721	0.9889	0.9959	0.9986	0.9996	0.9999
4.4	0.9214	0.9642	0.9851	0.9943	0.9980	0.9993	0.9998
4.6	0.9049	0.9549	0.9805	0.9922	0.9971	0.9990	0.9997
4.8	0.8867	0.9442	0.9749	0.9896	0.9960	0.9986	0.9995
5.0	0.8666	0.9319	0.9682	0.9863	0.9945	0.9980	0.9993
5.2	0.8449	0.9181	0.9603	0.9823	0.9927	0.9972	0.9990
5.4	0.8217	0.9027	0.9512	0.9775	0.9904	0.9962	0.9986
5.6	0.7970	0.8857	0.9409	0.9718	0.9875	0.9949	0.9980

TABLE T8   Cumulative Poisson Probabilities (Continued)

λ \ x	7	8	9	10	11	12	13
5.8	0.7710	0.8672	0.9292	0.9651	0.9841	0.9932	0.9973
6.0	0.7440	0.8472	0.9161	0.9574	0.9799	0.9912	0.9964
6.2	0.7160	0.8259	0.9016	0.9486	0.9750	0.9887	0.9952
6.4	0.6873	0.8033	0.8858	0.9386	0.9693	0.9857	0.9937
6.6	0.6581	0.7796	0.8686	0.9274	0.9627	0.9821	0.9920
6.8	0.6285	0.7548	0.8502	0.9151	0.9552	0.9779	0.9898
7.0	0.5987	0.7291	0.8305	0.9015	0.9467	0.9730	0.9872
7.2	0.5689	0.7027	0.8096	0.8867	0.9371	0.9673	0.9841
7.4	0.5393	0.6757	0.7877	0.8707	0.9265	0.9609	0.9805
7.6	0.5100	0.6482	0.7649	0.8535	0.9148	0.9536	0.9762
7.8	0.4812	0.6204	0.7411	0.8352	0.9020	0.9454	0.9714
8.0	0.4530	0.5925	0.7166	0.8159	0.8881	0.9362	0.9658
8.2	0.4254	0.5647	0.6915	0.7955	0.8731	0.9261	0.9595
8.4	0.3987	0.5369	0.6659	0.7743	0.8571	0.9150	0.9524
8.6	0.3728	0.5094	0.6400	0.7522	0.8400	0.9029	0.9445
8.8	0.3478	0.4823	0.6137	0.7294	0.8220	0.8898	0.9358
9.0	0.3239	0.4557	0.5874	0.7060	0.8030	0.8758	0.9261
9.2	0.3010	0.4296	0.5611	0.6820	0.7832	0.8607	0.9156
9.4	0.2792	0.4042	0.5349	0.6576	0.7626	0.8448	0.9042
9.6	0.2584	0.3796	0.5089	0.6329	0.7412	0.8279	0.8919
9.8	0.2388	0.3558	0.4832	0.6080	0.7193	0.8101	0.8786
10.0	0.2202	0.3328	0.4579	0.5830	0.6968	0.7916	0.8645
10.5	0.1785	0.2794	0.3971	0.5207	0.6387	0.7420	0.8253
11.0	0.1432	0.2320	0.3405	0.4599	0.5793	0.6887	0.7813
11.5	0.1137	0.1906	0.2888	0.4017	0.5198	0.6329	0.7330
12.0	0.0895	0.1550	0.2424	0.3472	0.4616	0.5760	0.6815
12.5	0.0698	0.1249	0.2014	0.2971	0.4058	0.5190	0.6278
13.0	0.0540	0.0998	0.1658	0.2517	0.3532	0.4631	0.5730
13.5	0.0415	0.0790	0.1353	0.2112	0.3045	0.4093	0.5182
14.0	0.0316	0.0621	0.1094	0.1757	0.2600	0.3585	0.4644
14.5	0.0239	0.0484	0.0878	0.1449	0.2201	0.3111	0.4125
15.0	0.0180	0.0374	0.0699	0.1185	0.1848	0.2676	0.3632
15.5	0.0135	0.0288	0.0552	0.0961	0.1538	0.2283	0.3171
16.0	0.0100	0.0220	0.0433	0.0774	0.1270	0.1931	0.2745
16.5	0.0074	0.0167	0.0337	0.0619	0.1041	0.1621	0.2357
17.0	0.0054	0.0126	0.0261	0.0491	0.0847	0.1350	0.2009
17.5	0.0040	0.0095	0.0201	0.0387	0.0684	0.1116	0.1699
18.0	0.0029	0.0071	0.0154	0.0304	0.0549	0.0917	0.1426
18.5	0.0021	0.0052	0.0117	0.0237	0.0438	0.0748	0.1189
19.0	0.0015	0.0039	0.0089	0.0183	0.0347	0.0606	0.0984
19.5	0.0011	0.0028	0.0067	0.0141	0.0273	0.0488	0.0809
20.0	0.0008	0.0021	0.0050	0.0108	0.0214	0.0390	0.0661

TABLE T8   Cumulative Poisson Probabilities (Continued)

λ \ x	14	15	16	17	18	19	20
4.2	1.0000						
4.4	0.9999	1.0000					
4.6	0.9999	1.0000					
4.8	0.9999	1.0000					
5.0	0.9998	0.9999	1.0000				
5.2	0.9997	0.9999	1.0000				
5.4	0.9995	0.9998	0.9999	1.0000			
5.6	0.9993	0.9998	0.9999	1.0000			
5.8	0.9990	0.9996	0.9999	1.0000			
6.0	0.9986	0.9995	0.9998	0.9999	1.0000		
6.2	0.9981	0.9993	0.9997	0.9999	1.0000		
6.4	0.9974	0.9990	0.9996	0.9999	1.0000		
6.6	0.9966	0.9986	0.9995	0.9998	0.9999	1.0000	
6.8	0.9956	0.9982	0.9993	0.9997	0.9999	1.0000	
7.0	0.9943	0.9976	0.9990	0.9996	0.9999	1.0000	
7.2	0.9927	0.9969	0.9987	0.9995	0.9998	0.9999	1.0000
7.4	0.9908	0.9959	0.9983	0.9993	0.9997	0.9999	1.0000
7.6	0.9886	0.9948	0.9978	0.9991	0.9996	0.9999	1.0000
7.8	0.9859	0.9934	0.9971	0.9988	0.9995	0.9998	0.9999
8.0	0.9827	0.9918	0.9963	0.9984	0.9993	0.9997	0.9999
8.2	0.9791	0.9898	0.9953	0.9979	0.9991	0.9997	0.9999
8.4	0.9749	0.9875	0.9941	0.9973	0.9989	0.9995	0.9998
8.6	0.9701	0.9848	0.9926	0.9966	0.9985	0.9994	0.9998
8.8	0.9647	0.9816	0.9909	0.9957	0.9981	0.9992	0.9997
9.0	0.9585	0.9780	0.9889	0.9947	0.9976	0.9989	0.9996
9.2	0.9517	0.9738	0.9865	0.9934	0.9969	0.9986	0.9994
9.4	0.9441	0.9691	0.9838	0.9919	0.9962	0.9983	0.9992
9.6	0.9357	0.9638	0.9806	0.9902	0.9952	0.9978	0.9990
9.8	0.9265	0.9579	0.9770	0.9881	0.9941	0.9972	0.9987
10.0	0.9165	0.9513	0.9730	0.9857	0.9928	0.9965	0.9984
10.5	0.8879	0.9317	0.9604	0.9781	0.9885	0.9942	0.9972
11.0	0.8540	0.9074	0.9441	0.9678	0.9823	0.9907	0.9953
11.5	0.8153	0.8783	0.9236	0.9542	0.9738	0.9857	0.9925
12.0	0.7720	0.8444	0.8987	0.9370	0.9626	0.9787	0.9884
12.5	0.7250	0.8060	0.8693	0.9158	0.9481	0.9694	0.9827
13.0	0.6751	0.7636	0.8355	0.8905	0.9302	0.9573	0.9750
13.5	0.6233	0.7178	0.7975	0.8609	0.9084	0.9421	0.9649
14.0	0.5704	0.6694	0.7559	0.8272	0.8826	0.9235	0.9521
14.5	0.5176	0.6192	0.7112	0.7897	0.8530	0.9012	0.9362
15.0	0.4657	0.5681	0.6641	0.7489	0.8195	0.8752	0.9170
15.5	0.4154	0.5170	0.6154	0.7052	0.7825	0.8455	0.8944

TABLE T8   Cumulative Poisson Probabilities (Continued)

λ \ x	14	15	16	17	18	19	20
16.0	0.3675	0.4667	0.5660	0.6593	0.7423	0.8122	0.8682
16.5	0.3225	0.4180	0.5165	0.6120	0.6996	0.7757	0.8385
17.0	0.2808	0.3715	0.4677	0.5640	0.6550	0.7363	0.8055
17.5	0.2426	0.3275	0.4204	0.5160	0.6089	0.6945	0.7694
18.0	0.2081	0.2867	0.3751	0.4686	0.5622	0.6509	0.7307
18.5	0.1771	0.2490	0.3321	0.4226	0.5156	0.6061	0.6898
19.0	0.1497	0.2148	0.2920	0.3784	0.4695	0.5606	0.6472
19.5	0.1257	0.1840	0.2550	0.3364	0.4246	0.5151	0.6034
20.0	0.1049	0.1565	0.2211	0.2970	0.3814	0.4703	0.5591

λ \ x	21	22	23	24	25	26	27
7.8	1.0000						
8.0	1.0000						
8.2	1.0000						
8.4	0.9999	1.0000					
8.6	0.9999	1.0000					
8.8	0.9999	1.0000					
9.0	0.9998	0.9999	1.0000				
9.2	0.9998	0.9999	1.0000				
9.4	0.9997	0.9999	1.0000				
9.6	0.9996	0.9998	0.9999	1.0000			
9.8	0.9995	0.9998	0.9999	1.0000			
10.0	0.9993	0.9997	0.9999	1.0000			
10.5	0.9987	0.9994	0.9998	0.9999	1.0000		
11.0	0.9977	0.9990	0.9995	0.9998	0.9999	1.0000	
11.5	0.9962	0.9982	0.9992	0.9996	0.9998	0.9999	1.0000
12.0	0.9939	0.9970	0.9985	0.9993	0.9997	0.9999	0.9999
12.5	0.9906	0.9951	0.9975	0.9988	0.9994	0.9997	0.9999
13.0	0.9859	0.9924	0.9960	0.9980	0.9990	0.9995	0.9998
13.5	0.9796	0.9885	0.9938	0.9968	0.9984	0.9992	0.9996
14.0	0.9712	0.9833	0.9907	0.9950	0.9974	0.9987	0.9994
14.5	0.9604	0.9763	0.9863	0.9924	0.9959	0.9979	0.9989
15.0	0.9469	0.9673	0.9805	0.9888	0.9938	0.9967	0.9983
15.5	0.9304	0.9558	0.9730	0.9840	0.9909	0.9950	0.9973
16.0	0.9108	0.9418	0.9633	0.9777	0.9869	0.9925	0.9959
16.5	0.8878	0.9248	0.9513	0.9696	0.9816	0.9892	0.9939
17.0	0.8615	0.9047	0.9367	0.9594	0.9748	0.9848	0.9912
17.5	0.8319	0.8815	0.9193	0.9468	0.9661	0.9791	0.9875

TABLE T8 Cumulative Poisson Probabilities (Continued)

λ \ x	21	22	23	24	25	26	27
18.0	0.7991	0.8551	0.8989	0.9317	0.9554	0.9718	0.9827
18.5	0.7636	0.8256	0.8755	0.9139	0.9424	0.9626	0.9765
19.0	0.7255	0.7931	0.8490	0.8933	0.9269	0.9514	0.9687
19.5	0.6854	0.7580	0.8196	0.8697	0.9087	0.9380	0.9591
20.0	0.6437	0.7206	0.7875	0.8432	0.8878	0.9221	0.9475

λ \ x	28	29	30	31	32	33	34
12.0	1.0000						
12.5	1.0000						
13.0	0.9999	1.0000					
13.5	0.9998	0.9999	1.0000				
14.0	0.9997	0.9999	0.9999	1.0000			
14.5	0.9995	0.9998	0.9999	1.0000			
15.0	0.9991	0.9996	0.9998	0.9999	1.0000		
15.5	0.9986	0.9993	0.9997	0.9998	0.9999	1.0000	
16.0	0.9978	0.9989	0.9994	0.9997	0.9999	0.9999	1.0000
16.5	0.9967	0.9982	0.9991	0.9995	0.9998	0.9999	1.0000
17.0	0.9950	0.9973	0.9986	0.9993	0.9996	0.9998	0.9999
17.5	0.9928	0.9959	0.9978	0.9988	0.9994	0.9997	0.9999
18.0	0.9897	0.9941	0.9967	0.9982	0.9990	0.9995	0.9998
18.5	0.9857	0.9915	0.9951	0.9973	0.9985	0.9992	0.9996
19.0	0.9805	0.9882	0.9930	0.9960	0.9978	0.9988	0.9994
19.5	0.9739	0.9838	0.9902	0.9943	0.9967	0.9982	0.9990
20.0	0.9657	0.9782	0.9865	0.9919	0.9953	0.9973	0.9985

λ \ x	35	36	37	38	39	40
17.0	1.0000					
17.5	0.9999	1.0000				
18.0	0.9999	0.9999	1.0000			
18.5	0.9998	0.9999	1.0000			
19.0	0.9997	0.9998	0.9999	1.0000		
19.5	0.9995	0.9997	0.9999	0.9999	1.0000	
20.0	0.9992	0.9996	0.9998	0.9999	0.9999	1.0000

TABLE T9   Random Digits

28107	23829	13740	59690	65965	80655	53273	48623	16912	20123
50179	90739	81831	88170	95842	25385	26129	51780	17253	36766
94329	52632	11650	76830	62609	89095	84995	96798	04199	41783
11038	44621	60120	22812	34922	75266	70153	95530	10385	21930
40275	78047	00219	98206	14361	21459	58445	22424	16904	56288
01407	15739	86875	27277	66380	54303	46812	71528	75392	53372
33485	09310	78412	12599	36321	76476	18465	68494	22133	39115
64003	97882	10983	78924	61198	05541	84906	65151	04056	81600
72209	69736	67471	07054	53616	03400	47623	92598	36099	99977
92444	01439	61619	38022	78580	58452	05612	07682	00020	35056
20481	76598	98683	06887	19218	04472	41412	84196	34060	17247
74451	72762	91374	65761	18931	82318	47935	51770	88436	57068
41810	97698	82483	78213	02049	92007	48955	34845	27216	29929
73736	06684	47983	42743	30193	78648	01717	55762	18614	49529
56219	73820	50028	23601	52930	22098	53181	24804	96945	41994
35651	96994	69889	14775	16638	71529	04335	17809	72839	99971
90015	62765	46776	11723	88073	80962	40307	87170	96383	35410
71648	49083	84586	90980	06756	71653	83500	36297	39554	18877
72121	92018	35819	06932	01742	18787	74188	23331	88798	94539
71866	07566	54229	16549	68590	55928	29054	08020	78337	18128
61407	09540	43035	10941	37845	15873	52926	77178	97812	22016
44836	60799	80686	70336	48114	51114	82624	40834	28800	45653
22308	51951	02803	52252	70559	83171	39270	86578	52416	31001
69771	54986	37522	77021	30129	18268	26660	34602	76987	37078
07583	12769	48235	66179	73152	50174	29099	25237	26217	37106
01247	14343	69520	06099	13087	36017	65033	04224	22213	82990
13377	46751	90029	55223	47966	33843	09890	82526	51728	01467
69729	01789	72712	36686	30352	56203	97231	62825	65920	31397
49200	94278	49210	12656	55502	19098	69749	09463	42938	00804
05472	98080	70729	29331	71968	86350	28529	82860	03210	35420
70222	16409	07546	87578	72821	70512	03620	41893	39173	76940
59120	07466	17840	44983	44351	21136	10543	89311	18485	17444
57668	81415	21605	73704	37336	42646	32923	03462	96683	53500
31101	46134	16540	18097	81542	30660	01622	43568	00443	74086
61313	31302	68332	72992	47828	19029	64349	47294	57543	57638
34908	17008	19318	07045	72182	98160	55747	30191	79625	24711
42390	42929	58140	25168	06308	93504	69082	22753	24843	55796
37797	52368	21421	41523	22310	15194	23329	92819	85177	31402
31897	76918	21886	61485	48511	68674	21658	08420	24749	61067
27635	95249	25370	44988	16765	69314	58719	61311	97687	87988

TABLE T9 Random Digits (Continued)

54525	63237	57130	41930	33575	90830	04157	40830	46305	01831
77754	23138	99867	61382	02612	29308	08626	00969	73065	92311
18544	69488	62907	25813	01041	25187	12778	46710	74923	13331
04027	74203	46847	40954	77897	27317	84280	77669	09836	27500
43174	40997	07193	12507	62016	05869	06285	60790	81711	31919
61249	48504	62027	70876	59326	86829	29225	01202	60010	90416
43474	47213	13520	28247	27349	27940	65598	37917	54316	92088
90400	05567	07580	54073	04566	31990	65458	64288	23279	09233
24962	22942	29093	32645	95343	90314	97185	53453	62096	55447
35962	96580	41847	75323	09840	68072	64374	67871	46788	94038
12109	11756	02464	31220	80955	04725	66136	86029	95407	94614
84055	35598	51486	28531	72752	26527	26993	95692	59468	02619
52925	08903	60644	13859	74022	63359	94094	97940	11839	19957
59701	60724	37419	44420	34981	52175	01366	03915	32137	25522
62289	18280	89699	50247	21969	62561	58926	26974	17683	29397
44851	44530	98588	34953	80901	55828	95273	06194	53489	77769
95651	88962	95620	67850	48944	18384	93207	85619	16869	87362
71420	14244	68227	90777	97587	94469	90438	92050	78986	60891
38581	37450	09858	72573	04676	37988	53479	85549	28508	54707
84138	17742	67152	12264	80272	54063	11869	98161	25803	09443
46148	84191	67119	05327	45834	34596	85050	52929	17309	67918
51148	38221	91188	28479	84281	76305	04811	82885	97959	04585
20998	14642	76158	20439	10482	77499	97294	22869	73300	78849
01292	45688	12506	98292	18031	11983	33766	19699	39380	27525
34110	03411	40343	61261	20610	76358	61675	62862	03529	35029
43476	70714	70719	15845	79646	20279	95151	06813	80117	70605
46131	42067	96483	67611	79656	21038	98884	49770	06716	22423
88939	13222	74938	21599	61640	61872	26199	55576	89224	27083
59245	32185	02071	39607	31614	99589	24507	77876	59044	38750
11960	86939	77870	67033	56174	15524	64076	81109	48958	27039
06351	01216	95624	91486	88620	11118	89963	01896	64532	87732
91952	14279	93610	81697	54005	47796	81467	55133	32469	03787
00790	44087	86987	28712	14192	55746	61014	67132	81164	26784
31617	99782	41789	09758	01347	24402	92852	53862	46227	90025
47413	71411	76091	01917	08256	77870	77659	75650	14242	39700
07225	76048	37409	48226	47235	83703	27383	59730	46361	90765
92493	18753	69027	58506	04243	99958	97090	92105	20051	54357
12212	05646	19614	77393	56686	19894	75641	71546	18245	44669
25188	28499	39348	13656	11307	28761	87926	81271	73195	57646
78481	55007	54545	25265	25573	93444	97012	66406	06833	92913

TABLE T10 Critical Values of D

Sample Size n	$D_{.10}$	$D_{.05}$	$D_{.01}$
1	0.950	0.975	0.995
2	0.776	0.842	0.929
3	0.642	0.708	0.828
4	0.564	0.624	0.733
5	0.510	0.565	0.669
6	0.470	0.521	0.618
7	0.438	0.486	0.577
8	0.411	0.457	0.543
9	0.388	0.432	0.514
10	0.368	0.410	0.490
11	0.352	0.391	0.468
12	0.338	0.375	0.450
13	0.325	0.361	0.433
14	0.314	0.349	0.418
15	0.304	0.338	0.404
16	0.295	0.328	0.392
17	0.286	0.318	0.381
18	0.278	0.309	0.371
19	0.272	0.301	0.363
20	0.264	0.294	0.356
25	0.24	0.27	0.32
30	0.22	0.24	0.29

Adapted from F.J. Massey, Jr. (1951) "The Kolgomorov–Smirnov Test for Goodness of Fit," *Journal Amer. Stat. Assoc.*, Vol. 46, p. 70, with the kind permission of the author and publisher.

TABLE T11   Random Standard Normal Deviates

0.115	-1.189	0.102	-2.597	0.559	-0.320	-1.153	1.060	0.106	0.427
-1.894	-0.189	1.930	-0.590	0.569	1.952	-0.138	0.372	0.203	0.987
-0.857	1.795	0.160	1.213	1.852	-0.364	1.409	0.804	0.059	-0.546
-0.840	0.504	0.234	0.046	-2.316	1.150	-0.609	-0.040	-0.025	0.503
-0.303	-0.489	0.884	-0.637	1.083	-0.070	-1.666	-0.157	0.905	-0.473
-0.969	0.047	1.239	1.218	-0.021	0.848	-2.356	0.994	0.127	1.064
-0.789	-0.211	1.372	1.726	-2.268	-0.220	-0.940	0.195	-0.004	-0.036
-0.563	0.265	0.536	0.419	-2.059	0.631	-1.498	-1.575	-0.376	0.211
0.806	-2.371	0.053	0.746	0.302	1.141	-0.633	-1.367	-0.511	1.559
0.239	0.889	0.785	1.883	-0.388	0.226	1.409	-0.105	1.032	0.762
-1.317	-0.502	0.707	-1.310	-0.097	0.464	-0.537	-0.504	-0.471	0.051
-0.168	0.356	1.896	1.289	-1.128	-0.720	2.345	-0.683	0.068	0.114
-0.722	-0.158	-1.408	0.059	1.591	-1.078	1.535	-0.281	-0.328	-1.575
0.044	-0.095	0.403	-0.308	0.524	0.028	1.986	1.751	-0.766	1.091
-0.605	-0.285	2.419	0.761	0.082	-0.027	0.031	0.288	0.631	0.825
-0.431	0.088	-0.714	1.353	-0.048	1.583	1.423	1.048	0.135	0.191
-2.230	0.889	-1.815	0.362	0.281	0.239	0.786	-0.077	1.561	1.543
-0.512	2.044	-1.697	0.332	-0.563	1.864	1.633	2.914	-0.508	-0.340
1.883	0.303	0.465	-0.286	0.555	1.078	-0.640	-1.027	0.217	0.566
0.416	-1.289	1.407	0.244	-1.885	-0.779	-2.295	-0.570	-0.774	-1.374
0.426	-1.136	-0.898	0.285	-0.381	0.232	0.364	0.663	-0.463	-0.992
-0.084	-1.551	1.045	-0.394	0.254	-0.009	1.587	0.875	-0.623	1.399
-1.265	0.530	-0.968	-0.617	-2.072	-1.041	-1.794	2.624	0.098	0.655
0.116	0.499	-0.571	-0.370	-1.880	-1.464	0.237	1.968	-0.128	-1.587
0.295	0.409	0.756	-1.207	0.615	-0.863	-0.804	2.102	1.624	-2.029
-1.464	1.623	0.273	0.038	0.860	-0.600	0.116	-0.807	0.117	-1.263
-0.199	-2.538	0.012	1.502	-1.165	-1.327	0.933	0.776	0.658	-0.810
-0.158	0.796	-0.970	-0.510	-1.805	0.537	-0.099	1.894	-0.205	-1.389
0.299	0.135	-0.385	1.105	1.341	-0.615	0.350	-0.179	-0.678	-0.219
0.734	-0.658	-1.038	0.559	-1.680	0.795	0.758	0.488	-0.427	1.889
-0.134	-0.083	-1.412	-1.545	-0.304	-1.553	-0.359	2.782	-0.121	0.257
0.612	-1.043	1.681	0.094	0.030	1.399	-2.020	-0.640	-0.413	0.780
0.720	-0.063	1.347	0.180	0.606	0.253	-0.448	-1.129	0.547	-0.002
0.042	-0.227	0.145	-0.081	1.864	-0.445	0.622	-0.068	-0.598	-0.171
-0.140	0.255	0.150	-0.036	-1.821	-1.131	-0.167	-0.578	0.171	-1.817
-0.055	0.190	0.550	-0.574	1.464	-0.237	1.435	0.655	0.409	0.021
-0.156	-0.105	0.084	-0.403	1.589	-1.153	1.941	1.231	-0.691	-0.357
0.240	-0.550	-1.188	-1.116	-0.528	0.789	2.036	-1.234	-0.390	0.420
-1.535	0.330	0.904	0.471	1.534	1.577	-0.796	1.388	-0.930	0.544
1.420	-1.823	0.533	-0.455	0.096	-0.194	0.387	0.219	-0.260	0.674
1.607	0.046	1.282	0.137	-2.095	0.393	1.375	-0.643	-0.427	-1.560
-0.777	-0.211	0.126	0.192	0.996	-0.556	0.022	-1.321	0.520	1.795
1.539	-1.312	0.660	-0.097	0.212	0.413	-0.699	-0.948	2.444	-0.272
1.280	-1.274	3.198	-1.111	0.466	0.983	-0.650	0.913	-0.269	-0.141
1.399	-0.537	1.429	0.303	1.103	0.613	-0.737	0.206	-0.983	-0.095
0.170	-0.742	0.316	-0.744	0.822	-0.578	-1.640	-1.438	-0.346	0.204
0.104	0.931	0.038	0.163	-1.000	-1.271	-0.052	1.447	1.327	0.178
1.725	1.170	0.417	-1.458	-0.875	-1.399	-0.691	0.929	0.761	0.071
-1.262	-0.087	1.126	0.177	-0.168	0.387	0.438	0.569	1.087	-0.407
-0.492	0.543	0.312	-0.929	1.330	0.330	-0.493	0.437	1.326	-0.863

References

Abramowitz, Milton, and Stegun, Irene A. (1968), *Handbook of Mathematical Functions*, Applied Mathematics Series No. 55, National Bureau of Standards.

Afzali, Nader (1979), "Albert's Reliability Allocation Technique with Relaxed Assumption on Effort Function," MS Report, Kansas State University, Manhattan, KS.

Albert, A. (1958), "A Measure of the Effort Required to Increase Reliability," Applied Mathematical and Statistical Library, Technical Report No. 43, May, Stanford University.

Albert, A. (1981), Private communication.

AT&T Reliability Manual (1983), Bell System Information Publication IP 10475.

Aitchison, J., and Brown, J. A. C. (1957), *The Lognormal Distribution*, Cambridge University Press, London.

Bain, Lee J. (1978), *Statistical Analysis of Reliability and Life Testing Models*, Marcel Dekker, New York.

Banerjee, S. K., and Rajamani, K. (1976), "Closed Form Solutions for Delta-Star and Star-Delta Conversions of Reliability Networks," *IEEE Trans. Reliability*, Vol. R-25, No. 2 (June), pp. 118–119.

Banerjee, S. K., and Rajamani, K. (1972), "Parametric Representation of Probability in Two Dimensions—A New Approach in System Reliability, Evaluation," *IEEE Trans. Reliability*, Vol. R21, No. 1 (February), pp. 56–60.

Barlow, Richard E., and Proschan, Frank (1975), *Statistical Theory of Reliability and Life Testing*, Holt, Rinehart and Winston, New York.

Bazovsky, Igor (1961), *Reliability Theory and Practice*, Prentice-Hall, Englewood Cliffs, NJ.

Chatterjee, Saumitra (1971), *Availability Models of Maintained Systems*, MS Thesis, Kansas State University, Manhattan, KS.

Davis, D. J. (1952), "The Analysis of Some Failure Data," *Journal Amer. Stat. Assoc.*, Vol. 47, pp. 113–150.

363

Dhillon, Balbir S. (1983), *Reliability Engineering in Systems Design and Operation*, Van Nostrand Reinhold, Princeton, NJ.

Dhillon, Balbir S., and Singh, Chanan (1981), *Engineering Reliability*, John Wiley, New York.

Epstein, Benjamin (1954), "Some Theorems Relevant to Life Testing from an Exponential Distribution," *Annals of Mathematical Statistics*, Vol. 25, pp. 373–381.

Epstein, Benjamin (1959), *Statistical Techniques in Life Testing*, Document PB171580, U.S. Dept. of Commerce.

Epstein, Benjamin (1960a), "Estimation from Life Test Data," *Technometrics*, Vol. 2, No. 4 (November), pp. 447–454.

Epstein, Benjamin (1960b), "Statistical Life Test Acceptance Procedures," *Technometrics*, Vol. 2, No. 4 (November), pp. 435–446.

Epstein, Benjamin (1960c), "Tests for the Validity of the Assumption that the Underlying Distribution of Life is Exponential. Part I," *Technometrics*, Vol. 2, No. 1 (February), pp. 83–101.

Epstein, Benjamin (1960d), "Tests for the Validity of the Assumption that the Underlying Distribution of Life is Exponential. Part II," *Technometrics*, Vol. 2, No. 4 (May), pp. 167–183.

Epstein, Benjamin, and Sobel, Milton (1953), "Life Testing," *Journal Amer. Stat. Assoc.*, Vol. 48, pp. 486–501.

Feller, William (1966), *An Introduction to Probability and Statistics*, Vol. II, John Wiley, New York, p. 11.

Gorski, Andrew G. (1968), "Beware of the Weibull Euphoria," *IEEE Trans. Reliability*, Vol. R-17, No. 4 (December), pp. 202–203.

Grosh, Doris L. (1982), "A Parallel System of CFR Units is IFR," *IEEE Trans. Reliability*, Vol. R-31, No. 4 (October), p. 403.

Grosh, Doris L. (1983), "Comments on the Delta-Star Problem," *IEEE Trans. Reliability*, Vol. R-32, No. 4 (October), pp. 391-394.

Gupta, Hariom, and Sharma, Jaydev (1978), "A Delta-Star Transformation for Reliability Evaluation," *IEEE Trans. Reliability*, Vol. R-27, No. 3 (August), p. 212.

Hamming, Richard W. (1962), *Numerical Methods for Scientists and Engineers*, McGraw-Hill, New York.

Henley, Ernest J., and Kumamoto, Hiromitsu (1981), *Reliability Engineering and Risk Assessment*, Prentice-Hall, Englewood Cliffs, NJ.

Hillier, Frederick S., and Lieberman, Gerald J. (1986), *Introduction to Operations Research*, 4th ed., Holden-Day, San Francisco.

Hwang, C. L., Tillman, Frank A., and Lee, M. H. (1981), "System–Reliability Evaluation Techniques for Complex/Large Systems—A Review," *IEEE Trans. Reliability*, Vol. R-3, No. 5 (December), pp. 416–422.

Jensen, Finn, and Peterson, Niels E. (1982), *Burn-In*, John Wiley, New York.

Johnson, Leonard G. (1951), "The Median Ranks of Sample Values in Their Population with an Application to Certain Fatigue Studies," *Industrial Mathematics*, Vol. 2, pp. 1–6.

Johnson, Leonard G. (1964), *The Statistical Treatment of Fatigue Experiments*, Elsevier, New York.

Kapur, K. C., and Lamberson, L. R. (1977), *Reliability in Engineering Design*, John Wiley, New York.

Kimball, Bradford F. (1960), "On the Choice of Plotting Positions on Probability Paper," *Journal Amer. Stat. Assoc.*, Vol. 55, No. 291 (September), pp. 546–560.

Lee, Eliza T. (1980), *Statistical Methods for Survival Data Analysis*, Lifetime Learning Publications, Division of Wadsworth, Belmont, CA.

Lloyd, David K., and Lipow, Myron (1962), *Reliability: Management, Methods and Mathematics*, Prentice-Hall, Englewood Cliffs, NJ.

Mann, Nancy R., Schafer, Ray E., and Singpurwalla, Nozer D. (1974), *Methods for Statistical Analysis of Reliability and Life Data*, John Wiley, New York.

Nelson, Lloyd S. (1967), "Weibull Probability Paper," *Industrial Quality Control*, March 1967.

Parzen, Emanuel (1962), *Stochastic Processes*, Holden-Day, San Francisco.

Ramamoorty, M. (1970), "Block Diagram Approach to Power System Reliability," *IEEE Trans. Power Apparatus and Systems*, Vol. 89 (May–June), pp. 801–811.

Rosenthal, Arnie (1978), "Note on 'Closed Form Solution for Delta-Star and Star-Delta Conversion of Reliability Networks,'" *IEEE Trans. Reliability*, Vol. R-27, No. 2 (January), pp. 110–111.

Rys, Malgorzata (1986), "An Evaluation of Various Plotting Positions," MS Report, Kansas State University, Manhattan, KS.

Saaty, T. L. (1957), "Resume of Useful Formulas in Queueing Theory," *Operations Research*, Vol. 5, No. 2, pp. 161–200.

Sandler, G. H. (1963), *System Reliability Engineering*, Prentice-Hall, Englewood Cliffs, NJ.

Shooman, Martin L. (1968), *Probabilistic Reliability*, McGraw-Hill, New York.

Singh, Balbir, and Proctor, C. L. (1976), "Reliability Analysis of Multistate Device Networks," *Proceedings 1976 Annual Reliability and Maintainability Symposium*, Institute of Electrical and Electronic Engineers, pp. 31–35.

Singh, C., and Kankam, M. D. (1976), "Comments on 'Closed Form Solutions for Delta-Star and Star-Delta Conversion of Reliability Networks,'" *IEEE Trans. Reliability*, Vol. R-25, No. 5 (December), p. 336.

Smith, Charles O. (1976), *Introduction to Reliability in Design*, McGraw-Hill, New York.

Spiegel, Murray R. (1961), *Statistics*, Schaum's Outline Series, McGraw-Hill, New York.

Spiegel, Murray R. (1965), *Laplace Transforms*, Schaum's Outline Series, McGraw-Hill, New York.

U.S. Army (1970), *Elements of Reliability and Maintainability*, U.S. Army Management Engineering Training Agency.

U.S. Department of Defense (1957), *Reliability of Military Electronic Equipment*, Advisory Group on Reliability of Electronic Equipment, Office of the Assistant Secretary of Defense (June), pp. 52–27.

von Alven, William, ed. (1964), *Reliability Engineering*, Prentice-Hall, Englewood Cliffs, NJ.

Wald, Abraham (1947), *Sequential Analysis*, John Wiley, New York.

Weibull, W. (1951), "A Statistical Distribution Function of Wide Applicability," *Journal of Applied Mechanics*, Vol. 18, pp. 293–297.

Wilks, Samuel (1962), *Mathematical Statistics*, John Wiley, New York.

Sources for Special Graph Papers

Codex Book Company, 74 Broadway, Norwood, ME 02062.

TEAM (Technical and Engineering Aids in Management), Box 25, Tamworth, NH 03886.

Notations and Abbreviations

$$
\begin{aligned}
\text{r.v.} \;&:\; \text{random variable} \\
T \;&:\; \text{lifetime of a unit, a r.v. (Chapters 1--4)} \\
T, \tilde{T} \;&:\; \text{total life of all components (Chapters 5--6)} \\
\mathbf{T} \;&:\; \text{Markov transition matrix (Chapter 7)} \\
\tilde{t} \;&:\; \text{lifetime of a unit, a r.v. (Chapters 5--6)} \\
f(t) \;&:\; \text{density function for length of life} \\
h(t) \;&:\; \text{hazard function} \\
H(t) \;&:\; \text{cumulative hazard function} \\
F(t) \;&:\; \text{CDF or unreliability function} \\
R(t) \;&:\; \text{reliability function} \\
\text{MVT} \;&:\; \text{mean value theorem} \\
g(x;p) \;&:\; \text{geometric distribution (or r.v.)} \\
f^{*}(t), h^{*}(t), F^{*}(t), R^{*}(t) \;&:\; \text{empirical counterparts to } f(t), \text{ etc.} \\
\text{IFR} \;&:\; \text{increasing failure rate} \\
\text{CFR} \;&:\; \text{constant failure rate} \\
\text{DFR} \;&:\; \text{decreasing failure rate} \\
\text{pdf} \;&:\; \text{probability density function} \\
\text{CDF} \;&:\; \text{cumulative distribution function} \\
ABC \;&:\; \text{short for } A \cap B \cap B, \text{ the intersection} \\
&\quad\; \text{of events (sets) } A, B, C.
\end{aligned}
$$

pmf	:	probability mass function
p.e.	:	probability element
CUD	:	continuous uniform distribution
DUD	:	discrete uniform distribution
Ex. 2.3.1	:	Exercise 1 of Section 2.3
NGEX(λ)	:	negative exponential distribution (or r.v.) with parameter λ
$N(\mu, \sigma^2)$	:	normal distribution (or r.v.) with mean μ and variance σ^2
poi($x; \lambda$)	:	Poisson probability (mass) function
POI($x; \lambda$)	:	cumulative Poisson probability
$\Gamma(\alpha)$	:	gamma function (a constant)
$G(\alpha, \beta)$	:	gamma distribution (or r.v.) with shape parameter α and scale parameter β.
WEI(θ, β)	:	Weibull distribution (or r.v.) with scale parameter θ and "slope" β
$B(m, n)$	:	beta function (a constant)
be(m, n)	:	beta distribution or variable with parameters m, n
MLE	:	maximum likelihood estimator
OC	:	operating characteristic
TC	:	total cost
E(TC)	:	expected total cost
ETT$_{r,n}$	:	expected waiting time for r th failure with n items on test, with no replacement
ETT$^*_{r,n}$	:	expected waiting time for r th failure with n items on test, with replacement
SPRT	:	sequential probability ratio test
bi($x; n, p$)	:	binomial pmf
Bi($x; n, p$)	:	binomial CDF
$\Lambda(\mu, \sigma^2)$	:	lognormal distribution or variable
α	:	significance level or gamma distribution parameter
γ	:	significance level replacing α when confusion could result with gamma parameter
MTTF	:	mean time to failure
MTTR	:	mean time to repair
SAS	:	Statistical analysis system
IMSL	:	International Mathematical and Statistical Library
$P_a(\theta)$	:	probability of acceptance for value θ
$L(\theta)$	:	probability of acceptance for value θ

$\mathbf{I}$: identity matrix

$\mathbf{T}$: Markov transition matrix

$P_i(t)$: probability system in state i at time t

$\mathbf{P}(t)$: column vector $[P_0(t), P_1(t), \ldots]^t$ (superscript t means transpose)

$\mathbf{P}^t(t)$: row vector, transpose of $\mathbf{P}(t)$

$\mathscr{L}[f(t)]$: Laplace transform of $f(t)$

$\mathscr{L}^{-1}[f(s)]$: Inverse Laplace transform of $f(s)$

$Z_i(s) \equiv Z_i$: Laplace transform of $P_i(t)$

$\mathbf{C}$: stands for $(T^t - I)/h$

$\mathbf{C}^*$: stands for $sI - C$

Index